Limnology of Parakrama Samudra - Sri Lanka

Developments in Hydrobiology 12

Series editor
H. J. Dumont

DR W. JUNK PUBLISHERS   THE HAGUE-BOSTON-LONDON   1983

Limnology of Parakrama Samudra – Sri Lanka

A case study of an ancient man-made lake in the tropics

Edited by
F. Schiemer

DR W. JUNK PUBLISHERS THE HAGUE-BOSTON-LONDON 1983

Distributors:
for the United States and Canada

Kluwer Boston, Inc.
190 Old Derby Street
Hingham, MA 02043
USA

for all other countries

Kluwer Academic Publishers Group
Distribution Center
P.O. Box 322
3300 AH Dordrecht
The Netherlands

Library of Congress Cataloging in Publication Data
Main entry under title:

Limnology of Parakrama Samudra, Sri Lanka.

(Developments in hydrobiology ; 12)
1. Limnology--Sri Lanka--Parakrama Samudra Reservoir. 2. Parakrama Samudra Reservoir (Sri Lanka) I. Schiemer, F. II. Series.
QH183.5.L55 1983 574.92'9'5493 82-18645
ISBN-13: 978-94-009-7283-4 e-ISBN-13: 978-94-009-7281-0
DOI: 10.1007/978-94-009-7281-0

'Let not even a drop of rain water go to the sea, without benefitting man'

Parakrama Bahu The Great
(in Mahavansa, the old
historical chronicle of Sri Lanka)

Colossal statue at Polonnaruwa, considered to represent King Parakrama Bahu I (1153–1186 AD), the founder of important irrigation systems.

Contents

Floristic and faunistic surveys

1. The Parakrama Samudra Project: scope and objectives

F. Schiemer

Keywords: tropics, limnology, training, hydrography

Abstract

The demand for environmental control and aquatic resource management necessitates the promotion of limnological training and research in many tropical countries. Sri Lanka has high potential for the development of fisheries in its several thousand of irrigation reservoirs. On the other hand, intensification of land use in the 'dry zone' of the country is creating new problems connected with environmental deterioration. The Parakrama Samudra Project was planned as a cooperative venture of European and Sri Lankan scientists with the object of analyzing the main components and interrelationships of a reservoir ecosystem. The project combined training and research interests: in the course of the two field-research phases (August/September 1979 and March/April 1980), a number of post-graduates from Sri Lankan universities joined the project and received training within the specific research programmes.

Parakrama Samudra is a large (25.5 km^2), shallow (z_{max} = 12.7 m), man-made irrigation reservoir situated within the dry zone of Sri Lanka. Due to the seasonal cycle in rainfall and the management of water for irrigational purposes, it exhibits strong fluctuations in water level, volume and water exchange rates during the course of the year. Research concentrated on topics of particular relevance to fish production: availability of nutrients, primary productivity, the characteristics of the fish fauna and trophic interactions between phytoplankton, zooplankton, fish and fish-eating birds.

1. Limnological training and research in tropical countries

The demand for proper management of inland water resources in tropical countries is increasing as a result of acute population growth and gradual industrialization. The urgent need for drinking water, food and recreational areas on one hand, and the dangers of increasing eutrophication and pollution on the other, are problems well known in the temperate zone. Although not yet as serious as in many industrial, temperate countries, the latter problems may develop quickly and may – for various reasons – become more serious in the tropics.

In most tropical countries, there is a dearth of qualified limnologists and fishery biologists to tackle these problems and many organizations have repeatedly emphasized the urgent need for better training and education within this sphere in developing nations (see e.g. Mori & Ikusima 1981).

The successful application of limnological techniques and the value of environmental recommendations come only from a good background in basic limnological and ecological theory. It is facile to suppose that a general ecological 'know-how' of temperate lakes can solve the complex problems arising in tropical waters. Most of our limnological knowledge at present comes from temperate-zone research and is not directly applicable to tropical aquatic ecosystems. The structure and annual cycles of the tropical aquatic biota, the thermal conditions and nutrient situation are too different for any

Schiemer, F. (ed.), Limnology of Parakrama Samudra – Sri Lanka
 ISBN 90 6193 763 9

immediate transfer of knowledge obtained in the temperate zone. Many well-known failures of recommendations put forward in tropical resource management, which have been based upon insufficient knowledge of the functioning of tropical ecosystems, are clear warnings in this respect.

The scope is very wide for developing new technologies in aquaculture and for management of water bodies for their economic exploitation. The possibilities range from increasing fish production, establishing new fisheries and exploiting macrophyte crops for animal husbandry to the development of integrated land-water systems (such as the Chinese 'mulberry-fish pond system' or an intensified use of exposed littoral areas of shallow tanks for live-stock and water-fowl) (see Löffler, in press).

Various attempts have been made recently to promote training and research in tropical limnology. An annual post-graduate course in limnology for students from developing countries has been organized since 1975 in Austria with the cooperation of Unesco. A workshop to promote the study of limnology in developing countries was held in 1980 at the SIL Congress in Kyoto and further workshops and training courses are planned for the next few years to be held in the tropical regions. The ultimate goal of these efforts is to establish self-sustaining research and training centres in the tropics. Stimulation of and participation in such endeavours offer a challenging opportunity to ecologists from developed countries.

2. The Sri Lankan example

Sri Lanka is an island situated 6°–10° N of the equator, and covers an area of 66 000 km². The morphological structure is characterized by a flat coastal area and a central highland (Fig. 1a). The

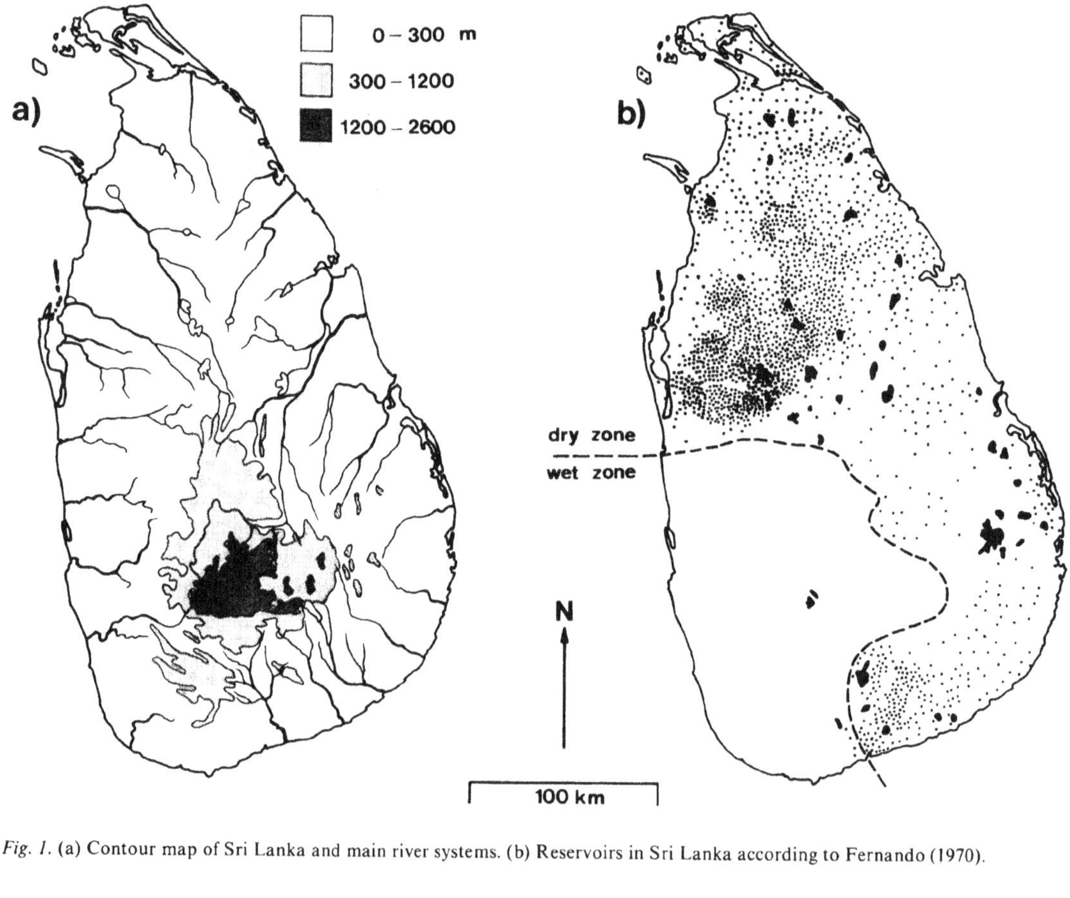

Fig. 1. (a) Contour map of Sri Lanka and main river systems. (b) Reservoirs in Sri Lanka according to Fernando (1970).

latter form a climatic barrier dividing the island into a 'wet zone' with continuously high precipitation, and a 'dry zone' with high rainfall only during the NE monsoon (October–January) (Dobesch 1983). About 100 major river systems drain the island radially to the sea (Fig. 1a).

Although Sri Lanka has no natural lakes, its dry zone possesses a sophisticated irrigation system of reservoirs and canals (Fig. 1b), built mainly for rice cultivation. The origins of this system reach back approximately 4 000 years (Brohier 1934). These reservoirs have been maintained and enlarged to a varying extent throughout the history of Sri Lanka. At present, there are approximately 3 500 reservoirs in existence, representing a surface area of 1 250 km^2.

A major enlargement of the irrigation system will be achieved by the Mahaweli River Diversion Scheme. This programme aims to improve and expand rice cultivation in the dry zone as well as increase hydro-electric power and control flooding. The scheme will make available an additional 200 km^2 of reservoir area (Wijeyaratne & Costa 1981) and change the hydrological regime of some of the major existing reservoirs, e.g. Parakrama Samudra. The environmental impact of the planned changes upon the aquatic biota will be profound. Due to the increase in the number of settlements and intensified land cultivation, a significant increase in nutrient and pesticide accumulation and in water-borne diseases (malaria) as well as a siltation of reservoirs by land erosion can be envisaged. The change in flow regimes could result in a reduction of existing fisheries output in some of the lakes. On the other hand, availability of aquatic habitats for fish production will be increased, as will be the local demand for fishery products (see TAMS report 1980).

Attention has been focussed recently on reservoirs as a protein resource. For various reasons, the development of freshwater fisheries in Sri Lanka is now receiving high priority:

1) inland fisheries have a high potential for development,
2) the development of inland fisheries is less costly than marine fisheries, and
3) inland fisheries are of great socio-economic importance within the dry zone, in regard to food supply and employment opportunities.

Landings of inland fisheries increased enormously following the introduction of *Sarotherodon* ('Tilapia') *mossambicus* in 1952. The reasons for the spectacular success of this species in some of the reservoirs (lack of lacustrine competitors among the indigenous fauna) and the low yield of indigenous species have been aptly discussed (C. H. Fernando & Indrasena 1969; C. H. Fernando 1971, 1973, 1977, 1982; De Silva & C. H. Fernando 1980). The varying success of *S. mossambicus* in different reservoir types has led to a subsequent search for other exotic species that would allow the yields from inland fisheries to be further increased. This search has not been successful (C. H. Fernando 1973) and inland fisheries yields have been stagnant for some years, despite a considerable investment of funds (Oglesby 1979). The main reason for this failure is the lack of knowledge about the biological structure and functioning of the reservoir ecosystems, their carrying capacity for different trophic forms, the effects of the hydrological regime imposed by irrigational management and the impact of the very large populations of fish-eating birds so characteristic of these water bodies.

In order to master the forthcoming problems of eutrophication and pollution and to allow a successful integration of the multiple uses of the reservoirs (irrigation, fisheries, recreation, washing facilities, pastures for live-stock etc.), a profound insight into how they are functioning is indispensable. The aim of the project was to gain knowledge of the structure and functioning of Parakrama Samudra, as an example reservoir, in order to provide a basis for effective management along these lines.

3. Objectives of the Parakrama Samudra project

The project is investigating ecosystem properties that are especially relevant to fisheries management. The Parakrama Samudra reservoir was selected as an example of a large shallow impoundment in the dry zone with high fish production (C. H. Fernando & Indrasena 1969; C. H. Fernando 1979; De Silva & C. H. Fernando 1980). By carrying out a detailed investigation of one characteristic reservoir type – including basic limnological parameters, productivity, nutrient flow and major trophic interactions – it was hoped to provide a model for further limnological studies of man-made bodies of water within the South-East Asian zone.

A comprehensive study of this nature was feasible only as a cooperative effort of Sri Lankan and foreign scientists. Several scientists from Sri Lankan universities (Univ. Colombo, Univ. Kelaniya, Ruhuna Univ. Campus) and other scientific institutions (Ceylon Institute for Scientific and Industrial Research, CISIR) cooperated in the project with limnologists from Austria, the Netherlands, England and West Germany.

The project was further designed to train postgraduate students of Sri Lankan Universities in limnological techniques in the course of the field research periods.

4. A general description of the lake

4.1. Geographical description and origin

Parakrama Samudra, the 'Sea of King Parakrama' as the name literally means, is one of the larger reservoirs of an ancient irrigation system. The reservoir is situated in the North Central province at the historical capital Polonnaruwa within the dry zone (7°55'N, 81°E, 58.5 m above sea level) (see Fig. 2). The natural catchment area of about 75 km^2 is located on the western side of the lake and is bordered by the Sudukanda ridge. To a large extent this watershed is covered by jungle. There are only a few settlements, mainly fishermen dwellings (about 100 families; see also C. H. Fernando & Indrasena 1969). The drainage of the watershed occurs by surface run-off via small rivulets that dry out during the 'dry season'.

The main water supply to the reservoir is through a channel from the river Amban Ganga. The inflow is regulated at the anicut at Angamedilla, situated about 8 km southwest of the lake. This inflow enters the southern basin (PSS) at its west shore. Amban Ganga, which has its origin near Kandy (Fig. 2), is a major tributary of Mahaweli, the largest river of Sri Lanka. The geology of the Amban Ganga catchment and that of the Mahaweli are described as the 'highland series' and consist of Pre-Cambrian crystalline rocks (charnockites and quartzites with some crystalline limestone) (Cooray 1967). Amban Ganga passes through agricultural areas.

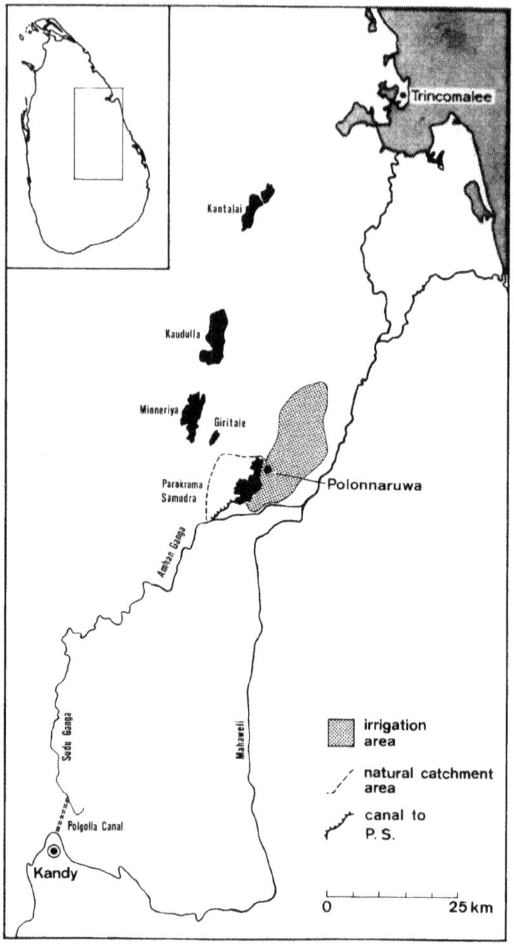

Fig. 2. Location of Parakrama Samudra with its sources of water supply, natural catchment area and the existing and planned irrigation area.

Within the Mahaweli Development Program – a long-term project to improve irrigation within the Mahaweli basin – water from the Mahaweli is diverted via the Polgolla tunnel (see Fig. 2) to Amban Ganga. Thus the actual catchment area comprises a broad spectrum of geographical regions.

Parakrama Samudra has a long history. The present reservoir is a result of connecting three original reservoirs (Fig. 3). The oldest one is the northernmost part, Topa Wewa (hereafter referred to as PSN), which was built around 386 AD (Brohier 1934). The middle part, Eramudu Wewa (PSM),

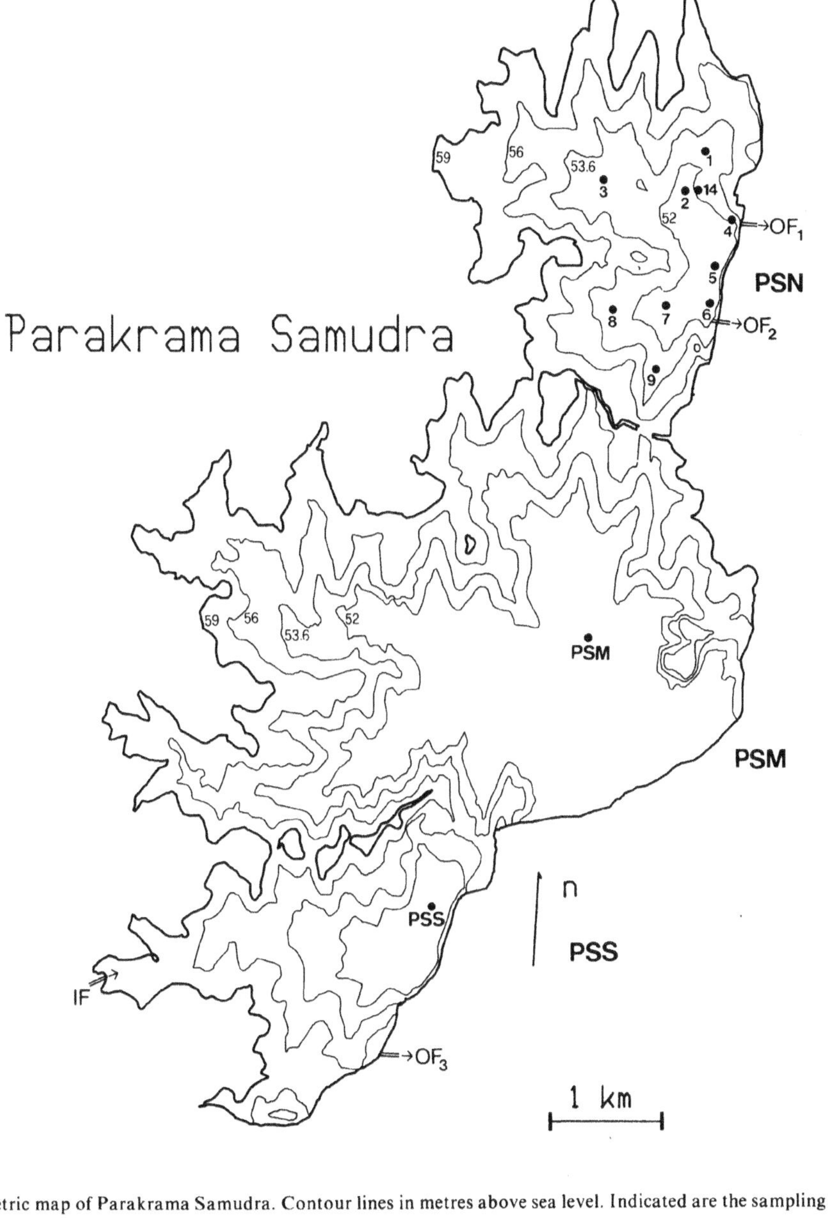

Fig. 3. Bathymetric map of Parakrama Samudra. Contour lines in metres above sea level. Indicated are the sampling stations referred to in different papers in this volume.

and the southern part, Dumbutulla Wewa (PSS), were constructed during the reign of King Parakrama Bahu (1 153–1 183 AD), in which period the ancient irrigation system attained its maximal development.[1] During this period, Polonnaruwa was a very densely populated area (see e.g. A. D. N. Fernando 1979) that was later abandoned due to invasions. The dam of PSM was damaged by a hurricane in 1848 and subsequently the lake bed was covered by jungle. In 1945 the dam was reconstructed and the lake was again filled. The area of PSM is still covered with dead trees (see Fig. 7).

4.2. Morphometry

The following description is based on a map provided by the Irrigation Department, Sri Lanka, and on a series of depth soundings in August/September 1979. At full supply level (194 feet, 59.13 m a.s.l.), the reservoir has a length of 10.7 km, a width of 5.6 km and a surface area of 25.5 km². The basin is shallow, with a maximal depth of 12.7 m. Mean depth at full supply level is 5.0 m. The water depth increases on the whole from west to east, i.e. towards the embankment (see Fig. 3). The shoreline is irregular except for the dam on the eastern shore. Consequently, the index for littoral development (according to Hutchinson 1957; see Table 1) is high.

Table 1. Morphometric parameters of the three basins of Parakrama Samudra referring to a water level of 59.13 m a.s.l. (194 f a.s.l.).[a]

	Area	z_m	$\bar{z}$	Volume	D_L
PSN	6.522	8.23	3.95	25.772	2.00
PSM	15.376	12.73	5.50	84.563	1.99
PSS	3.626	8.03	4.68	16.978	2.04

[a] Surface area in km²; maximum and mean depths (z_m, $\bar{z}$) in m; volume in 10^6 m³; littoral development (D_L) according to Hutchinson (1957).

The reservoir is divided by chains of islands into three basins. At low water level, these basins are connected only by narrow channels and form fairly distinct lakes. Even at a higher water level, when the lake forms a closed water surface, the three basins maintain a certain limnological identity. Morphological data and hypsographic curves of the three basins are compared in Table 1 and Figure 4, respectively. These data have been derived by planimetry of the contour lines shown in Figure 3. There is no complete agreement in the water-level–lake-volume relationship between the data obtained from the Irrigation Department and from the sum of volumes calculated for the three basins from planimetry. The latter are about 10×10^6 m³ lower, independent of the water level from 51.8 m a.s.l. onwards. This discrepancy can not be clarified on the basis of the available information. In the following description, the official values from the Irrigation Department of Sri Lanka are used for data on hydrology referring to the whole lake, while hydrological calculations for the individual basins refer to the hypsographic curves shown in Figure 4.

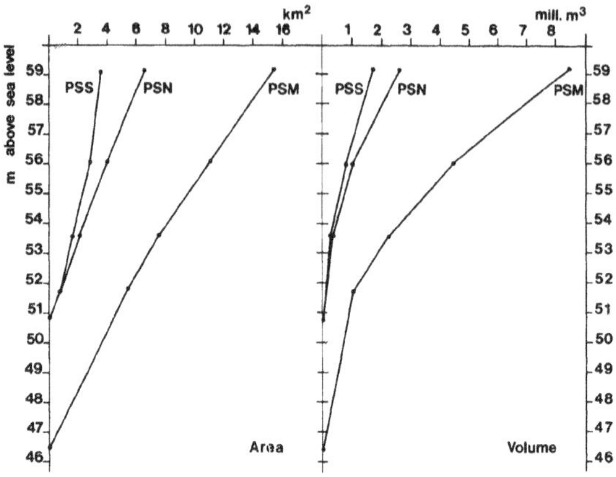

Fig. 4. Depth–area and depth–volume curves of the three basins. Depth in m a.s.l. (Volumes to be multiplied by a factor of ten).

PSM is by far the largest and deepest (z_{max} = 12.7 m) basin, while PSS and PSN are similar both in surface area and water depth (z_{max} = 8.0 and 8.2 m). The ratio in volumes of PSN–PSM–PSS at full supply level is 1:3.4:0.7. The relative differences in maximum and mean water depth and in water volume of PSM compared to the smaller basins increase with decreasing water level; e.g. at a water level of 52.5 m a.s.l. (the mean water level of July, see Table 2), the ratio in volumes between the three basins is 1:10.4:0.7.

Several fjord-like bays cause considerable horizontal heterogeneities in limnological characteristics even within the three separate lake basins. This

[1] It is said that Parakrama Bahu constructed or restored 165 dams, 3 910 canals, 163 major tanks and 2 376 minor tanks during his reign of 30 years.

was obvious during the period of low water level in August/September 1979, when PSN was subdivided by a large peninsula into two regions exhibiting distinct differences in the biomass of phytoplankton, zooplankton and zoobenthos (see Dokulil et al. 1983; Duncan & Gulati 1981; Schiemer 1983).

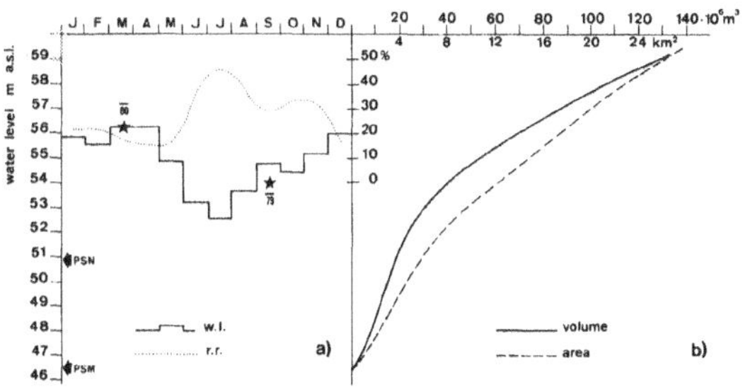

Fig. 5. (a) Seasonal water level fluctuations (full line = metres above sea level) and water renewal rates (r.r., pointed line = monthly inflow rates in % of the monthly mean lake volumes). Mean values of the period 1974–1977 according to Nedeco (1979). Asterisks mark the water level at the period of the two research phases, Aug/Sept 1979 and March/April 1980. Arrows indicate the maximum depth of PSN and PSM. (b) Depth-volume and depth-area curves of the total lake, according to data from the Irrigation Department of Sri Lanka.

4.3. Hydrography

The seasonal hydrological regime reflects the monsoon cycle and the irrigation demands for rice cultivation in the Polonnaruwa district. Inflow, outflow and storage are strictly managed for irrigation needs. In 1978, when a study was performed by Nedeco for the 'Mahaweli Development Board', there were 77 km^2 of irrigated area. An expansion to 114 km^2 is planned within the 'Accelerated Mahaweli Development Program'.

The water gains comprise precipitation, periodical inflow from the natural watershed and the supply from the Amban Ganga. The contribution of superficial and groundwater inflow from the catchment area to the total water gain of the lake is highest in the period of the NE monsoon, between October and December (Fig. 5), and comprises 16% of the total water gain during one year (based on water balance data for the years 1974–1977; Nedeco 1977); 84% of the water gain derives from the Amban Ganga. The water supply rates are kept at similar levels throughout the year, except for lower rates in December and May (Table 2).

The water release is minimal in March, August/September and December. These reflect the harvesting periods when paddy fields have to fall dry for seed maturation (March and August/September) and the periods when monsoonal precipitation is sufficient to flood the rice fields and an additional release of water from the reservoir is not necessary. Maximal discharge is from May to July when the fields are flooded for the second harvesting cycle.

Table 2. Annual cycle of water balance components: monthly means of the data published by Nedeco (1979) for 1974–1977.[a]

Month	Water gains			Water losses			STOR	AREA	ELEV	RR
	IF	CM	TOT	OF	EVAP	TOT				
Jan	13.00	1.75	14.75	13.50	2.00	15.50	67.50	16.8	55.80	21.9
Feb	13.00	1.00	14.00	17.75	1.75	19.50	62.50	16.0	55.48	22.4
Mar	12.25	1.25	13.50	0.50	2.25	2.75	73.25	18.0	56.20	18.4
Apr	9.50	2.25	11.75	11.50	0.50	12.00	73.25	17.8	56.23	16.0
May	7.75	1.25	9.00	28.50	2.00	30.50	51.75	14.2	54.83	17.4
Jun	12.25	0.25	12.50	27.25	2.50	29.75	35.00	11.1	53.58	35.7
Jul	10.00	0.75	10.75	20.25	1.75	22.00	23.75	8.5	52.53	45.3
Aug	13.00	0.75	13.75	0.25	1.25	1.50	36.00	11.1	53.63	38.2
Sep	13.00	1.25	14.25	0.00	1.25	1.25	49.25	13.8	54.65	28.9
Oct	13.00	2.25	15.25	19.50	-1.00	18.50	46.25	13.1	54.38	33.0
Nov	13.00	6.00	19.00	10.00	-3.50	6.50	58.75	15.2	55.13	32.3
Dec	5.75	6.50	12.25	2.25	-0.75	1.50	69.75	17.1	55.93	17.6

[a] IF, inflow from Amban Ganga; CM, gains from the natural catchment area; OF, outflow by sluices; E, evaporation; STOR, water capacity; AREA, lake area in km^2; ELEV, water level in m a.s.l.; RR, monthly renewal rate, IF/STOR. 100. Water volumes in 10^6 m^3.

The overall seasonal cycle of water-level fluctuations is outlined in Figure 5 and Table 2, based on data from the years 1974–1977. Water storage is highest from December to April and lowest from June to August. The average annual amplitude in water level changes was 52.2–56.2 m a.s.l., i.e. 3.7 m, representing a change in storage from 23.8 to 73.3×10^6 m³. As a consequence of the seasonality of inflow, outflow and water storage, the values for water renewal rates exhibit strong seasonal differences. In Table 2, renewal rates are expressed as inflow in percentage of the lake volume. Due to the low storage, the maximal renewal occurs in July. A second peak is at the end of November. This is the overall hydrological pattern. In November 1978 the dam of PSM was destroyed by a cyclone and the water level was drastically reduced. Repair work continued for approximately two years during which time the water level was kept at a lower level (see monthly values for the period January 1978 to June 1980 in Table 3).

In order to understand the biological processes in the different basins, the hydrology – especially water renewal and flushing rates – has to be considered separately. The lake forms a flow-through system from south to north (Fig. 6). The only permanent inflow, the canal from the Amban Ganga, enters PSS at its southwest end. The water for irrigation is released via three sluices. One is situated at PSS (OF_3). The remaining water flows through PSM (which has the highest water capacity, but no outflow at the dam) towards PSN, where one sluice is

Table 3. Hydrographic data for 1978–1980 from the records of the Irrigation Department of Sri Lanka. Monthly means for water level (m a.s.l.), lake volume and surface area. Water gains and water losses per month. Losses by evaporation are based on data from Dobesch (this volume).

		Elev m	Vol 10^6 m³	Area km²	Inflow 10^3 m³	Rain 10^3 m³	Σ Gains 10^3 m²	Outflow 10^3 m³	Evap 10^3 m³	Σ Losses 10^3 m³
1978	1.	58.7	122.4	24.9	29.1	1.7	30.8	26.8	2.0	28.8
	2.	58.9	128.5	25.8	33.8	0.7	34.5	33.2	2.9	36.1
	3.	59.1	132.7	26.0	15.2	2.3	17.5	6.2	3.0	9.2
	4.	59.0	131.9	25.8	8.4	0.3	8.7	14.4	2.9	17.3
	5.	58.4	116.2	24.2	25.5	0.4	25.9	37.3	4.4	41.7
	6.	57.8	103.2	22.7	26.7	0	26.7	28.3	4.8	33.1
	7.	57.2	92.2	20.8	22.0	0	22.0	29.9	4.9	34.8
	8.	56.7	78.3	18.9	21.2	0	21.2	27.8	4.6	32.4
	9.	56.5	77.6	18.5	16.1	0	16.1	4.6	3.9	8.5
	10.	57.3	93.1	20.8	21.2	7.9	29.1	12.2	2.6	14.8
	11.	57.9	105.3	22.7	22.5	5.9	28.5	7.6	1.8	9.4
	12.	56.7	79.9	18.9	0	10.2	10.2	29.5	1.3	30.8
1979	1.	56.6	72.2	18.7	14.7	2.2	16.9	30.8	1.5	32.3
	2.	56.4	68.6	18.2	17.7	0.7	18.4	24.1	2.1	26.2
	3.	56.0	67.3	17.2	8.3	1.0	9.3	11.2	2.0	13.2
	4.	55.8	63.8	16.7	15.8	1.5	17.3	20.7	1.9	22.6
	5.	54.9	49.8	14.3	13.3	1.4	14.7	29.8	2.6	32.4
	6.	54.2	40.9	12.4	18.0	0	18.0	21.8	2.6	24.4
	7.	53.6	34.7	10.9	12.0	0.1	12.1	23.0	2.6	25.6
	8.	53.5	33.3	10.7	12.8	0.3	13.1	23.5	2.6	26.1
	9.	53.9	37.2	11.6	1.5	1.9	3.4	4.8	2.4	7.2
	10.	54.7	47.8	13.7	25.6	2.6	28.2	15.6	1.7	17.3
	11.	55.0	50.8	14.5	15.4	4.5	19.9	31.3	1.2	32.5
	12.	55.5	57.9	15.9	20.1	5.2	25.3	21.7	1.1	22.8
1980	1.	56.6	78.3	18.7	37.8	0.1	37.9	31.8	1.5	33.3
	2.	56.7	80.2	18.9	11.9	0	11.9	29.8	2.1	31.9
	3.	56.2	71.5	17.8	8.7	0.6	9.3	11.7	2.0	13.7
	4.	56.4	74.6	18.2	14.7	5.2	19.9	7.4	2.1	9.5
	5.	56.8	82.0	19.1	19.2	0.7	19.9	30.7	3.5	34.2
	6.	55.9	65.7	16.9	9.4	0	9.4	24.6	3.6	28.2

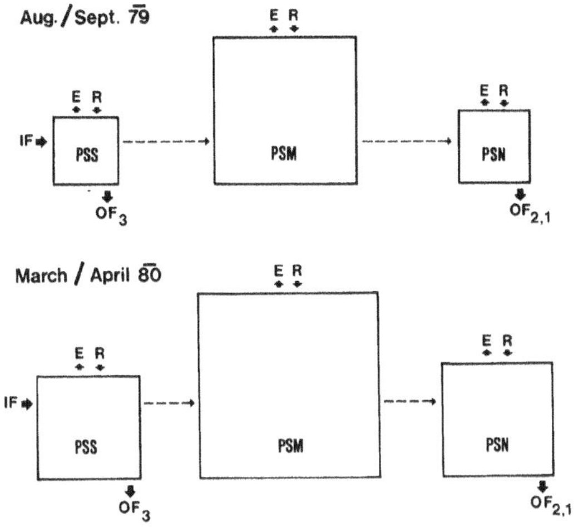

Fig. 6. Scheme of the hydrological regime and the through-flow system of the PS reservoir. The sides of the squares representing the three basins at the two research phases, Aug/Sept 1979 and March/April 1980, equal the cubic root of their volume. Rates of inflow (IF), outflow ($OF_{1,2,3}$) and rainfall are recorded by the Irrigation Department of Sri Lanka. Broken lines represent the water flow between the basins, which is not recorded but can be calculated from water budget parameters (see text).

situated near Polonnaruwa New Town (OF_2) and one near the Polonnaruwa Resthouse (OF_1). Additionally there is a spill at PSS for water release in the case of an emergency. The main release occurs at the northernmost sluice (OF_1) in PSN; this means that most of the water passes from the inflow in PSS to the north of PSN.

The water exchange in the three basins depends mainly on their water capacity, i.e. is much higher in PSS and PSN compared to PSM (roughly according to their capacity ratios).

The Irrigation Department of Sri Lanka keeps daily records of the water inflow Amban Ganga, the outflow at the three sluices, water level and precipitation. The flow rates from PSS into PSM and from PSM into PSN can be calculated as parameters of a budget of the individual basins, e.g.

$$OF_{PSM} (= IF_{PSN}) = \Delta\, Vol_{PSN} - R + OF_{1+2} + E$$

OF_{PSM} = outflow from PSM; $OF_{1,2}$ = see Figure 3; R = gain by precipitation; calculated as precipitation times lake area, assuming that, except for the monsoon period, rain-water from the catchment area does not enter the lake; E = losses by evaporation.

The hydrological data for the period of the two research visits are summarized in Table 4. The following comments refer to the northern basin of the reservoir (PSN), which was the main object of the study. Water level and flow-through rates differed strongly at the two visits (Fig. 6). In August and September 1979, the water level was particularly low (53.2–54.0 m a.s.l.; volume of PSN $2.2\text{–}3.8 \times 10^6\ m^3$). Water through-put rates were high shortly before and at the beginning of our visit (10–22 August 1979) with inflow and outflow rates of 40% of the PSN volume per day. Slightly lower gains than losses in this period led to a decrease in water level. From 24 August onwards until the end of our visit, the outflow rates were kept at a lower level (approximately 3% per day). Inflow rates ranged from 0 to 17%.

During the second research phase (25 February to 15 April 1980), the water level of the reservoir was higher (56.0–56.4 m a.s.l.; PSN volume $9.6\text{–}11.6 \times 10^6\ m^3$). The inflow rates throughout the visit ranged from 1%–4% per day ($\phi = 2.2\%$), except on 10 March, when the inflow was 13.2% d^{-1} of the PSN volume. Outflow rates were 5%–10% d^{-1} in the period between 25 February and 8 March (decreasing water level), 1%–2% d^{-1} until the end of March (± constant water level) and <1% d^{-1} in the first half of April (increasing water level).

5. Working plan

The topics studied during the two research periods were selected according to their presumable relevance for fishery management. The choice of topics was furthermore based on a scheme of system interrelationships that has been discussed in a previous paper (Schiemer 1981).

One part of the investigation was devoted to factors controlling the production of phytoplankton and its accessibility for economically important fish, especially *Sarotherodon mossambicus*. This is a complex question necessitating various lines of research, e.g. (a) availability and recycling of inorganic nutrients (Gunatilaka & Senaratne 1981; Gunatilaka 1983); (b) nutrient versus light limitation of primariry production, effects of grazing and hydrological flushing on the phytoplankton biomass, the taxonomic differentiation and size structure of primary productivity (Dokulil et al. 1983); (c) the

Table 4. Elements of daily water budget for the two research periods (10 August to 30 September 1979; 25 February to 10 April 1980), based on data from the Irrigation Department of Sri Lanka.[a]

Date	Water level	Area	Volume	IF	Rain	OF_1	OF_2	OF_3	Σ OF	IF/vol $\times$ 100
Aug 10	53.7	11.15	35.3	881		989	174	98	1261	2.50
11	53.7	11.15	34.9	930		976	171	0	1147	2.60
12	53.7	11.15	34.8	795		973	171	0	1144	2.28
13	53.7	11.15	34.7	220		969	169	0	1138	0.63
14	53.6	10.8	34.4	147		949	169	56	1174	0.43
15	53.6	10.8	33.9	73		949	166	56	1171	0.22
16	53.5	10.65	33.1	404		925	161	54	1140	1.22
17	53.4	10.4	32.6	355		903	159	54	1116	1.09
18	53.4	10.4	32.1	245		891	154	0	1045	0.76
19	53.3	10.15	31.7	232		874	120	0	994	0.73
20	53.3	10.15	31.2	245		874	120	0	994	0.78
21	53.2	9.8	30.7	232		844	115	49	1008	0.76
22	53.4	10.4	30.3	232		844	115	15	974	0.76
23	53.4	10.4	32.2	196	4	0	0	12	12	0.60
24	53.5	10.65	33.3	0	12	49	154	0	203	0
25	53.6	10.8	33.9	0		51	159	0	210	0
26	53.6	10.8	34.0	0		51	166	0	217	0
27	53.4	10.4	34.3	0	13	51	81	0	132	0
28	53.6	10.8	34.4	318		54	83	0	137	0.92
29	53.7	11.15	34.7	196		54	37	0	91	0.57
30	53.7	11.15	34.8	220		54	37	0	91	0.63
31	53.7	11.15	34.9	220		54	37	0	91	0.63
Sept 1	53.7	11.15	34.9	245	36	54	37	0	91	0.70
2	53.8	11.4	35.5	346	15	54	37	0	91	0.98
3	53.8	11.4	36.0	881	7	54	37	0	91	3.4
4	53.9	11.65	36.9	0		54	37	24	115	0
5	53.9	11.65	37.1	0		54	39	24	117	0
6	53.9	11.65	37.3	0		56	39	24	119	0
7	53.9	11.65	37.3	0		56	39	24	119	0
8	53.9	11.65	37.4	0		56	39	24	119	0
9	53.9	11.65	37.6	0	40	56	39	24	119	0
10	53.9	11.65	37.8	0		56	39	24	119	0
11	53.9	11.65	38.0	0		56	39	24	119	0
12	53.9	11.65	38.0	0	25	56	39	24	119	0
13	54	11.8	38.2	0	15	56	39	24	119	0
14	54	11.8	38.2	0		56	39	24	119	0
15	54	11.8	38.6	0		59	42	24	125	0
16	54	11.8	38.7	0	10	59	42	24	125	0
17	54	11.8	39.0	0		59	42	24	125	0
18	54	11.8	39.1	0		59	42	24	125	0
19	54	11.8	39.8	0		10	17	24	51	0
20	54	11.8	39.8	0		59	42	24	125	0
21	54	11.8	39.3	0		184	83	24	291	0
22	54	11.8	39.3	0		12	83	24	119	0
23	54	11.8	39.0	0		184	10	24	218	0
24	54	11.8	39.0	0		184	83	24	291	0
25	54	11.8	39.0	0		181	83	22	286	0
26	54	11.8	38.2	0		176	81	22	279	0
27	53.9	11.65	38.0	0		176	81	22	297	0
28	53.9	11.65	37.6	0		174	78	20	272	0
29	53.9	11.65	37.6	0		174	78	20	272	0
30	53.9	11.65	37.6	0		174	78	20	272	0

Table 4. (Continued).

Date	Water level	Area	Volume	IF	Rain	OF_1	OF_2	OF_3	Σ OF	IF/vol × 100
Feb 25	56.4	18.3	75.6	318		327	161	0	488	0.42
26	56.4	18.3	75.1	281		327	254	0	581	0.37
27	56.4	18.3	75.1	281		327	254	0	581	0.37
28	56.4	18.3	74.6	318		918	0	28	946	0.43
29	56.4	18.3	74.0	318		913	0	28	941	0.43
Mar 1	56.3	18.0	73.5	306		913	0	28	941	0.42
2	56.3	18.0	73.0	306		905	264	28	1 197	0.42
3	56.2	17.8	72.2	391		902	264	0	1 166	0.54
4	56.2	17.8	71.5	293		675	141	0	816	0.41
5	56.2	17.8	70.9	269		673	141	0	814	0.38
6	56.2	17.8	70.6	220		669	0	28	697	0.31
7	56.1	17.5	70.0	220		669	0	28	697	0.31
8	56.1	17.5	69.4	245	20	665	0	28	693	0.35
9	56.1	17.5	69.4	245		39	93	28	160	0.35
10	56.0	17.3	69.1	1 272		39	93	0	132	1.84
11	56.2	17.8	70.9	514		40	94	0	134	0.72
12	56.2	17.8	71.8	318		40	94	0	134	0.44
13	56.2	17.8	71.5	245		132	0	13	145	0.34
14	56.2	17.8	71.5	269		132	0	13	145	0.37
15	56.2	17.8	71.5	269		132	0	0	132	0.37
16	56.2	17.8	72.2	318		132	79	0	211	0.44
17	56.3	18.0	72.5	318		132	79	0	211	0.44
18	56.3	18.0	72.5	392		132	79	0	211	0.54
19	56.3	18.0	72.8	245		132	79	13	224	0.34
20	56.3	18.0	72.8	245		132	0	13	145	0.34
21	56.3	18.0	72.8	281		132	0	0	132	0.38
22	56.3	18.0	72.8	196		371	0	0	371	0.27
23	56.3	18.0	72.5	196		371	79	7	457	0.27
24	56.3	18.0	72.3	179		371	79	0	450	0.25
25	56.2	17.8	72.3	179		132	79	0	211	0.25
26	56.2	17.8	72.2	122		132	79	20	231	0.17
27	56.2	17.8	72.2	122		132	0	20	152	0.17
28	56.2	17.8	72.2	122		132	0	3	135	0.17
29	56.2	17.8	72.2	98		132	0	3	135	0.14
30	56.2	17.8	72.2	147		132	79	3	214	0.20
31	56.2	17.8	71.9	147		132	79	3	214	0.20
Apr 1	56.2	17.8	71.9	147	4	40	0	7	47	0.20
2	56.2	17.8	72.0	73	6	40	0	7	47	0.10
3	56.2	17.8	72.4	73	20	40	0	7	47	0.10
4	56.3	18.0	73.3	220	6	40	64	7	111	0.30
5	56.3	18.0	73.6	391		40	64	7	111	0.53
6	56.3	18.0	74.0	257	55	40	64	0	104	0.35
7	56.4	18.3	74.0	501		40	64	0	104	0.68
8	56.4	18.3	74.0	416		40	64	0	104	0.56
9	56.4	18.3	74.6	1444		40	64	0	104	1.94
10	56.4	18.3	75.6	612		188	64	0	252	0.81

[a] Water level in metres; lake volume in 10^6 m^3; surface area in km^2; inflow and outflow at the three sluices (OF_1, OF_2, OF_3) in 10^3 m^3; rain in mm; daily renewal rate (IF/vol. 100). Evaporation is 7.03 mm day^{-1} in August/September and 3.8 mm day^{-1} in February/April.

biology of feeding and digestive physiology of algivorous fish (Hofer & Schiemer 1981, 1983; Dokulil 1983); and (d) food consumption and competition by herbivorous zooplankton (Duncan & Gulati 1983a).

A second area of research concentrated on the distribution pattern and feeding ecology of fishes (including economically unexploited mass fishes) with the aim of discovering economically exploitable food niches and predation effects of fishes on zoobenthos (Schiemer & Hofer 1983) and zooplankton (Duncan 1983). The presence of large populations of fish-eating birds, especially cormorants, made it necessary to clarify their impact on the ecosystem (a) with regard to predation effects on fish and (b) with regard to nutrient recycling (Winkler 1983).

A lack of limnological information on the PS reservoir made it necessary to include in our programme an assessment of basic limnological parameters, e.g. on thermics (Bauer 1983), light (Dokulil et al. 1983) and chemistry (Gunatilaka & Senaratne 1981). It soon became clear that several ecosystem processes are strongly influenced by the hydrological regime imposed on the lake by irrigation management (see above). This situation necessitated a monitoring of limnological parameters (especially those pertaining to the limnetic zone) throughout our two visits (see especially Duncan & Gulati 1981; Duncan 1983; Dokulil et al. 1983). The diel pattern of several biological processes was analysed in respect to nutrient recycling (Gunatilaka 1983), primary production (Dokulil et al. 1983), zooplankton dynamics (Duncan & Gulati 1983b), feeding and migrations of fish (Schiemer & Hofer 1983; Hofer & Schiemer 1983) and birds (Winkler 1983).

6. Operation

The planning phase extended over approximately two years. A preliminary discussion on the feasibility of a joint project took place between the author and Prof. H. H. Costa and Dr. S. De Silva at the University of Kelaniya in January 1977. Subsequently, grants were requested from the Austrian Chancellor's office, Unesco and the National Science Foundation in Sri Lanka.

In spring 1979, Mr. A. Gunatilaka, a Sri Lankan post-graduate working at the University of Vienna, left for Sri Lanka to make arrangements for the first research visit, due to begin six months later. Field research was carried out during two visits of the European group, the first from 8 August to 5 October 1979 and the second from 25 February to 15 April 1980. The research parties included scientists from Austria (6), England (1), West Germany (1) and the Netherlands (1, only in 1979) and scientists and post-graduates from Sri Lanka (7).

A field laboratory was established in the spacious rooms of the Polonnaruwa Resthouse, situated on the dam of PSN. In addition, we were allowed to use the laboratory facilities of the Fisheries Station Polonnaruwa, which also provided a large and very convenient motor-boat.

The subsequent chapters refer to common sampling stations in the lake, which are indicated in Figure 3. The limnological monitoring programme included weekly recording of light, chemistry, chlorophyll and zooplankton at stations IF, PSS, PSM, PSN 3, PSN 8 and OF 1. Studies on diurnal patterns of temperature stratification, chemistry, primary production, and zooplankton dynamics were for convenience carried out at PSN 14, a station that is in easy reach of the Polonnaruwa Resthouse.

Acknowledgements

The project was sponsored by Unesco and the Austrian Federal Chancellor's office. Additional financial contributions were made to participants by the Municipality of Vienna, the Dutch Organization of Tropical Research, the British Council and the German Science Foundation. Prompt assistance was rendered to us by various institutions in Sri Lanka: we are especially obliged to the National Science Council, the Ministry of Fisheries, the Irrigation Department, the Government Agent of Polonnaruwa and the local officers of UNDP and FAO.

We thank Professor Abeywickrama, Professor Arudpragasam, Professor De Fonseka (University of Colombo), Professor Costa (University Kelaniya), Professor De Silva (Ruhuna University College), Mr. Jeyaraj (Deputy Director of CISIR) and Dr. Dharmawardene (Director, Radio-Isotope Centre, Colombo University) for their kind support.

Fig. 7. PSM with dead trees, Aug/Sept 1979 (phot. P. Newrkla).

Fig. 8. PSN at low water level (Aug/Sept 1979) with dry fallen littoral zones used as pastures. Sudukanda ridge in the background (phot P. Newrkla).

Fig. 9. Water buffalos in the littoral zone of PSM (phot. P. Newrkla).

Fig. 10. Washing laundry in the littoral zone of PSN.

Fig. 11. Cast net fisherman in PSN (phot. P. Newrkla).

Fig. 12. Gill net fishermen selling their catch on the shore of PSM (phot. P. Newrkla).

In Polonnaruwa we received continuous help from the officers at the Freshwater Fisheries Station and we gratefully acknowledge the hospitality of our friends Mr. Setunga, Mr. Sumanaweera and Mr. Samarakoon (Condensed Milk Factory, Polonnaruwa).

The project was administrated by the Institute for International Cooperation, Vienna, to whom we are greatly obliged for their efficient cooperation.

I am grateful to Dr. P. Newrkla, who helped with the figures and Mrs. Joy Wieser for correcting the English manuscripts. Finally I would like to express our gratitude to Professor H. Löffler and Professor C. H. Fernando for their continuous support in this venture.

References

Bauer, K., 1983. Thermal stratification, mixis and advective currents in Parakrama Samudra Reservoir, Sri Lanka. In: Schiemer, F. (ed.) Limnology of Parakrama Samudra - Sri Lanka: a case study of an ancient man-made lake in the tropics. Developments in Hydrobiology (this volume). Dr W. Junk, The Hague.

Brohier, R. L., 1934. Ancient Irrigation Works in Ceylon. Ceylon Government Press, Colombo, Ceylon. 79 pp.

Cooray, P. G., 1967. An introduction to the geology of Ceylon. Spolia Zeylan 31: 1-324.

De Silva, S. S. & Fernando, C. H., 1980. Recent trends in the fishery of Parakrama Samudra, an ancient man-made lake in Sri Lanka. In: Furtado, J. (ed.) Tropical Ecology and Development, pp. 927-937. Society for Tropical Ecology, Kuala Lumpur, Malaysia.

Dobesch, H., 1983. Energy and water budget of a tropical man-made lake. In: Schiemer, F. (ed.) Limnology of Parakrama Samudra - Sri Lanka: a case study of an ancient man-made lake in the tropics. Developments in Hydrobiology (this volume). Dr W. Junk, The Hague.

Dokulil, M., 1983. Aspects of gut passage of algal cells in *Sarotherodon mossambicus* (Pisces, Cichlidae). In: Schiemer, F. (ed.) Limnology of Parakrama Samudra - Sri Lanka: a case study of an ancient man-made lake in the tropics. Developments in Hydrobiology (this volume). Dr W. Junk, The Hague.

Dokulil, M., Bauer, K. & Silva, I., 1983. An assessment of the phytoplankton biomass and primary productivity of Parakrama Samudra, a shallow man-made lake in Sri Lanka. In: Schiemer, F. (ed.) Limnology of Parakrama Samudra - Sri Lanka: a case study of an ancient man-made lake in the tropics. Developments in Hydrobiology (this volume). Dr W. Junk, The Hague.

Duncan, A., 1983. The composition, density and distribution of the zooplankton in Parakrama Samudra. In: Schiemer, F. (ed.) Limnology of Parakrama Samudra - Sri Lanka: a case study of an ancient man-made lake in the tropics. Developments in Hydrobiology (this volume). Dr W. Junk, The Hague.

Duncan, A. & Gulati, R. D., 1981. Parakrama Samudra (Sri Lanka) Project, a study of a tropical lake ecosystem. III. Composition, density and distribution of the zooplankton in 1979. Verh. Internat. Verein. Limnol. 21: 1001-1008.

Duncan, A. & Gulati, R. D., 1983a. A diurnal study of the planktonic rotifer populations in Parakrama Samudra, Sri Lanka. In: Schiemer, F. (ed.) Limnology of Parakrama Samudra - Sri Lanka: a case study of an ancient man-made lake in the tropics. Developments in Hydrobiology (this volume). Dr W. Junk, The Hague.

Duncan, A. & Gulati, R. D., 1983b. Feeding studies with natural food particles on tropical species of planktonic rotifers. In: Schiemer, F. (ed.) Limnology of Parakrama Samudra - Sri Lanka: a case study of an ancient man-made lake in the tropics. Developments in Hydrobiology (this volume). Dr W. Junk, The Hague.

Fernando, A. D. N., 1979. Major ancient irrigation works of Sri Lanka. Proc. Royal Asiatic Soc. (Sri Lanka Branch) 1979: 1-24.

Fernando, C. H., 1971. The role of introduced fish in fish production in Ceylon's freshwaters. In: Duffey, D. & Watt, A. S. (eds.) The Scientific Management of Animal and Plant Communities for Conservation, pp. 285-310. Blackwell, Oxford.

Fernando, C. H., 1973. Man-made lakes of Ceylon: a biological resource. In: Ackerman, W. C., White, G. F. & Worthington, E. B. (eds.) Man-made lakes: their problems and environmental effects. American Geophysical Union Monograph Washington 17: 664-671.

Fernando, C. H., 1977. Reservoir fisheries in South East Asia: past, present and future. In: Proceedings of the Indo-Pacific Fisheries Council Symposium on the Development and Utilization of Inland Fishery Resources, Colombo, Sri Lanka, October 1976, IPFC/76/SYM/49, pp. 27-29.

Fernando, C. H., 1982 (in press). Lake and reservoir ecosystems in South East Asia (oriental region). In: Taub, F. B. (ed.) Lake and reservoir ecosystems. Elsevier, Amsterdam.

Fernando, C. H. & Indrasena, H. H. A., 1969. The freshwater fisheries of Ceylon. Bull. Fish. Res. Stat. Ceylon 20: 101-134.

Gunatilaka, A., 1983. Phosphorus and phosphatase dynamics in a tropical man-made lake based on diurnal observations. In: Schiemer, F. (ed.) Limnology of Parakrama Samudra - Sri Lanka: a case study of an ancient man-made lake in the tropics. Developments in Hydrobiology (this volume). Dr W. Junk, The Hague.

Gunatilaka, A. & Senaratna, C., 1981. Parakrama Samudra (Sri Lanka) Project, a study of a tropical lake ecosystem. II. Chemical environment with special reference to nutrients. Verh. Internat. Verein. Limnol. 21: 994-1000.

Hofer, R. & Schiemer, F., 1981. Proteolytic activity in the digestive tract of several species of fish with different feeding habits. Oecologia (Berl.) 48: 342-345.

Hofer, R. & Schiemer, F., 1983. Feeding ecology, assimilation efficiencies and energetics of two herbivorous fish: *Sarotherodon* (Tilapia) *mossambicus* (Peters) and *Puntius filamentosus* (Cuv. et Val.). In: Schiemer, F. (ed.) Limnology of Parakrama - Sri Lanka: a case study of an ancient man-made lake in the tropics. Developments in Hydrobiology (this volume). Dr W. Junk, The Hague.

Hutchinson, G. E., 1957. Treatise on Limnology, vol. 1. John Wiley, New York. 1115 pp.

Löffler, H., (in press). Limnological aspects of shallow lakes. Scope Symp. Wetland Mod. 1981 Minsk.

Mori, S. & Ikusima, J. (eds.), 1981. Proceedings of the first workshop on the promotion of limnology in the developing countries. Organizing Committee, 21st SIL Congress. 72 p.

Nedeco, 1979. Mahaweli Ganga Development Program Implementation Strategy Study: main report, Annex A, Annex B. Ministry of Mahaweli Development, Colombo.

Oglesby, R. T., 1979. Development of fisheries in the man-made lakes and reservoirs (FAO report, unpubl.).

Schiemer, F., 1981. Parakrama Samudra (Sri Lanka) Project, a study of a tropical lake ecosystem. I. An interim review. Verh. Internat. Verein. Limnol. 21: 987–993.

Schiemer, F. & Duncan, A., 1983. Parakrama Samudra project: a summary of main results. In: Schiemer, F. (ed.) Limonology of Parakrama Samudra – Sri Lanka: a case study of an ancient man-made lake of the tropics. Developments in Hydrobiology (this volume). Dr W. Junk, The Hague.

Schiemer, F. & Hofer, R., 1983. A contribution to the ecology of the fish fauna of the Parakrama Samudra reservoir. In: Schiemer, F. (ed.) Limnology of Parakrama Samudra – Sri Lanka: a case study of an ancient man-made lake in the tropics. Developments in Hydrobiology (this volume). Dr W. Junk, The Hague.

TAMS mid-term report, 1980. Environmental assessment; Accelerated Mahaweli Development Program. Tippets, Abbet, McCarthy, Stratton.

Wijeyaratne, M. J. S. & Costa, H. H., 1981. Stocking estimations of *Tilapia mossambicus* fingerlings for some inland reservoirs of Sri Lanka. Int. Revue ges. Hydrobiol. 66: 327–333.

Winkler, H., 1983. The ecology of cormorants (genus *Phalacrocorax*). In: Schiemer, F. (ed.) Limnology of Parakrama Samudra – Sri Lanka: a case study of an ancient man-made lake in the tropics. Developments in Hydrobiology (this volume). Dr W. Junk, The Hague.

Author's address:
E. Schiemer
Institute of Zoology
University of Vienna
Althanstr. 14
A-1090 Vienna
Austria

2. Energy and water budget of a tropical man-made lake

H. Dobesch

Keywords: energy budget, water budget, tropical lake

Abstract

A simple model is applied for the calculation of the water budget of a tropical man-made shallow lake. The lacking climatological data are deduced from the general climatic conditions of Sri Lanka. This one-layer model assumes that the lake is well mixed, so that the concept of equilibrium temperature with time constant and simple heat transfer can be applied.

1. Introduction

A review of the Parakrama Samudra Project, Sri Lanka, is given in Schiemer (1981, 1983). Beside the interrelationship and dynamic properties of the reservoir's ecosystem, the working plan includes the evaluation of the water budget of Parakrama Samudra Reservoir (PS), which is the main topic of this chapter. PS is a large, shallow impoundment situated at 7°55′N, 81°E, at an elevation of 55–60 m above sea level in the 'dry zone' of Sri Lanka. The lake consists of three basins with a total surface area of 25.5 km^2 at full supply level and a catchment area of 75 km^2. At low water levels during summer, the lake is divided into three basins connected only by narrow channels. At present, PS is used for the irrigation of approximately 70 km^2 of rice fields and is supplied by the artificial channel of Ambanganga, which is regulated by an anicut at Angamedilla, some 8 km from the reservoir. In the new Mahaveli Development scheme, Ambanganga receives a definite quota of water from the Mahaveli (Nedeco 1979). Data available for the Polonnaruwa area are air temperature values from 1966 to 1976, rainfall data for the standard period 1931–1960, and the data needed for the water management of the dam and irrigation system. In order to obtain significant data for the evaluation of the lake energy budget, especially evaporation, the general climate of Sri Lanka has to be considered.

2. Short summary of the general climatic conditions in Sri Lanka

2.1. Monsoon regime and precipitation

The dominant characteristic of the climate of Sri Lanka is the monsoon, which occurs twice a year as the southeast monsoon and the northeast monsoon, according to the predominant wind direction. Together with the 'intermonsoon seasons', the year can thus be divided into four seasons: March–mid-May (the first intermonsoon season), mid-May–September (SW monsoon season), October–November (second intermonsoon season) and December–February (NE monsoon season). According to this classification, the monsoon seasons occupy 7½ months, i.e. the greater part of the year, and the intermonsoon seasons only 4½ months. Because of the entirely different structures of the SW and NE monsoons and the effect of the central highlands, there are considerable spatial and seasonal differences in rainfall distribution. A

Schiemer, F. (ed.), Limnology of Parakrama Samudra – Sri Lanka
 ISBN 90 6193 763 9

pattern of humid and arid months in Ceylon with the 'aridity index' (De Martonne 1926; Lauer 1952) can be derived, and is well presented by Domrös (1974). In this pattern, the high regional variation of the number of humid or arid months becomes obvious, reaching from a maximum of 12 humid months in the southwest quarter of Sri Lanka to a minimum of four humid months is the Jaffna Peninsula. The northwestern parts of the islands are the most arid zones. The island can therefore be divided into a 'wet zone' and a 'dry zone'. The 'driest' station with any precipitation measurements is Maha Lewaya Saltern, near Hambantota on the southeast coast, with an average annual rainfall of 929 mm during the 30-year period of observation from 1931–1960. On the western and southwestern slopes of the highlands, precipitation in the wet zone reaches more than 5 000 mm in the Crinigathena region.

Over the major part of the island, the wettest and driest months differ by 250–400 mm rainfall with significant deviation only in the driest and in the wettest areas: In the dry southeast and northwest, the absolute annual rainfall range amounts to only about 150–200 mm (Hambantota: November 187 mm, August 42 mm). The largest absolute annual rainfall occurs on the slopes of the highlands, i.e. Hendon 862 mm: December 938 mm, June 76 mm during the NE monsoon season – and Watawala 802 mm: June 911 mm, February 109 mm during the SW monsoon season.

2.2. Air temperature

The temperature conditions of Sri Lanka can be summarized by the term 'thermic diurnal climate', which means that the temperatures are characterized predominantly by differences in their diurnal course. The changes of temperature between day and night are greater than the seasonal differences, making a division of the year into temperature seasons scarcely possible.

In the lowlands up to an altitude of 150 m, the monthly mean temperature varies between 24.5 °C and 28.5 °C. The annual temperature range, which is represented by the difference between the average temperature of the warmest and coldest months, amounts to a maximum of 4.3 °C (Trincomalee, east coast) and a minimum of 1.5 °C (Radnapura and Talawakele in the southwestern quarter and in the highlands). The warmest months are April–May and the coldest those of the NE monsoon period (December–February); both, however, are poorly defined. Therefore a well-balanced annual course of temperature can be considered typical in Sri Lanka.

Consideration of the diurnal temperature ranges reveals considerable spatial differences. The coastal stations have the smallest amplitudes (Galle 4.7 °C and Jaffna 4.8 °C) whilst the stations situated further towards the interior (Badulla 9.8 °C and Talawakele 8.9 °C) have the highest. The largest diurnal temperature ranges occur, for all except the eastern parts of Sri Lanka, in February/March, for the east during the SW monsoon season (i.e. Batticaloa 8.2 °C in July).

The temperature observations indicate a relationship between the diurnal temperature range and the monsoon regime, so that the areas affected by the SW monsoon have their smallest diurnal ranges during the SW monsoon, while those influenced by the NE monsoon experience theirs during this season. The highest diurnal ranges occur during the intermonsoon seasons.

2.3. Atmospheric humidity

Like the monthly mean temperature, the monthly mean values of relative humidity vary but little over Sri Lanka. The spatial differences in atmospheric humidity are greatest in the SW monsoon period, and lowest in the NE monsoon and intermonsoon seasons. Thus the greatest spatial differences occur in June and July, whereas the lowest fluctuations and highest air humidity occur in October and November. The smallest annual range exists on the south and southwest coasts of Sri Lanka, where there are no distinct seasonal differences (about 3%) in humidity. The highest annual range of humidity (about 15%) is found on the east coast and the eastern slopes of the highlands, with a significant annual pattern in which the SW monsoon from May/June to September is characterized by low humidity values (Trincomalee 70%). In the NE monsoon season, humidity climbs to 85%–88%. For the southwest, the values are between 80% and 83% during February–April and 86% during October–December.

2.4. Surface wind

The surface-wind conditions of Sri Lanka are very complex. Three components have to be considered: the monsoon winds (all winds that are linked with the monsoon system), the land and sea breezes, which are well established (the sea breeze can reach across the whole island) and the mountain and valley winds in the highlands. The SW monsoon determines the wind conditions over Sri Lanka for 4½ months of the year and that of the NE monsoon for three months. The annual course of wind direction and velocity is strongly influenced by the seasonal monsoon structure, with highest wind speed during the SW monsoon months. The NE monsoon winds blow only a little stronger than the intermonsoon winds; the weakest winds occur in the first intermonsoon season. These observations fit into the general pattern of monsoon circulation; generally the SW monsoon blows with a considerable higher wind speed than the NE monsoon. Prevailing wind directions are SW and W from May to October, and NE and E from December to February. Wind directions are changeable in the intermonsoon seasons.

2.5. Sunshine duration

Over Sri Lanka, sunshine duration varies mainly between 7.9 and 3.8 h on an average per day. Great differences exist between the highlands and the lowlands. In the lowlands, not only is the annual sunshine duration greater, but the annual course is also more balanced and the extreme values between the sunniest and least sunny months are less significant. For almost all parts of the island, February or March is the sunniest month of the year. The occurrence of the least sunny month differs regionally. On the western and southwestern slopes of the highlands, and in the lowlands of the southwest sector, the lowest sunshine duration normally occurs during the SW monsoon months in June and July, but in December on the eastern slopes of the highlands and in the eastern and northern lowlands.

There is a drop in the sunshine duration during the rainy monsoon season, whereas the intermonsoon season has a longer duration of sunshine. Two types of sunshine that differ fundamentally in their duration can be distinguished according to the two major climatic regions (wet and dry zones) of Sri Lanka in both the monsoon seasons. The wet-zone type has a distinct minimum of sunshine in the SW monsoon season and a comparatively long duration in the NE monsoon season although the absolute sunshine maximum is in February/March. The dry-zone type is characterized by a NE monsoonal sunshine minimum and a SW monsoonal maximum; the absolute maximum is partly in June and partly in February/March. Sunshine conditions during the two intermonsoon seasons differ fundamentally: the maximum sunshine duration occurs in the first intermonsoon season, the minimum mostly in the second intermonsoon season.

3. The climatic regime in the Polonnaruwa region

The foregoing information can be drawn upon in order to obtain climatological data for the Polonnaruwa area, for which no long-term climatological observations of humidity, wind and sunshine duration are available. The catchment area of Parakrama Samudra has nine humid months a year. The annual mean rainfall is (according to Table 1) 1 850 mm with the maximum in December (430 mm) and the minimum in June (16 mm). The area should have an annual temperature range about 4 °C, with the warmest months in May/June (maximum of monthly mean temperature of about 29 °C). The maximum monthly mean relative humidity can be set at 85% in December and (with an annual range of about 15%) the minimum at 70% in August/September, which is fairly near the data for Anuradhapura. The wind speed was assumed to be nearly the same as in Anuradhapura (with slightly higher values in July–September), as well as the ratio of actual maximum possible sunshine duration (with slightly lower values in December–February). The extra-terrestrial radiation has been evaluated from Dogniaux (1976). The short-wave radiation budget was calculated according to Glover & McCulloch (1958a, b) with an albedo of 6%.

4. The model used to estimate the energy budget of Parakrama Samudra

The model applied here has been derived with success for shallow lakes in temperate climatic zones for the summer season (Dobesch 1980b), where the whole water body is assumed to be well mixed across the depth $\bar{d}$. The energy-balance equation in this model is expressed as

$$dU/dt = S + E + H + Q + B + P \quad (1)$$

where S denotes the net radiation, E the flux of latent heat, H the flux of sensible heat, Q the 'hydrological term' ie.e. heat advection due to in- or outflow, B the heat flux into the ground, P the energy change due to precipitation and dU/dT the change of energy stored in the water body. The different terms in equation (1) are given by:

$$S = Gl(1-\alpha) + L\downarrow - L\uparrow \quad (2a)$$

where Gl is the global radiation, α the reflectance of the underlying surface, $L\downarrow$ the incoming atmospheric counter-radiation, and $L\uparrow$ the outgoing long-wave radiation;

$$W \equiv dU/dt = \bar{d}\rho c \partial T_w/\partial t \quad (2b)$$

where ρ is the water density, c the specific heat of water, T_w the water temperature and t the time;

$$E = -\frac{1}{\gamma} f(u)(e_s - e) \quad (2c)$$

where γ is the parameter constant, f(u) an empirical function of wind speed, e_s the saturation vapour pressure at the surface temperature T_s and e the vapour pressure of the air;

$$H = -f(u)(T_s - T) \quad (2d)$$

where T is the air temperature.

The evaluation of the surface temperature of the lake has been made using the 'time constant' of the water body and the heat transfer across the water surface according to

$$T_{s,i} = T_i + (T_{s,i-1} - T_i)\exp(-\Delta t/\tau) \quad (3)$$

with

$$\tau = (c\rho/\alpha')(V/F) \quad (4)$$

and T_i the air temperature and time t_i, $T_{s,i}$ and $T_{s,i-1}$ the water surface temperature at time t_i and t_{i-1}, Δt the difference $t_i - t_{i-1}$, F the area, V the volume of the water body ($V/F \equiv \bar{d}$) and α' a heat transfer coefficient.

With the concept of equilibrium temperature (= the water temperature at which no heat transfer across the surface exists), which means that a certain temperature (T_e) the short- and long-wave radiation heat inputs are exactly balanced by the heat losses from evaporation, convection and back-radiation, T_e can be calculated by setting W = O and $T_e = T_s$ in equation (1) and solving iteratively for T_e.

A linearized form of equation (1) is

$$W = -K(T_e - T_s) \quad (5)$$

where the surface heat exchange coefficient K (which is $\approx \alpha'$ as it can be shown) can be calculated with

$$K = 4\epsilon\sigma T^3 + f(u)(1 + s/\gamma) \quad (6)$$

In equation (6), ϵ is the emissivity of long-wave radiation, σ the Boltzmann constant and $s = de_s/dT_s$ the slope of saturation water vapour (at the temperature T_s) versus T_s. It can be shown that there is a connection between τ and κ in the form $\tau/\bar{d} = \rho c/K$. With the relationship (2a) to (2d), (3), (6), a parametrization of $L\downarrow$ in the form $L\downarrow = f(T,e)\sigma T^4$ (Dobesch 1980a) and with neglect of the last two terms on the right side of equation (1), the balance equation (1) can be written as (with $K_w \equiv Gl(1-\alpha)$)

$$K_w + \epsilon\sigma T^4\{[f(T,e) - 1] - 4(T_s/T - 1)\} -$$
$$(1 + s/\gamma)f(u)(T_s - T) + K(T_s - T_e) + Q = O \quad (7)$$

The unknown temperature T_s can be calculated by solving equation (7) iteratively.

5. Results

5.1. The annual course of the components of the heat balance equation

The results achieved by the model (7) from monthly mean climatological data of Table 1 are shown in Table 2. Besides the surface temperature (in °C), sensible heat H and net radiations (in W m^{-2}), the amount of evaporated water E (in mm $month^{-1}$) from the model and from other authors is given. The water surface temperature of Parakrama Samudra is only slightly higher than the air temperature, due to the mixing effect of the relatively high wind speeds during June–September and the high evaporation rates; during the rest of the year the water temperature is 1°–1.5° higher, with a maxi-

Table 1. Data for the evaluation of surface temperature and evaporation from Parakrama Samudra (albedo = 6%).

Element	Month 1	2	3	4	5	6	7	8	9	10	11	12	Year
Rainfall	218	141	99	140	63	16	57	64	86	211	326	430	1 849 mm
Air temperature	24.9	26.2	28.1	28.7	28.6	28.8	28.7	28.7	28.4	27.4	26.6	25.0	27.5 °C
Relative humidity	82	76	73	78	77	73	71	70	70	78	84	85	76.4%
Wind speed	6.0	6.3	5.6	5.2	10.3	13.5	13.1	12.9	11.8	7.7	4.5	5.6	8.5 km h $^{-1}$
Short-wave radiation budget	204	209	254	248	242	230	238	238	237	219	201	190	226 W m $^{-2}$
Mean depth of Parakrama	4.9	5.8	7.1	7.0	5.2	3.5	2.5	3.8	5.0	4.4	4.9	6.1	5.0 m

mum in April. Though the data are mostly not for the Polonnaruwa area, a considerable similarity exists between E, E_2 and E_4. However, E_1 – the Penman estimate of evaporation for Anuradhapura – seems high, especially in the winter months when air humidity is high (and so the saturation deficit is low) and wind speed low. On the other hand, these evaporation values seem to be possible if one assumes that this is the evapotranspiration from a wet vegetation surface. The values for potential evapotranspiration rate E_3 calculated after Papadakis (1961) are too low, mostly because wind dependency is not taken into account.

A water balance equation has been used in the form (Gray 1970)

$$I + P + O + \Delta S + E = O_g - I_s = A \quad (8)$$

where I is the inflow, P is the precipitation, O the outflow, ΔS the change in storage, O_g the seepage and I_s the surface inflow from the banks of the lake. The data for the left side of equation (8) are given in Nedeco (1979) and by the model (7), so equation (8) can easily be established as is shown in Table 3 for 1974–1977 (in mm water column; to get the real amount of water for the tank operation, the values in Table 3 have to be multiplied with the area of the reservoir). Because no values are available for the seepage and the surface inflow, these two properties are summarized in A of (8). Further on, Table 3 shows the values of the water balance for the exist-

Table 2. The annual course of water surface temperature T_s and elements of the energy budget (E, S, H) calculated with model (7) for Parakrama Samudra and evaporation data from other sources (E_1, E_2, E_3, E_4).

Element	Month 1	2	3	4	5	6	7	8	9	10	11	12	Year
T_s	25.8	27.6	28.5	30.2	29.5	28.8	28.8	28.9	28.9	28.1	27.3	26.0	°C
E	81	114	116	114	183	213	234	243	211	126	81	70	1 786 mm
E_1	155	165	205	204	217	216	223	226	222	183	150	143	2 309 mm
E_2	122	139	154	142	136	128	125	133	138	134	122	116	1 600 mm
E_3	99	126	153	144	124	116	126	132	141	125	106	92	1 484 mm
E_4	124	127	162	151	176	176	185	189	200	154	123	118	1 885 mm
S	124	132	174	164	158	153	158	158	159	137	122	108	146 W m $^{-2}$
H	-63	-92	-27	-81	-95	0	-13	-26	-11	-60	-39	-68	-48 W m $^{-2}$

E_1, Penman estimate of evaporation for Anuradhapura.
E_2, from Atlas of World Water Balance (1976).
E_3, potential evapotranspiration from Domrös (1974), computed after Papadakis (1961) for Anuradhapura.
E_4, free water surface evaporation from 'Class A Pan' (Domrös 1974) for Kalawewa.

Table 3. The annual course of the components of the water balance of Parakrama Samudra (period 1974–1977).

	Month 1	2	3	4	5	6	7	8	9	10	11	12	Year
I	890	880	790	670	630	1 170	1 290	1 230	1 010	1 150	1 250	700	11 660 mm
P	133	74	97	146	75	13	59	65	104	156	438	461	1 821 mm
O	950	1 250	170	670	2 150	2 700	2 590	180	70	1 450	460	120	12 760 mm
Δs	-45	-344	607	-14	-1 493	-1 559	-1 329	1 079	935	-992	822	632	-1 701 mm
E	81	114	116	114	183	213	234	243	211	126	81	70	1 786 mm
A	-53	-754	1 208	18	-3 121	-3 289	-2 804	1 951	1 768	-1 262	1 969	1 603	-2 766 mm

ing water management as outlined in Nedeco (1979), which will make it possible to enlarge the irrigation area by 50%.

The cause of the fairly high deficit seems to be in the choice of the four years 1974–1977. But these are the only years for which monthly values of the water balance components relevant for Table 3 are available. If the period from 1950–1977 – for which only annual totals exist, at least in Nedeco (1977) – is considered the deficit becomes very low, so that the seepage is almost balanced by the surface inflow, as is shown in Table 4.

Table 4. The annual totals of the water balance components (1950–1977) in mm.

I	(inflow)	9 933
O	(outflow)	10 067
P	(precipitation)	1 849
E	(evaporation)	1 786
ΔS	(change in storage)	-133

5.2. Daily course of the components of the heat balance of Parakrama Samudra

The daily course of the components of the heat balance has been evaluated with the model described by equations (1)–(6) for the few days for which data are available. One of these days, 1 September 1979, is shown in Table 5 (data) and in Table 6 (results). It was a characteristic day at the end of the SW monsoon season, with a thunderstorm at about 3 p.m. that caused a rapid cooling of the surface layer of the water body (see Table 5). In the morning, the water temperature increased rapidly due to the high short-wave radiation income, and air temperature rose to as high as 39.5 °C at noon. Light winds were established from 11 p.m., raising the evaporation from 0.06 to 0.01 mm h^{-1} water column. During the thunderstorm, when

Table 5. Data for the evaluation of the daily course of the heat budget for 1 September 1979.

Element	Hour 6	7	8	9	10	11	12	13	14	15	16	17	18	
T	24.5	27.0	30.5	33.5	36.0	37.8	39.0	39.5	39.5	39.0	32.5	29.8	26.0	°C
RF	76	69	65	62	60	55	53	53	52	55	69	73	78	%
u	0.2	0.2	0.2	0.3	0.3	1.8	2.0	2.1	2.0	2.0	3.8	0.5	0.2	m s^{-1}
Kw	21	188	623	1 127	1 155	1 040	607	867	1 289	286	432	124	21	W m^{-2}
Ts	28.5	28.8	29.0	30.1	30.4	30.8	31.3	31.7	32.6	32.7	31.6	31.5	31.4	°C
T_{20}	28.8	28.8	28.9	29.6	30.3	30.6	31.3	31.7	32.4	33.2	31.7	31.6	31.4	°C
T_{40}	28.8	28.8	28.8	29.2	30.0	30.5	31.2	31.5	31.9	32.5	31.8	31.7	31.4	°C
T_{80}	28.7	28.7	28.7	28.8	28.9	29.6	30.8	31.2	31.4	31.8	31.7	31.5	31.3	°C
T_{120}	28.4	28.4	28.4	28.4	28.4	28.5	30.0	30.8	31.2	31.3	31.3	31.0	30.2	°C
T_{160}	28.4	28.4	28.4	28.4	28.4	28.4	28.4	28.4	28.5	28.5	28.4	28.4	28.4	°C
T_{180}	28.4	28.4	28.3	28.2	28.1	28.0	27.9	27.9	27.9	27.9	28.1	28.2	28.3	°C

T = air temperature, RF = relative humidity, u = wind speed, Kw = short-wave radiation budget, Ts = water surface temperature and T_{20}–T_{180} = water temperature between 20-cm and 180-cm depth.

Table 6. The daily course for 1 September 1979 of evaporation (E), net radiation (S), long-wave radiation of the water surface (L↑), sensible heat (H) and change in net heat content of the water body ($\frac{dU}{dt}$) evaluated with the data of Table 5 (in W m^{-2}), and the evaporated water column (E′) (in mm h^{-1}). Reference level is water surface.

Element	Hour 6	7	8	9	10	11	12	13	14	15	16	17	18
E	-59	-57	-44	-44	-33	-88	-98	-102	-134	-124	-243	-79	-74
S	-260	-89	353	859	891	779	345	604	1 002	16	151	-163	-275
L↑	-463	-465	-466	-473	-474	-477	-480	-482	-488	-489	-482	-481	-481
H	-10	-5	3	9	16	50	59	63	53	49	12	-6	-14
$\frac{dU}{dt}$	238	-35	-87	-296	-401	-413	-1 005	-488	-517	-447	436	169	413
E′	0.09	0.08	0.06	0.06	0.05	0.13	0.14	0.15	0.19	0.18	0.35	0.12	0.11

winds rose to an hourly average of nearly 4 m s^{-1}, evaporation reached 35 mm h^{-1}. Because of the high surface temperature, low wind speed and high air humidity, evaporation is low, giving on an average about 6 mm day^{-1}, which agrees well with the monthly data from Table 2. Table 6 does not contain values for the heat flux into the bottom of the lake (B) nor for the hydrological term (Q) and precipitation term (P) according to equation (1). The main reason why equation (1) is only balanced by the values of Table 6 in the early morning and late afternoon is that there is no estimate of Q throughout the day (B and P are normally small). Another reason is that the energy fluxes contained in equation (1) are on different time scales, so only averaging the data (at least over one day) gives satisfying results. In fact, this model can be said to give good estimates for evaporation and change of heat storage for hourly values. The other terms in equation (1) should be considered with caution.

6. Conclusion

The model applied here gives a good estimate of monthly data for the energy budget and for the water balance. The estimate of hourly values of the components of these equations should be considered with caution, mainly because there is no information about the heat advections due to water inflow and the different time scales of these components. Later, the radiation budget should be measured for use in hourly values to overcome the uncertainty in determining the atmospheric counter-radiation by parametrization procedures.

References

Atlas of World Water Balance, 1976. USSR Committee for the International Hydrological Decade, Hydrometeorological Publishing House, Unesco Press.

De Martonne, E., 1926. Arésime et indice d'aritidé. C. Roy. Acad. Sci. 182: 1395–1398.

Dobesch, H., 1980a. Die Parametrisierung der atmosphärischen Gegenstrahlung im Ostalpenraum. Arch. Met. Geoph. Biokl. Ser. B 28: 365–371.

Dobesch, H., 1980b. On the estimation of energy budgets in stagnant bodies of water. Arch. Met. Geoph. Biokl. Ser. A 29: 363–372.

Dogniaux, R., 1976. Computer procedure for accurate calculation of radiation data related to solar energy utilization. Solar Energy, proceedings of the Unesco/WMO Symposium Genf. 30.8–3.9.1976, pp. 191–197.

Domrös, M., 1974. The Agroclimate of Ceylon. Geoecological Research, vol. 2. Franz Steiner Verlag, Wiesbaden.

Glover, J. & McCulloch, J. S. G., 1958a. The empirical relation between solar radiation and hours of bright sunshine in the high altitude tropics. Quart. J. R. Met. Soc. 84: 56–60.

Glover, J. & McCulloch, J. S. G., 1958b. The empirical relation between solar radiation and hours of sunshine. Quart. J. R. Met. Soc. 84: 172–175.

Gray, D. M. (ed.), 1970. Handbook of the Principles of Hydrology. Canadian National Committee for the International Hydrological Decade, Ottawa.

Lauer, W., 1952. Humide and aride Jahreszeiten in Afrika und Südamerika und ihre Beziehung zu den Vegetationsgürteln. Bonner Geograph. Abh. 9: 15–98.

Nedeco, 1979. Mahaweli Ganga Development Program Implementation Strategy Study 2. Annex B: Irrigation and Drainage Requirements.

Papadakis, J., 1961. Climatic Tables of the World. Buenos Aires.

Schiemer, F. (ed.), 1981. Parakrama Samudra (Sri Lanka) Limnology Project, Interim Report. Institute for International Cooperation, Vienna.

Schiemer, F., 1983. Parakrama Samudra Project – scope and objectives. In: Schiemer, F. (ed.) Limnology of Parakrama Samudra – Sri Lanka: a case study of an ancient man-made lake in the tropics. Developments in Hydrobiology (this volume). Dr W. Junk, The Hague.

Author's address:
H. Dobesch
Central Institute of Meteorology and Geodynamics
Hohe Warte 38
A-1190 Vienna
Austria

3. Thermal stratification, mixis, and advective currents in the Parakrama Samudra Reservoir, Sri Lanka

K. Bauer

Keywords: stability, advection, convection, tropical, reservoir

Abstract

Thermal stratification in the northern basin of Parakrama Samudra (PSN) was investigated in diurnal cycles at a point as well as in horizontal surveys. The data were used to calculate Birgean mechanical work (Birge 1916) and derived values. Although similar temperature ranges were recorded (28.0°–34.5°C) for August/September 1979 and March 1980, fluctuations in mixing work (Eckel 1948) (up to 4 J m^{-2} and 20 J m^{-2}, respectively) were observed, which is in accordance with the changing water levels of 1.8 m and 3.8 m, respectively. Deviations of 1.7–2.8 J m^{-2} from mean potential energy can drive currents to equalize local inhomogeneities of the stratification. The main source of these inhomogeneities may be the stronger temperature effect of heating and cooling in the shallower areas. It obviously drives advective currents towards the deeper areas where it amplifies the apparent heat budget about three times. Losses of Birgean energy incurred by cooling (1.7–5.9 J m^{-2} in 1979, 27.1 J m^{-2} in 1980) are transformed into convective energy. This leads to erosion processes in the order of a few millimetres of sediment depth daily and hence settled algae can be redistributed every night. The density effects caused by heat exchange are obviously the predominant factor in the mechanics of such shallow tropical lakes.

1. Introduction

Parakrama Samudra is a shallow irrigation reservoir in the dry zone of Sri Lanka. Its geographical and hydrographical properties are described in Schiemer (1983). The lake was investigated during August/September 1979 and March 1980.

The main objective of this study was to investigate the physical factors responsible for the mixing rhythm of the lake. As the water temperatures are high, strong density gradients are linked with temperature gradients. Whilst this leads to high stability against mixis, surface cooling brings about energy-rich convective currents, which can cause strong mixing. Furthermore, heat fluxes are comparatively greater in the tropics. For these reasons such water bodies should, hypothetically, be governed by density currents of thermal origin rather than by wind influences.

2. Methods

Depth profiles of water temperature were measured at 10-cm intervals with an NTC thermistor of 0.1 °C accuracy. Diurnal cycles of temperature profiles were taken at the station PSN 14 (see Schiemer 1983). Horizontal surveys on local differences in the pattern of stratification were undertaken in March 1980.

Calculations of the heat content were done by trapezoidal integration of the temperature profiles. Estimations of the 'expected changes in heat content' were calculated according to Dobesch (1983).

A set of analogous formulae (see the Appendix)

Schiemer, F. (ed.), Limnology of Parakrama Samudra – Sri Lanka
 ISBN 90 6193 763 9

adapted from Schmidt (1914) and Birge (1916) were used for the following evaluations:

1) energy necessary for mixis (as defined in the Appendix) and sediment erosion,
2) energy content of horizontal patchiness in stratification, and
3) kinetic energy gain of convective currents produced by the loss of potential energy due to surface cooling.

3. Results

3.1. Temperature ranges

In 1979 the lowest surface temperature at sunrise 0600 hours was 28.5 °C. Temperatures as low as 28 °C sometimes occurred at the bottom of the lake. In the morning and evening, when the heat balance was negative, temperature inversions of 0.1°–0.3 °C were found regularly in the uppermost half metre.

In March 1980 the morning temperatures were close to 30 °C, with inversions at the surface and distinct layers of cooler water near the bottom. In both periods, maximum surface temperatures occurred in the early afternoon. The range was 33.0°–34.5 °C in 1979 and 33.2°–33.9 °C in 1980. Temperature data from diurnal cycles that were selected for further evaluation are given in Table 1 (1979) and Table 2 (1980).

3.2. Horizontal inhomogeneity

Local differences in thermal stratification were expected due to the shallowness of the lake. Two surveys on horizontal differences in temperature profiles were conducted in the afternoon (9 March 1980: 1325–1410 hours; 16 March 1980: 1420–1610 hours), where an equilibrium in heat gains and losses could be assumed to exist. During the surveys, observed differences in the local heat content of the water column exceeded by far the possible differences due to heat exchange through the water surface. This implies that the observed local differences reflect a horizontal patchiness.

Table 3 shows the means and standard deviations of heat content, mixing energy (M) and potential energy (B) for the 2-m and 3-m water columns (see the Appendix), and also the mean absolute deviation of B from $\bar{B}$. Heat content varies in the order of 4 MJ m^{-2}, while the empirical daily heat budget is

Table 1. Diurnal cycles of temperature stratification at PSN 14 in 1979.

Depth	30.8.79					1.9.79										
	Daytime															
(m)	0600	0930	1230	1515	1750	0615	0640	0810	0854	0945	1145	1245	1305	1345	1605	1750
	Temperatures (°C)															
0.0	28.5	30.4	34.0	30.3	29.3	28.6	28.5	29.1	30.5	30.4	31.4	31.4	32.1	32.9	31.5	31.1
0.1	28.6	30.7	34.0	30.0	29.4	28.6	28.6	29.1	30.1	30.3	31.2	31.4	32.1	32.9	31.5	31.0
0.2	28.7	30.4	34.3	30.0	29.5	28.6	28.6	29.1	30.0	30.3	31.2	31.5	32.1	32.8	31.6	31.0
0.3	28.7	29.8	32.0	30.0	29.5	28.6	28.6	29.1	29.7	30.1	31.2	31.5	32.0	32.3	31.6	31.1
0.4	28.7	29.3	31.8	30.0	29.6	28.6	28.6	28.9	29.6	30.0	31.2	31.3	31.9	32.3	31.5	31.0
0.5	28.7	29.1	30.8	30.0	29.6	28.6	28.6	28.8	29.4	29.8	31.1	31.2	31.6	31.5	31.6	31.0
0.6	28.8	29.0	30.0	29.8	29.6	28.6	28.6	28.7	29.2	29.6	31.1	31.2	31.4	31.4	31.5	31.0
0.7	28.7	28.9	29.8	29.7	29.6	28.6	28.6	28.7	29.1	29.0	30.1	31.2	31.4	31.4	31.5	31.0
0.8	28.7	28.9	29.7	29.7	29.5	28.6	28.6	28.6	28.8	28.8	31.0	31.1	31.1	31.3	31.6	31.0
0.9	28.7	28.9	29.7	29.7	29.3	28.6	28.6	28.5	28.8	28.7	31.0	31.0	31.1	31.3	31.5	31.0
1.0	28.8	28.8	29.4	29.7	29.1	28.6	28.6	28.5	28.8	28.5	30.9	31.0	31.0	31.2	31.6	31.0
1.1	28.8							28.4	28.7	28.3	30.9	31.0	31.0	31.2	31.5	30.0
1.2	28.8					28.5		28.4	28.5	28.2	30.7	31.0	31.0	31.2	31.4	27.7
1.3	28.8						28.5	28.4	28.4	28.2	30.5	30.9	30.9	31.2	30.0	29.5
1.4	28.8					28.4		28.2	28.4	28.1	30.1	30.5	30.8	31.2	29.9	28.9
1.5	28.7	28.7	28.8	28.7	28.9		28.5	28.1	28.3	28.1	28.1	29.4	30.0	30.2	28.3	28.8
1.6						28.4		28.1	28.1	27.9	28.0	28.6	28.8	28.7	28.3	28.7
1.7								28.1	28.1	27.9	27.9	28.2	28.3	28.2	28.2	28.7
1.8	28.5	28.7	29.1	28.5	28.6	28.4	28.1	28.1	28.2	28.1	27.9	28.1	28.0	28.1	28.6	28.6

Table 2. Diurnal cycles of temperature stratification at PSN 14 in 1980.

Depth (m)	8.3.1980 Daytime 0600	0850	1230	1520	1830	2340	18/19.3.1980 0945	1245	1545	1805	0745
0.0	30.0	31.0	33.8	34.2	31.6	30.6	30.5	32.2	33.1	31.3	30.4
0.1	30.1	30.7	33.6	34.1	31.6	30.6	30.5	32.2	33.1	31.4	30.5
0.2	30.1	30.7	33.4	34.1	31.7	30.7	30.5	32.2	33.1	31.5	30.5
0.3	30.2	30.6	33.0	34.1	31.9	30.7	30.5	32.2	33.1	31.7	30.5
0.4	30.2	30.6	32.6	34.0	31.8	30.7	30.5	32.1	33.1	31.7	30.5
0.5	30.2	30.5	32.3	34.0	31.8	30.7	30.5	32.1	33.0	31.7	30.5
0.6	30.2	30.5	32.3	34.0	31.7	30.7	30.5	32.1	33.0	31.9	30.5
0.7	30.2	30.4	32.1	33.3	31.7	30.7	30.5	32.0	33.0	31.9	30.5
0.8	30.2	30.4	32.1	32.8	31.7	30.7	30.5	32.0	33.0	31.9	30.5
0.9	30.2	30.3	32.0	32.8	31.6	30.7	30.5	32.0	33.0	31.9	30.5
1.0	30.2	30.3	32.0	32.1	31.6	30.7	30.5	32.0	32.9	31.9	30.5
1.1	30.2	30.3	32.0	32.0	31.5	30.7	30.5	31.9	32.9	31.7	30.5
1.2	30.2	30.3	31.7	32.0	31.4	30.7	30.5	31.7	32.9	31.7	30.5
1.3	30.2	30.3	31.6	32.0	31.4	30.7	30.5	31.3	32.8	31.6	30.5
1.4	30.2	30.3	31.3	31.9	31.4	30.7	30.5	31.1	32.0	31.1	30.5
1.5	30.2	30.2	31.2	31.8	31.3	30.7	30.5	31.0	31.4	30.9	30.5
1.6	30.2	30.2	31.1	31.6	31.3	30.7	30.5	30.9	31.2	30.9	30.5
1.7	30.2	30.2	31.0	31.6	31.3	30.6	30.5	30.9	31.1	30.7	30.5
1.8	30.2	30.2	31.0	31.6	31.2	30.6	30.5	30.8	31.0	30.4	30.5
1.9	30.2	30.2	30.9	31.6	31.2	30.6	30.5	30.7	31.0	30.3	30.5
2.0	30.2	30.2	30.8	31.2	31.0	30.6	30.5	30.7	30.9	30.2	30.5
2.1	30.2	30.1	30.8	31.1	31.0	30.6	30.5	30.7	30.9	30.1	30.5
2.2	30.2	30.1	30.8	31.0	31.0	30.6	30.5	30.7	30.7	30.1	30.5
2.3	30.2	30.0	30.8	30.6	30.9	30.6	30.5	30.6	30.7	29.9	30.5
2.4	30.1	29.9	30.8	30.5	30.9	30.6	30.5	30.6	30.7	29.9	30.5
2.5	30.1	29.6	30.6	30.4	31.1	30.6	30.5	30.6	30.6	29.9	30.5
2.6	30.0	29.5	30.5	29.9	31.1	30.6	30.4	30.5	30.6	29.9	30.5
2.7	29.8	29.5	30.5	29.6	31.1	30.6	30.4	30.5	30.6	29.9	30.5
2.8	29.7	29.5	30.5	29.6	31.1	30.6	30.4	30.5	30.6	29.9	30.5
2.9	29.4	29.4	30.3	29.3	31.1	30.6	30.4	30.5	30.5	29.8	30.5
3.0	29.3	29.4	30.3	29.2	31.1	30.6	30.3	30.5	30.5	29.8	30.4
3.1	29.3	29.4	30.2	29.2	31.1	30.6	30.3	30.4	30.4	29.3	30.4
3.2	29.2	29.3	30.2	29.2	31.0	30.6	30.1	30.4	30.4	29.3	30.4
3.3	29.2	29.3	30.1	29.2	30.9	30.6	30.1	30.4	30.3	29.2	30.3
3.4	29.2	29.3	30.0	29.2	30.9	30.6	30.0	30.3	30.0	29.1	30.3
3.5	29.1	29.3	30.0	29.2	30.6	30.6	30.0	30.3	30.0	29.1	30.2
3.6	29.1	29.3	30.1	29.2	30.5	30.6	29.9	30.0	30.0	29.1	30.0
3.7	29.1	29.3	30.1	29.2	30.3	30.6	29.9	30.3	29.9	28.9	30.3
3.8	29.0	29.3	30.0	29.2	29.9	30.6	29.9	30.0	29.9	28.9	30.0

about 20 MJ m^{-2} (see Table 4). Such a degree of patchiness in horizontal heat distribution must be an important source of energy for advective currents. The possible kinetic energy of these currents is equal to the mean absolute deviation of the local potential energy (B) from the mean potential energy ($\bar{B}$):

$$E_{kin} = \frac{\Sigma|B - \bar{B}|}{n}$$

This energy amounts to 1.7–2.8 J m^{-2} (Table 3), i.e. between 21% and 84% of the respective M values.

3.3. Diurnal cycles of heat content and advective mechanisms

The importance of horizontal currents can also be seen from the changes in heat content throughout the day (Table 4). On days where the development of thermal stratification was registered at sta-

Table 3. Local differences in stratification. Means ($\bar{x}$) and standard deviations (SD) of the potential energy (B), mixing work (M), and heat content (U); for B also the mean absolute deviation of B(z) from the mean $\overline{B(z)}$, i.e. the available energy content of the thermal patchiness. Numbers of samples: $n = 9$ for 2-m and 3-m depth on 9.3.1980, $n = 15$ for 2-m and $n = 8$ for 3-m depth on 16.3.1980.

Date	Parameter (depth)	Unit	$\bar{x}$	SD	$x-\bar{x}$
9.3.80	B (2 m)	J m^{-2}	92.26	2.21	1.69
	B (3 m)	J m^{-2}	201.43	3.50	2.77
	M (2 m)	J m^{-2}	3.36	0.56	
	M (3 m)	J m^{-2}	7.77	1.37	
	U (2 m)	MJ m^{-2}	266.52	3.72	
	U (3 m)	MJ m^{-2}	393.80	4.48	
16.3.80	B (2 m)	J m^{-2}	92.21	3.86	2.73
	B (3 m)	J m^{-2}	196.67	2.13	1.85
	M (2 m)	J m^{-2}	3.25	1.14	
	M (3 m)	J m^{-2}	8.77	1.48	
	U (2 m)	MJ m^{-2}	265.94	4.14	
	U (3 m)	MJ m^{-2}	390.28	2.55	

tion PSN 14 (30 August 1979, 1 September 1979, 8 March 1980, 18–19 March 1980), the heat content for each measured profile was calculated. The change in heat content between every two measurements was compared with the expected change of heat content for the same time interval, calculated from Dobesch's (1983) average diurnal course of heat-balance parameters. In the absence of any horizontal transport of water and heat, the expected changes in heat content ($\Delta U_{exp.}$) and those actually observed ($\Delta U_{act.}$) should be equal. However, a regression of the values given in Table 4 results in:

$$\Delta U_{act.} = -4.2 \text{ MJ m}^{-2} + 3.36 \times \Delta U_{exp.}$$

($n = 22$, $r = 0.66$)

This means that the actual gains and losses of heat between the temperature measurements are in general about 3.36 times higher than those predicted from Dobesch's climatological calculations. The following mechanism could account for this discrepancy. The lake has a large shallow area and heat gains and losses can both lead to higher amplitudes in water temperature in the shallow zones than in the offshore areas (e.g. PSN 14). Warmer water from the periphery flows at the surface towards the middle of the lake at times of heat gain and in-

Table 4. Diurnal cycles of potential energy (B), mixing work (M), heat content (U), actual changes in heat content ($U_{act.}$), and theoretically expected changes in heat content ($U_{exp.}$). Calculated from the temperature readings in Tables 1 and 2.

Date, time	B(z) J m^{-2}	M(z) J m^{-2}	U(z) MJ m^{-2}	$U_{act.}$ MJ m^{-2}	$U_{exp.}$ MJ m^{-2}
30.8.1979 (z = 1.8 m)					
0600	62.65	0.03	216.10		
0930	63.35	1.36	219.45	+3.35	+0.25
1230	66.17	4.26	228.66	+9.21	+6.32
1515	68.84	1.47	222.04	-6.58	+4.60
17.50	64.34	0.78	220.12	-1.96	-0.42
1.9.1979 (z = 1.8 m)					
0.615	61.59	0.66	215.39		
0.640	61.98	0.10	214.85	-0.56	-0.50
0810	61.18	0.92	215.14	+0.29	-0.21
0854	62.20	1.66	218.07	+2.93	+0.17
0945	61.60	2.23	218.03	-0.04	+1.38
1145	67.91	2.60	228.78	+12.75	+4.18
1245	69.58	2.26	230.87	+2.09	+2.43
1305	70.41	2.70	232.88	+2.01	+0.75
1345	71.15	3.11	234.64	+1.80	+1.34
(thunderstorm)					
1605	69.47	2.73	231.46	-3.22	n. det.
1750	67.44	2.33	227.61	-3.85	-0.92
8.3.1980 (z = 3.8 m)					
0600	300.31	4.08	475.51		
0850	300.54	5.65	476.89	+1.38	+1.05
1230	320.35	12.01	496.43	+19.54	+7.09
1520	312.00	20.53	496.47	+0.08	+5.65
1830	327.38	4.72	496.22	-0.29	-1.84
2340	319.45	0.45	487.18	-9.04	-3.18
18/19.3.1980 (z = 3.8 m)					
0945	312.83	1.78	483.21		
1245	320.33	8.67	493.92	+10.71	+7.07
1545	324.16	13.91	500.57	+6.65	+5.52
1805	306.04	11.59	485.47	-15.10	-0.29
0745	314.57	1.04	483.96	-1.51	-3.77

creases the actual heat gain there. At times of negative heat balance, cooler water from the periphery flows along the bottom towards the deeper parts of the lake and increases the heat loss there. In both cases, there is a time lag between the origin of the advective water and its arrival at PSN 14. A model of this process with regular time intervals would result in a cyclic relationship of $\Delta U_{act.}$ and $\Delta U_{exp.}$. As most of our data were taken at times when heat gains were expected, the above linear regression is mainly influenced by the corresponding side of the cycle and thus the constant of -4.2 MJ m^{-2} reflects the time lag.

3.4. Convection

As pointed out above, many of the temperature profiles show inversions at the surface, indicating strong cooling, which results in convective currents. The heat gained during the day is completely lost during the night. Morning stratifications with cooler water at the bottom do not result from the remnants of the last day's stratification, but are explained by the advective mechanism discussed above. Actually the midnight stratification on 8 March 1980 (Table 2) was already completely homothermal at a much higher temperature than in the morning. There is clearly a total convective mixis down to the bottom every night.

The strong advections cast doubts about whether the stratification in PSN 14 is representative for the lake. An attempt is therefore made to calculate the kinetic energy of the convective currents that originates from the loss of potential.

By assuming that the evening stratification is the object of cooling and convective mixing, and that the morning stratification is the result of a similar process during the previous night, then the difference $\Delta B = B_{morn.} - B_{even.}$ equals the total potential energy that was transformed into kinetic energy of convective currents during one night. On 30 August 1979 the difference was 1.69 J m^{-2}, compared with a mixing energy of the evening stratification $M_{even.} =$ 0.78 J m^{-2}. For 1 September 1979 the figures are ΔB = 5.85 J m^{-2} and $M_{even.}$ = 2.33 J m^{-2} (from Table 4). On this day, however, the afternoon stratification had resisted a strong thunderstorm, which resulted in incomplete mixing.

As midnight temperatures are available for 8 March 1980, the diurnal changes in stratification can be discussed in more detail. The evening stratification had a mixing resistance of $M_{even.}$ = 4.72 J m^{-2} and it was dissipated by cooling. No mixing work was necessary except for a layer of 40–50 cm above the bottom, where a lower temperature existed before the homothermal midnight state. Thus a ΔB of 7.9 J m^{-2} was released as kinetic energy from evening until midnight.

3.5. Sediment erosion

The kinetic energy of the convective currents can have an erosive effect on the bottom sediments. Based on the data from the two research periods for PSN 14, the following considerations could be extended to illustrate the possible magnitude of the erosive effect. By assuming a sediment density of 1.1 g cm^{-3} (for very soft sediments; Schiemer, personal communication) and a water density of 1.0 g cm^{-3} (the density decrease to about 0.996 g cm^{-3} at 29 °C may be roughly compensated for by suspended matter), the formula for mixing energy M can be applied to this two-layer system.

To resuspend this sediment uniformly in the whole 1.8-m water column in 1979, 17.55 J m^{-2} were necessary for every centimetre of eroded sediment. In 1980, with a depth of 3.8 m, the corresponding value was 37.17 J m^{-2}. Thus the possible erosion effect would have been 0.96 mm on 30 August 1979, 1.33 mm on 1 September 1979, and 7.28 mm on 18 March 1980.

4. Discussion

Sly (1978) estimates that the relative importance of density effects caused by heating and cooling ranks third after wind and river inflow. Wind action was also estimated to be most important by Birge (1916) and Schmidt (1914, 1928) in their considerations of the potential energy content of lakes. A simple model calculation using their formulae (Bauer 1983) shows that in a temperate lake (hypolimnetic temperature, 4 °C) with an epilimnion of one-tenth of the total depth, a cooling of this epilimnion from 17 °C to 16 °C delivers enough convective energy to mix the lake down to the bottom. With the change from temperate to tropical zones, the higher temperature level should increase the importance of density effects, firstly by enhanced stability (mixing energy M) and secondly by greater energy of the convective currents (ΔB).

On 1 September 1979 a heavy thunderstorm did not suffice to mix the whole stratified water column down to 1.8-m depth. A 'sharpened thermocline' (Lewis 1973) remained between 1.2-m and 1.3-m depth with a temperature decrease of 1.4 °C within 10 cm. During the night, however, the stratification disappeared due to heat loss, without the necessity for mixing. The potential energy loss of the cooling water was transformed into convective energy. As there was no stratified water below to be mixed, the convective energy was available for sediment erosion.

The transformation of potential energy differences into kinetic energy for currents is incomplete and thus only a fraction of it would be considered as available. The transformation will be more or less complete in the case of cooling convection, but not for the horizontal inhomogeneities. Since there are steady sources for local differences of stratification (e.g. wind, water depth, through-flow), horizontal temperature equalization is never fully attained. However, the order of magnitude of the calculated values suggests that advective processes play an important role in lakes of this type.

Heating and cooling not only influence the local stratification and mixis, they also obviously drive a general circulation pattern in the lake. Local convective currents are mainly organized in Langmuir circulations and these seem to interfere with the lake-wide advection pattern. The northern basin of Parakrama Samudra (PSN) is subdivided at low water level into at least two bays between which the advective mechanism probably does not effect a complete exchange of water. There are significant differences in, for instance, chlorophyll content of these bays (Dokulil et al. 1983).

The relative importance of wind action and thermal density effects also depends on the strength and duration of the occurring winds. In August 1979 a prolonged period of strong wind led to water turbidity with extinction coefficients in the range of $e = 9\ m^{-1}$. The latter decreased in the calmer period of the end of August and in September. On 1 September the extinction decreased during the day from $4.3\ m^{-1}$ to $2.9\ m^{-1}$ (Dokulil et al. 1983). Probably sediment erosion during the night causes resuspension of much turbid material, which resettles during the day under conditions of stable stratification.

The sediment density of $1.1\ g\ cm^{-3}$ that was used for the calculation of erosion is a compromise. Soft sediments may have densities of $1.2\ g\ cm^{-3}$ and more, we must also take into consideration that there will be a layer of water that is enriched with settled algae and other light particulate matter. Wetzel (1975) gives a density range of 1.01–1.03 g cm^{-3} for most phytoplankton organisms. If distilled water of 4 °C were to contain settled phytoplankton in quantities equal to 50% of the volume, the maximum density of this suspension would thus be $1.015\ g\ cm^{-3}$. Due to smaller density differences, layers of such algal suspensions can be resuspended much more readily in the water column than the soft sediments.

The general conclusion may be drawn that, in shallow tropical lakes like Parakrama Samudra, the density effects of heating and cooling predominantly govern the daily mixis and recycling of particles and nutrients, except in seasons with a strong steady wind.

Appendix

The concept of static potential energy in the stratification of a lake: its alterations as driving and resisting forces in the dynamics of the stratification. Birge (1916) defines the 'work of the wind', which causes the distribution of heat in a stratified lake, by the formula:

$$B = \frac{g}{A_0} \int_0^{z_m} z\, A_z (1 - \rho_z)\, dz \qquad (1)$$

as quoted by Hutchinson (1957). The components are:

- B Birge's 'wind work', an expression assuming that the water is heated at the surface and pressed to its final depth z by wind action against the buoyancy in the initial 4 °C water, the density of which is set '1'.
- g is the earth gravitation constant. It amounts to $9.80665\ ms^{-2} = 9.80665\ N\ kg^{-1}$.
- z is the depth of the individual integrand.
- z_m is the maximum depth of the lake.
- A_0 is the surface area of the lake.
- A_z is the hypsographic area of the lake at depth z.
- dz is the 'thickness' of the integrand's layer.

In the present paper, this formula is applied to water columns with constant area $A_0 = A_z = 1\ m^2$. This is due to the patchiness of the stratification in a shallow lake.

A formula similar to Birge's is given by Schmidt (1914, 1928) for what he called 'stability'. This is the energy necessary for a total mixis of the stratified lake, assuming that the mixture has the mean density of the stratified body of water. Schmidt (1928) drew attention to the problem of the temperature anomaly of water, and Eckel (1948) solved this problem by a modified stability formula, which he called 'mixing work' (*Mischungsarbeit*). In contrast to Schmidt, he assumed a constant water mass and a mean temperature in the mixed water. Furthermore, Bauer (1978, 1983) showed that mixing work can be understood as an alteration of the Birgean potential energy (as it is termed in the new interpretation) caused by the mixing process. Therefore Eckel's mixing work is here written as a difference in the Birgean energies:

$$M(z_i) = \frac{g}{A_0} \left(\int_0^{z_i} z A_z \mu_m \, dz - \int_0^{z_i} z A_z \mu_z \, dz \right) \quad (2)$$

While Birge's, Schmidt's, and Eckel's energies are defined for the whole depth scale of the lake, our formulation is an integral depth function:

- z_i is the independent variable of the function. Physically it is the depth down to which a homogenizing mixing (as we define the term 'mixis' in this chapter) could proceed.
- $M(z_i)$ is the mixing work that was (or would be) necessary to homogenize the stratified water between the surface and the depth z_i.
- μ is a short form for the density difference $1 - \rho$.
- μ_z therefore equals $1 - \rho_z$ for the stratified water and
- μ_m is the μ value for the correctly determined density of the mixed water.

Most natural processes that influence the state of stratification (e.g. heating, cooling and wind) affect the lake through its surface. Alterations of the stratification and its Birgean energy content thus proceed from the surface towards the depth. A depth function of our type with the surface as a reference level is an adequate mathematical description and allows estimates of the possible depth of effect. The formula proposed by Idso (1973) cannot be used for such purposes for theoretical as well as practical reasons that will be discussed in a forthcoming paper (Bauer 1983).

Whenever the state of stratification is altered, the energetical analysis of this process can be carried out by a comparison of the Birgean energies according to the type-(2) formula, which can be rewritten in a generalized form:

$$\Delta B(z_i) = B(z_i)_2 - B(z_i)_1 \quad (3)$$

- $B(z_i)$ is the integral depth function of the Birgean potential energy and the indices 1 and 2 point, respectively, to the state of stratification before and after the process under consideration.

If $B(z_i)$ increases, the energy difference must have been supplied to the lake. Decreases of potential energy are transformed into other forms of energy, usually kinetic energy of currents.

Using formula (3) it is possible to analyse the effect of surface cooling. The loss of potential energy B caused by the cooling is transformed into the kinetic energy of Langmuir currents. These lead to further mixis, which can be calculated from the function $M(z_i)$ of an intermediate state (hypothetical or measured), when the currents are just running but have not yet brought about their mixing effect. This mixis can homogenize within the water body or can erode sediments, if more ΔB is available than is necessary to overcome the $M(z_i)$ down to the maximum depth $z_i = z_m$. The erosive effect can also be estimated by the $M(z_i)$ function of the layered sediment–water system.

Especially in large and shallow lakes important horizontal differences in stratification occur, which means that the B values differ locally. If the water were allowed to equalize, the mean value $\bar{B}$ would obtain in the whole lake. To reach this state there would be an increase of the local B in one half and an equal mean decrease in the other half of the lake sites. The mean change would be $\Sigma|B - \bar{B}|/n$, i.e. the mean absolute difference between the local B values and the lake mean $\bar{B}$. This amount of potential energy would be transformed into the kinetic energy of equalizing currents. It could lead to an internal mixing, partly with homogeneous layers within the stratification.

Acknowledgements

This work was supported by a travel grant from the DFG (Deutsche Forschungsgemeinschaft). Most measurements were performed with the help of other team members, especially A. Duncan in 1979 and I. L. Silva in 1980.

References

Bauer, K., 1978. A generalization of the lake stability concept. Abstract. Verh. Internat. Verein. Limnol. 20: 984–985.

Bauer, K., 1983 (in preparation). Die potentielle Energie in der Schichtung von Seen. Ph.D. thesis, University of Munich.

Birge, E. A., 1916. The work of wind in warming a lake. Trans. Wis. Acad. Sci. 18: 341–391.

Dobesch, H., 1983. Climatology and energy budget of a tropical man-made lake. In: Schiemer, F. (ed.) Limnology of Parakrama Samudra – Sri Lanka: a case study of an ancient man-made lake in the tropics. Developments in Hydrobiology (this volume). Dr W. Junk, The Hague.

Dokulil, M., Bauer, K. & Silva, I., 1983. An assessment of the phytoplankton biomass and primary productivity of Parakrama Samudra, a shallow man-made lake in Sri Lanka. In: Schiemer, F. (ed.) Limnology of Parakrama Samudra – Sri Lanka: a case study of an ancient man-made lake in the tropics. Developments in Hydrobiology (this volume). Dr W. Junk, The Hague.

Eckel, O., 1948. Über die Mischungsarbeit von stabil geschichteten Wassermassen. Arch. Meteorol. Geophys. Bioklim. (A) 1: 264–269.

Hutchinson, G. E., 1957. A Treatise on Limnology, vol. 1. John Wiley & Sons, New York.

Idso, S. B., 1973. On the concept of lake stability. Limnol. Oceanogr. 18: 681–683.

Lewis, W. M., Jr., 1973. The thermal regime of Lake Lanao (Philippines) and its theoretical implications for tropical lakes. Limnol. Oceanogr. 18: 200–217.

Schiemer, F., 1983. The Parakrama Samudra Project – scope and objectives. In: Schiemer, F. (ed.) Limnology of Parakrama Samudra – Sri Lanka: a case study of an ancient man-made lake in the tropics. Developments in Hydrobiology (this volume). Dr W. Junk, The Hague.

Schmidt, W., 1914. Über den Energiegehalt der Seen. Int. Rev. ges. Hydrogr. Hydrobiol. Suppl. 6: 1–25.

Schmidt, W., 1928. Über die Temperatur- und Stabilitätsverhältnisse in Seen. Geogr. Ann. 10: 145–177.

Sly, P. G., 1978. Sedimentary processes in lakes. In: Lerman, A. (ed.) Lakes – chemistry, geology, physics, pp. 65–90. Springer, New York.

Wetzel, R. G., 1975. Limnology. W. B. Saunders, Philadelphia.

Author's address:
Kurt Bauer
Fischgesundheitsdienst
Tiergesundheitsdienst Bayern e.V.
Senator Gerauer Str. 23
D-8011 Grub bei München
Federal Republic of Germany

4. Phosphorus and phosphatase dynamics in Parakrama Samudra based on diurnal observations

A. Gunatilaka

Keywords: hot-water-extractable-P, intracellular-P, dissolved organic-P, phosphatase activity, tropical

Abstract

Phosphate-deficiency indicators such as inducement of alkaline phosphatase activity, changes in the algal internal P pool (as hot-water-extractable phosphate) along with soluble reactive phosphate (SRP), soluble phosphate and particulate phosphate have been measured at a Parakrama Samudra North station on three days. Parallel observations were made on total and dissolved nitrogen fractions. The observations made indicate the existence of a diel SRP cycle. The average SRP concentrations ranged from 9 to 12 $\mu g\ l^{-1}$. Due to enhanced phosphatase activity during the afternoon, the values can go up to 19–26 $\mu g\ l^{-1}$. However, 80% of the SRP values observed were below 10 $\mu g\ l^{-1}$ and 94% below 20 $\mu g\ l^{-1}$. The level of the dissolved organic phosphate significantly decreased with the increase of phosphatase activity ($r = 0.522$, $P < 0.01$). There appears to be an inverse relationship between external inorganic P concentration and both P-deficiency indicators (phosphatase activity and hot-water-extractable P). Hot-water-extractable P per unit chlorophyll-a decreased during the day but increased gradually to reach a maximum at midnight. The above observations suggest that the algal population in Parakrama Samudra may be phosphorus limited at times of the day.

1. Introduction

Phosphorus has been implicated as a primary factor contributing to eutrophication in freshwaters (Sakamoto 1966; Vollenweider 1968; Thomas 1969; Lean 1973a; Schindler 1976; Dillon & Rigler 1974), but the supply of it in many lakes all over the world is generally low (Golterman 1973, 1975; Oglesby & Schaffner 1975; Rigler 1973; Wetzel 1975). Orthophosphate is undetectable in some natural waters during periods of maximal algal growth. Such a situation has been documented for Parakrama Samudra in August–September 1979 (Gunatilaka 1980; Gunatilaka & Senaratne 1981) and may be a common feature in tropical waters (Viner 1973). In temperate lakes during the summer stratification, phosphate turn-over time may be as short as 1–4 min (Rigler 1964; Lean 1973a and b; Lean & Nalewaiko 1979). Peters (1976) observed that in African lakes values may even drop below 1 min.

It is apparent that although inorganic orthophosphates (which is the main form of phosphate taken up by the algae) may be in short supply for phytoplankton, dissolved and colloidal organic phosphates and inorganic polyphosphates are present. The importance of organic phosphorus in lakes and in oceans has been recognized (Redfield et al. 1963; Ketchum & Corwin 1965; Hutchinson 1967), however, little is known about organic phosphorus compounds, and their kinetics and decomposition in natural waters.

Phosphatases are capable of hydrolysing dissolved organic phosphates (Rautanen & Karkkainen 1951; Overbeck 1962; Reichardt et al. 1967; Reichardt 1971; Berman 1969). Under phosphorus-limited conditions, phytoplankton release them to hydrolyse extracellular, dissolved organic phos-

Schiemer, F. (ed.), Limnology of Parakrama Samudra – Sri Lanka
 ISBN 90 6193 763 9

phates (Galloway & Krauss 1963; Kuenzler & Parras 1965; Healey & Hendzel 1975, 1976). Fitzgerald & Nelson (1966) demonstrated that in algal cultures (diatoms and blue-greens), phosphatase activity was 5–20 times greater when cells were phosphorus limited.

Phosphatase activity has been used extensively as an indicator of phosphate sufficiency or deficiency (Berman 1969, 1970; Jones 1972, 1973; Fuhs 1969, 1972; Rhee 1973; Smith & Kalff 1981; Wynne 1981b). Alkaline phosphatase activities have been studied in a number of temperate lakes as reviewed by Wetzel (1981) and correlated well with other signs of phosphorus shortage (Sproule & Kalff 1978; Healy & Hendzel 1980).

According to Fitzgerald & Nelson (1969), Fitzgerald (1972), Rother & Fay (1979), Wynne (1977), Wynne & Berman (1980) and Wynne (1981a and b), the fluctuation of the internal P pool in algae is indicative of their P status. Under high external P concentrations, luxury uptake of phosphorus takes place (Mackereth 1953; Fogg 1973), and excess phosphorus is stored intracellularly (Kuhl 1962, 1974), mainly as polyphosphates (Rhee 1973; Elgavish & Elgavish 1980). When external phosphate concentrations are limiting, algal growth can continue at the expense of intracellular P reserves. Fitzgerald & Nelson (1966) suggested that the amount of hot-water-extractable inorganic phosphate in cells ('surplus P') gives a measure of the available intracellular storage P, and hence the P status of the organisms. The work of Elgavish & Elgavish (1980) confirms the above findings. Rother & Fay (1979), Fitzgerald (1972) and Wynne & Berman (1980) have observed the slow depletion of internal P pools in laboratory cultures and in natural populations under P-limited conditions.

During three diurnal studies in March 1980, the fluctuations of alkaline phosphatase activity, hot-water-extractable phosphate pool (labile phosphate), and other phosphate fractions have been observed at a Parakrama Samudra North station. An attempt is made here to interpret these data in the light of the general nutrient situation in the lake to obtain an insight into phosphorus dynamics. This chapter also presents some evidence that the diel changes in soluble reactive phosphate (SRP) and dissolved P concentrations imply the existence of P limitation in the lake.

2. Study area and methods

Diurnal studies were carried out (3, 18 & 27 March 1980) at a Parakrama Samudra North station (PSN 14; see Schiemer 1983, Fig. 3). Water samples were collected at 3-h intervals throughout the day till midnight, at 0.5-m-depth intervals from the surface to 3.5 m, using a 2L Ruttner perspex sampler. Two successive samples were pooled to obtain a composite sample at each meter depth, and stored in pre-rinsed plastic bottles. The samples were processed immediately. Soluble fractions were collected by filteration through pre-washed 0.45-μm-membrane filters. Total phosphate, total nitrogen and phosphatase were determined in unfiltered samples.

Phosphate was analysed by the ascorbic acid modification of the molybdenum blue method (Murphy & Riley 1962, as outlined in Strickland & Parsons 1968). Inorganic phosphate (P_i, soluble reactive phosphate – SRP, after Strickland & Parsons 1968) measured as a phosphor molybdate complex may overestimate P_i under some circumstances (Rigler 1966, 1968, 1973). For total phosphate (TP), total nitrogen (TN), soluble phosphate (SP) and soluble nitrogen (SN), samples were digested under pressure (1.3 atm.) with persulphate in an autoclave (Raveh & Aunimeleck 1979) and TN and SN were determined after reduction with Deverda's alloy.

Particulate phosphorus (PP) and soluble unreactive phosphorus (SUP, viz. organic phosphorus) were determined by difference as (TP) – (SP) and (SP) – (SRP), respectively. Alkaline phosphatase was assayed according to a modified Jones (1972) procedure with *p*-nitrophenyl phosphate (PNPP) as substrate: 10-ml lake water samples were incubated wit 1 ml of 0.01 *M* PNPP (Sigma) in 0.2 *M* Tris–HCl buffer (pH 7.6) in a water-bath at 37 °C for 3 h. Two drops of chloroform were added (Berman 1970) to facilitate the assay. At the end of the incubation period, the reaction was terminated by the addition of 1 ml, 1 *M* NaOH, and the colour of the *p*-nitrophenol (PNP) was measured at 410 nm.

Hot-water-extractable phosphorus (labile phosphate, LP) was determined by a method similar to that of Rother & Fay (1979): 250-ml portions of lake water were filtered using Whatman GF/F glass-fibre filters under gentle suction. After rinsing them with distilled water, the filters were transferred into tubes containing 10 ml distilled water, and boiled for 1 h in a water-bath. The extract was filtered, and molybdate reactive phosphate was determined as above.

Primary productivity for the diurnal studies has been reported by Dokulil et al. (1983).

3. Results

The three diurnal series during 1980 showed complex patterns of phosphorus dynamics (Figs. 1–3; the data were analysed as by Gunatilaka & Senaratne [1981] using a modified Delhomme [1978] kriging technique (Winkler, unpublished). Fluctuations in SRP, SP, PA, PP and LP levels on 8, 18 and 27 March are given in Figures 1–3, respectively. Some main physical and chemical characteristics of the three basins of Parakrama Samudra are summarized in Table 1. Detailed information on SRP, dissolved organic phosphate (DOP), LP, chlorophyl-a concentrations and gross production during the three diurnals are given in Table 2.

During the post-monsoon period (February–April 1980), the average SRP concentration in the three basins was higher, 6–12 $\mu g\,l^{-1}$ (Table 1) than the pre-monsoon averages (0–6 $\mu g\,l^{-1}$; Gunatilaka & Senaratne 1981) and never dropped below undetectable levels. Early morning surface concentrations were always quite close to this average SRP concentration at the PSN 14 station (Table 2), but slightly higher values were recorded at 2–3 m in the water column. The general trend was an increase of the SRP concentration in the water column, from morning onwards, parallel with the increase of light intensity (Figs. 1a, 2a and 3a). Also, either at noon or in the afternoon, a distinct increase in the SRP concentration in the 2- to 3-m region was observed. The peak values ranged from 19 to 26 $\mu g\,l^{-1}$ and represent three- to fourfold increase over the lake SRP concentration. This increase in the SRP concentration was very pronounced on two of the three occasions (8 March, Fig. 1a; and 27 March, Fig. 3a).

During the diurnals, soluble phosphorus level fluctuated between 16 and 67 $\mu g\,l^{-1}$. On two instances during the day, SP indicated a gradual decrease at all depths (Figs. 1b and 2b). The highest concentrations for the day were observed either at midnight or in the early morning and were followed by a gradual decrease (Figs. 1b, 2b and 3b). On 18 March, SP concentrations showed a conspicuous decrease during the period of maximum SRP demand (ca. 1200–1500 hours, Fig. 2b). There was a

Table 1. Mean and range (in parentheses) of some physical parameters, phosphate fractions, total nitrogen and chlorophyll-a for the three main basins of Parakrama Samudra and PSN 14 for March–April 1980, based on weekly analysis of water from 0.5-m depth. PSS, PSM & PSN: Parakrama Samudra south, middle and north, respectively. PSS-IF: inflow at PSS, PSN 14: station where diel changes were measured.

	PSS-IF	PSS	PSM	PSN	PSN 14
Conductivity (μs 25 °C)	255 ± 140	255 ± 140	199 ± 50	233 ± 11	231 ± 18
	(155 - 354)	(155 - 354)	(163 - 234)	(218 - 243)	(218 - 243)
pH	8.3 ± 0.2	8.6 ± 0.4	8.4 ± 0.4	8.5 ± 0.2	8.6 ± 0.2
	(8.2 - 9.1)	(8.2 - 9.1)	(8.1 - 8.7)	(8.4 - 8.9)	(8.3 - 8.9)
Alkalinity (meq.)	2.84 ± 0.43	2.95 ± 0.25	2.35 ± 0.50	1.94 ± 0.18	2.04 ± 0.24
	(2.63 - 3.18)	(2.63 - 3.18)	(1.84 - 3.04)	(1.76 - 2.20)	(1.76 - 2.20)
SRP ($\mu g\,l^{-1}$)	9.8 ± 1.5	12.2 ± 8.8	6.3 ± 2.6	6.2 ± 2.4	6.4 ± 1.2
	(8.0 - 10.9)	(6.0 - 18.4)	(3.8 - 9.6)	(2.3 - 8.3)	(5.2 - 8.0)
DOP (($\mu g\,l^{-1}$)	36 ± 18	28 ± 18	36 ± 11	31 ± 11	27 ± 11
	(16 - 52)	(16 - 49)	(28 - 49)	(16 - 42)	(16 - 40)
TP ($\mu g\,l^{-1}$)	73 ± 20.9	67 ± 13.8	76 ± 21.9	74 ± 22.9	78 ± 26.0
	(51 - 108)	(51 - 47)	(48 - 91)	(63 - 107)	(48 - 107)
TN (mg l^{-1})	0.66 ± 0.09	0.51 ± 0.16	0.59 ± 0.34	0.97 ± 0.09	0.72 ± 0.34
	(0.59 - 0.72)	(0.39 - 0.62)	(0.35 - 0.83)	(0.84 - 1.08)	(0.49 - 0.95)
*Chlorophyll-a ($\mu g\,l^{-1}$)	2.4 - 6.1	20.0	14.1 ± 6.4	16.2 ± 6.4	16.3 ± 5.8
		(13.5 - 24.0)	(9.6 - 23.4)	(7.3 - 31.6)	(9.6 - 9.8)

* Data from Dokulil *et al.* (1982).

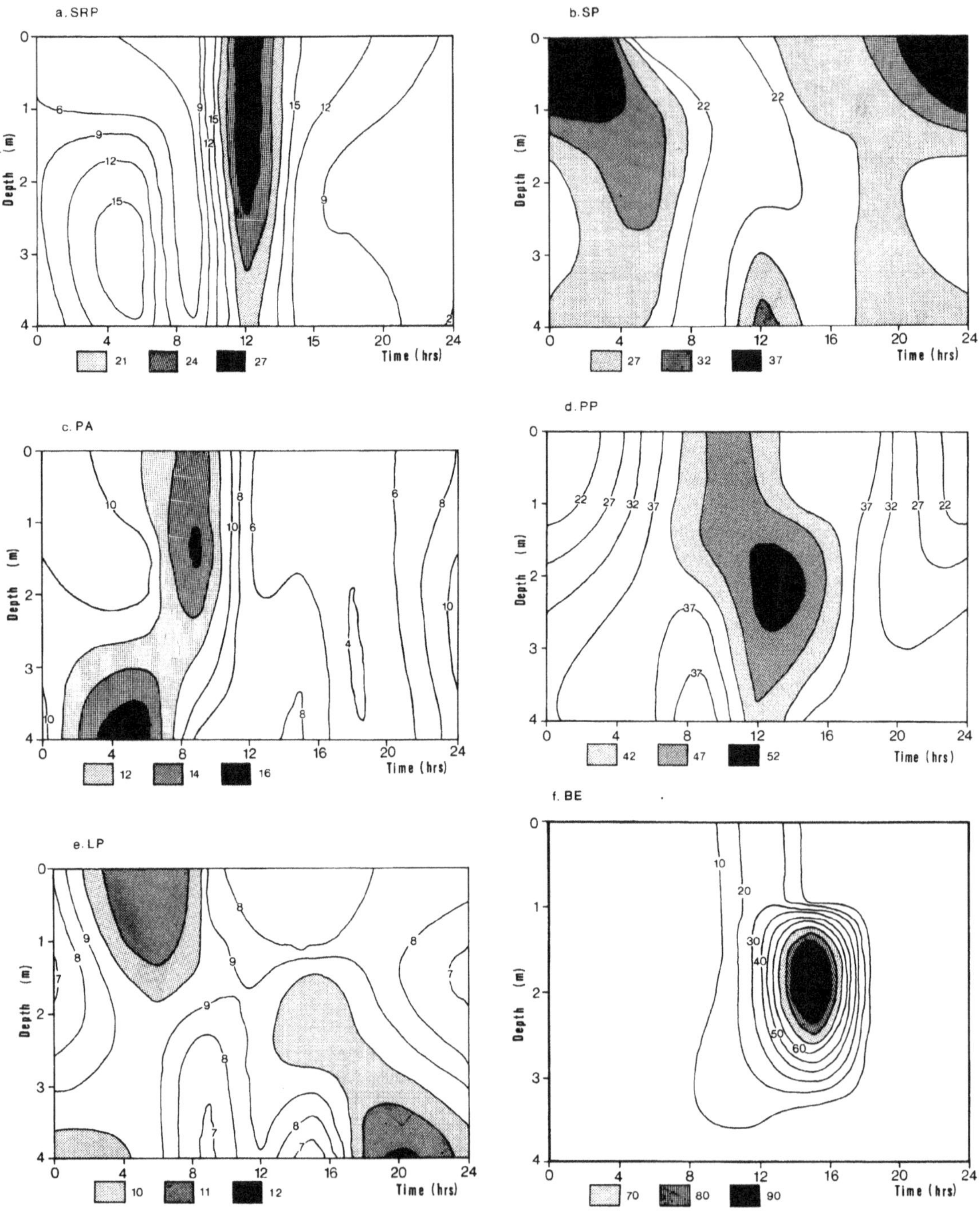

Fig. 1. Diel variations in (a) soluble reactive phosphate (SRP), (b) soluble phosphate (SP), (c) phosphatase activity (PA), (d) particulate phosphate (PP), (e) hot-water-extractable P as labile phosphate (LP) and (f) Birgean energy (BE) at PSN 14 on 8 March 1980. BE in relative units; PA: μmol PNPP hydrolysed per litre per hour and the rest in $\mu g\,l^{-1}$.

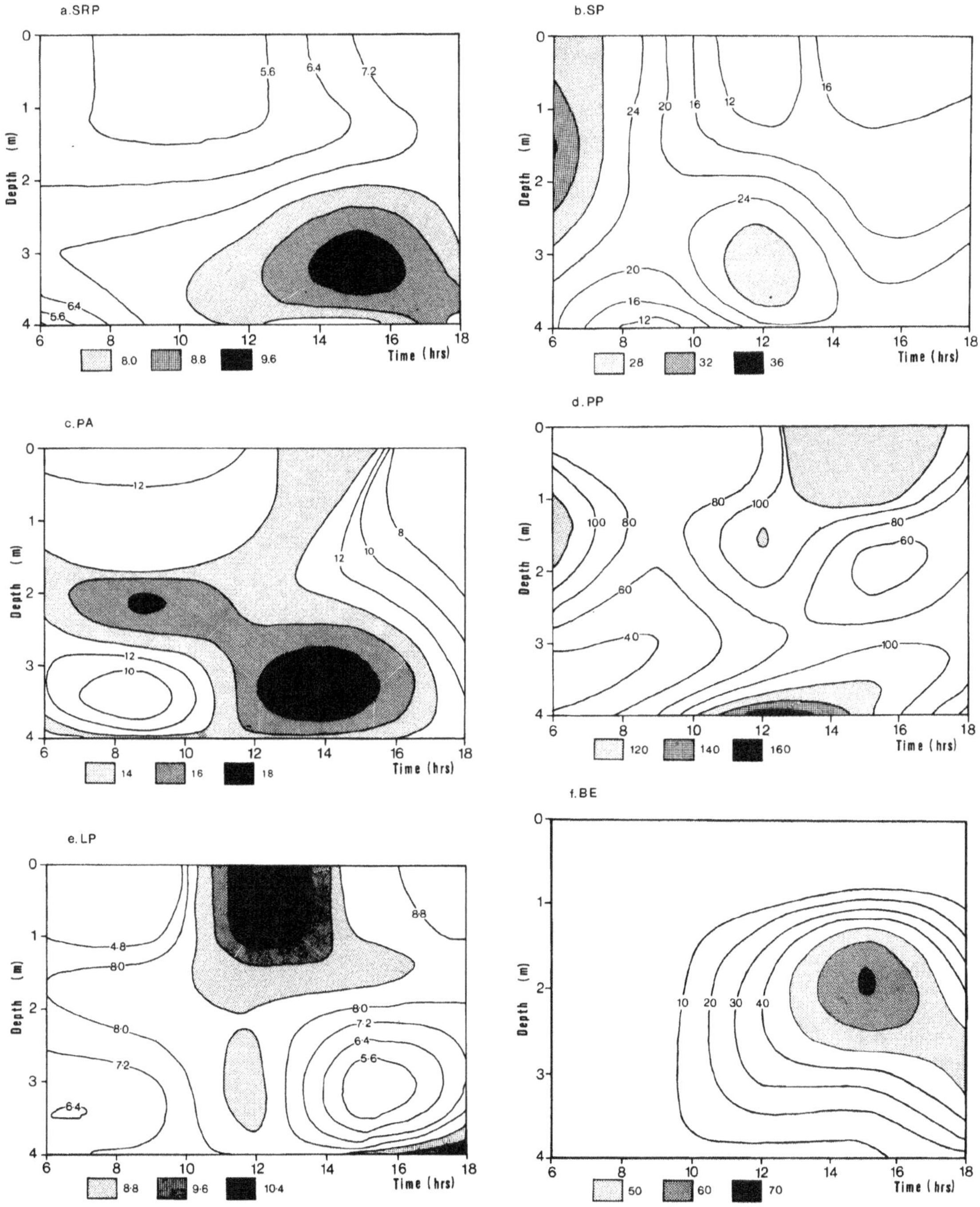

Fig. 2. Diel variations in (a) soluble reactive phosphate (SRP), (b) soluble phosphate (SP), (c) phosphatase activity (PA), (d) particulate phosphate (PP), (e) hot-water-extractable P as labile phosphate (LP) and (f) Birgean energy (BE) at PSN 14 on 18 March 1980. BE in relative units; PA: μmol PNPP hydrolysed per litre per hour and the rest in $\mu g\,l^{-1}$.

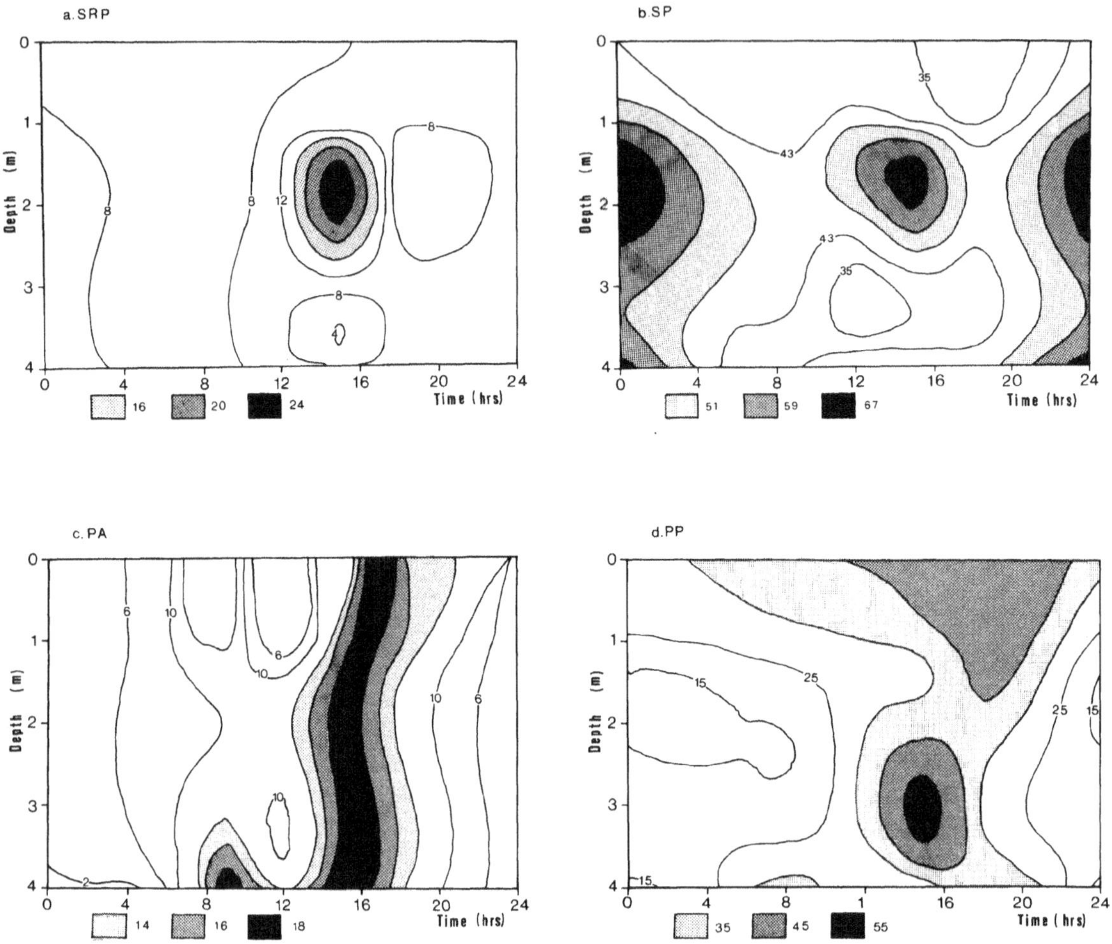

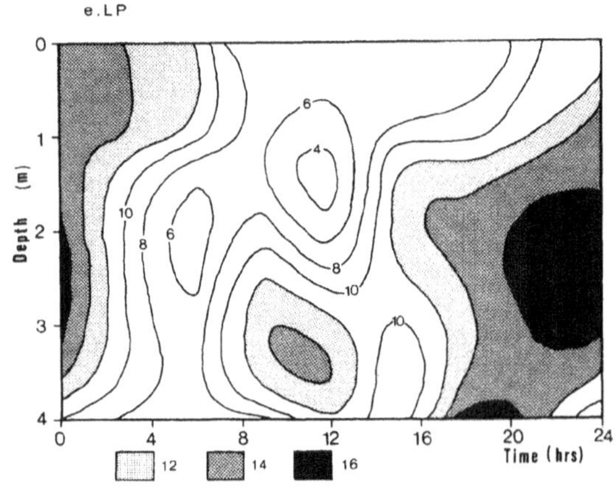

Fig. 3. Diel variations in (a) soluble reactive phosphate (SRP), (b) soluble phosphate (SP), (c) phosphatase activity (PA), (d) particulate phosphate (PP), (e) hot-water-extractable P as labile phosphate (LP) on 27 March 1980. PA: μmol PNPP hydrolysed per litre per hour and the rest in $\mu g\,l^{-1}$.

Table 2. Diel changes in soluble reactive phosphate (SRP), dissolved organic phosphate (DOP), labile phosphate (LP, measured as hot-water-extractable P), chlorophyll-a (Chl-a) concentrations and gross production (G. Prod.) at PSN 14 on 8, 18 and 27 March 1980.

Time (h)	Depth (m)	SRP ($\mu g\,l^{-1}$)	DOP ($\mu g\,l^{-1}$)	LP ($\mu g\,l^{-1}$)	Chl-a[a] ($\mu g\,l^{-1}$)	LP/Chl-a	G. Prod.[b] $mg\,O_2\,l^{-1}\,h^{-1}$
8. 03. 1980							
0630	0–1	5	28	11.6	12.1	0.96	0.26
	1–2	15	17	9.1	9.2	0.99	0.14
	2–3	14	12	9.1	12.5	0.73	0.05
	3–4	6	24	9.8	25.3	0.39	0.04
0930	0–1	6	16	9.4			0.23
	1–2	3	16	7.3			0.34
	2–3	6	11	6.6			0.27
	3–4						0.10
1230	0–1	26	8	8.0	19.1	0.42	0.20
	1–2	24	2	9.4	16.8	0.56	0.33
	2–3	19	15	8.9	16.6	0.54	0.21
	3–4	26	3	5.8	19.3	0.30	0.09
1530	0–1	14	14	7.4	13.3	0.56	0.12
	1–2	12	10	10.3	16.4	0.63	0.15
	2–3	12	13	6.8	13.9	0.49	0.02
	3–4	15	19	7.3	21.4	0.34	0.03
1830	0–1	12	15	8.5	17.6	0.48	
	1–2	9	19	9.9	14.4	0.69	
	2–3	12	16	11.1	12.0	0.93	
	3–4	9	13	10.1	11.3	0.89	
2400	0–1	6	33	7.0			
	1–2	8	15	7.7			
	2–3	6	24	10.4			
	3–4	5	26	9.1			
18. 03. 1980							
0630	0–1	6	28	7.1	16.0	0.44	0.28
	1–2	6	29	7.9	13.8	0.57	0.17
	2–3	7	20	6.7	10.6	0.63	0.05
	3–4	5	16	6.9	12.3	0.56	0.01
0915	0–1	5	16	6.8	13.1	0.52	0.37
	1–2	6	15	8.5	13.4	0.63	0.30
	2–3	8	13	7.1	11.9	0.60	0.15
	3–4	8	3	7.7	14.7	0.52	0.07
1230	0–1	5	5	11.1	11.9	0.93	0.47
	1–2	7	14	8.8	12.3	0.72	0.25
	2–3	8	23	9.1	16.4	0.55	0.18
	3–4	8	15	8.1	14.7	0.55	0.08
1540	0–1	7	12	8.7	9.8	0.89	0.21
	1–2	8	5	7.9	10.2	0.77	0.13
	2–3	10	8	5.0	10.8	0.46	0.03
	3–4	8	14	9.1	12.7	0.72	0.04
1830	0–1	8	6	7.9	10.6	0.74	
	1–2	8	6	7.9	10.4	0.76	
	2–3	8	13	7.3	15.8	0.46	
	3–4	10	13	11.0	22.0	0.50	
27. 03. 1980							
0630	0–1	5	33	10.8	26.7	0.30	0.41
	1–2	6	47	5.0	21.7	0.23	0.36
	2–3	5	43	6.8	21.9	0.31	0.08
	3–4	6	34	6.6	23.7	0.28	0.05

Table 2. (Continued).

Time (h)	Depth (m)	SRP ($\mu g\ l^{-1}$)	DOP ($\mu g\ l^{-1}$)	LP ($\mu g\ l^{-1}$)	Chl-a[a] ($\mu g\ l^{-1}$)	LP/Chl-a	G. Prod.[b] $mg\ O_2\ l^{-1}\ h^{-1}$)
1020	0–1	6	29	6.6	19.1	0.35	0.52
	1–2	7	38	8.9	20.7	0.43	0.41
	2–3	8	37	13.7	17.8	0.77	0.23
	3–4	7	33	7.9	12.5	0.63	0.89
1330	0–1	8	39	5.0	16.9	0.30	0.56
	1–2	11	42	5.0	14.5	0.34	0.29
	2–3	9	22	12.4	20.9	0.59	0.18
	3–4	9	43	11.8	13.4	0.88	0.05
1630	0–1	8	32	8.9	16.3	0.55	0.11
	1–2	27	13	12.7	25.7	0.49	0.09
	2–3	9	32	9.9	23.6	0.42	0.07
	3–4	9	41	8.9	22.2	0.40	0.01
1830	0–1	8	23	8.9	16.3	0.55	
	1–2	8	40	14.5	25.7	0.56	
	2–3	9	30	13.5	23.6	0.57	
	3–4	9	39	16.6	22.2	0.75	
2400	0–1	9	51	15.0	16.9	0.89	
	1–2	11	62	16.6	15.1	1.10	
	2–3	11	50	16.6	11.1	1.50	
	3–4	11	59	10.0	14.3	0.70	

[a,b] Data from Dokulil *et al.* (1982).

deviation from this general pattern on 27 March, when an increase in the whole water column occurred during mid-day and continued until late evening (Fig. 3b). The overall soluble phosphate concentration on this day was nearly doubled (27 March: 35–67; 18 March: 16–36; 8 March: 22–37 $\mu g\ l^{-1}$).

Phosphatase activity in the early morning as recorded in surface water ranged between 9 and 15 $\mu mol\ l^{-1}\ h^{-1}$. Slightly lower values were recorded at depths greater than 2 m. The phosphatase activity on two occasions increased gradually in the whole water column during the day (Figs. 1c and 3c), starting from 0900 hours onwards, reaching its maximum level between 1200 and 1600 hours. By noon, although the surface concentrations dropped to a minimum below 1 m, concentrations showed a conspicuous gain with maximum values reaching 16–19 $\mu g\ l^{-1}\ h^{-1}$. Subsequently these concentrations decreased by evening and continued to decrease till midnight. In contrast, on 18 March, neither a clear trend nor a clear distribution pattern of phosphatase activity was observed (Fig. 2c). Occasionally high values were recorded near the bottom. The level of organic phosphate significantly decreased with the increase of phosphatase activity ($n = 36$, $r = 0.522$, $P < 0.01$).

Although the labile phosphate (hot-water-extractable P) level fluctuated during the day, a slight decrease was apparent during mid-day (Figs. 1e, 2e and 3e). On 27 March, concentrations recorded for the early morning ranged between 10 and 11 $\mu g\ l^{-1}$ in the upper layers of the water column, but the values were nearly halved close to the bottom. On an other occasion (8 March, Fig. 1e), they differed by only a narrow margin (8–11 $\mu g\ l^{-1}$). Generally as the day progressed, the concentrations in the upper layers decreased, though slightly higher values were observed towards the bottom. By noon the surface values dropped to 5–6 $\mu g\ l^{-1}$, but a slight gain was seen in the afternoon (8–9 $\mu g\ l^{-1}$ at around 1500 hours). A positive recovery condition was approached by the evening and by midnight the values increased up to 11–13 $\mu g\ l^{-1}$. Also it was observed that, on 8 and 27 March, LP per unit chlorophyll-a decreased to a minimum during the day (Table 2), but increased gradually to reach a maximum at midnight.

The distribution of the particulate phosphorus appears to be close to the phytoplankton distribution pattern (Table 2 and Figs. 1d, 2d en 3d). From 0900 to 1500 hours, the surface particulate P concentrations showed a gradual increase that eventually extended to the bottom. However, the maximum accumulation was at the 1- to 2-m depth, and this coincides with the chlorophyll-a distribution pattern and gross production in the water column (Table 2; see also Dokulil et al. 1983). Between 1200 and 1500 hours, the average particulate phosphate concentration was comparatively high, which could in turn be related to the P accumulation in the phytoplankton biomass. Incidently this coincided with the primary production peak (Table 2). In the late afternoon the particulate P showed a gradual drop, at all depths, with a minimum at midnight.

On two occasions, early morning particulate P concentrations in the water column were between 31 and 41 $\mu g\,l^{-1}$. On 18 March, exceptionally high particulate P values were recorded during early morning and mid-day, with the highest values at 1500 hours (125 $\mu g\,l^{-1}$). In general, the PP concentrations could be considered rather low in comparison to SP values.

Figures 2f and 3f show the diurnal distribution of Birgean energy (in relative units) in the water column on 8 and 18 March. The energy calculations are based on temperature recordings by Bauer (1982). During these two days, the early morning energy differences in the water column were not large, but a slight gain occurred with depth, resulting in the development of a weak stratification. The stratification became stronger by noon, mostly between 1–3 m. After mid-day, the temperature-induced density differences became more pronounced, as indicated by steadily increasing energy levels, resulting in a stronger stratification between 1–3 m.

The total, daily radiation on both days was similar, but slight differences were recorded due to cloud cover and the lowest values were recorded on 8 March (see Dokulil et al. 1983).

4. Discussion

In the post-monsoon period, the SRP concentration in the lake was slightly higher than in the pre-monsoon phase, but this increase is not truely indicative of the available SRP level for the phytoplankton community. The overall concentration of dissolved phosphate in the lake is regulated by interrelated complex processes such as assimilation reactions, adsorption mechanisms, sedimentation and flushing rate.

The results presented here provide evidence for the existence of a diel SRP cycle. The variable patterns (Figs. 1–3) may reflect the changing external conditions as well as the varying physiological conditions of the phytoplankton population. The observed overall increase of the SRP level during the mid-day may be due to the hydrolysis of DOP in the lake water. The organic P fraction is comprised of organic phosphates brought into the lake via the inflow and excreted by phytoplankton. The ability of the phytoplankton to hydrolyse DOP compounds in laboratory experiments is well documented (Chu 1942; Harvey 1953; Kuenzler 1965, 1970). Berman (1970) and Wynne (1981a and b) reported that, in Lake Kinneret, *Peridinium* can utilize DOP under P-limited conditions. Also inorganic polyphosphates may form an important component of the algal excretory products (Lean & Nalewajko 1976) and they are to an extent susceptible to enzymatic hydrolysis. According to many workers (Torrani 1960; Kuenzler & Parras 1965; Healey 1973), phosphatases are induced or de-repressed when orthophosphate becomes growth limiting. In Parakrama Samudra, it was seen that SRP levels can increase during periods in the afternoon with a concomitant increase in phosphatase activity.

The hot-water-extractable phosphorus pool in algae (labile P) is mainly comprised of polyphosphates (Rhee 1973; Elgavish & Elgavish 1980). The study by Elgavish & Elgavish (1980) gave conclusive evidence to indicate that polyphosphates are the major intracellular phosphorus pool that is affected when orthophosphate is depleted from the growth medium. In Parakrama Samudra, there appears to be an inverse relationship between the hot-water-extracted P and phosphatase activity. The slight drop in the hot-water-extracted P content during mid-day corresponds to the depletion of the internal algal P pool at the time of maximum production (Table 2; see also Dokulil et al. 1983) and may supplement the algal inorganic P demand. This implies that some of the stored phosphates have been used by the algae.

The low ambient inorganic phosphate level associated with the concomitant increase in phosphatase activity and depletion in the algal internal-P pool as indicated by the low LP-chlorophyll-a ratios during mid-day implies that the algal population of Parakrama Samudra may be phosphorus limited at times of the day. This hypothesis is tested by a plot of dissolved nitrogen–SRP ratios at station PSN 14 in comparison to literature values for half-saturation constants (K_s) for P uptake by algae (Fig. 4). Based on Uehlinger's (1980) and Ahlgren's (1977) chemostat experiments for several algae, K_s for inorganic phosphate ranged between 1 and 10 $\mu g\,l^{-1}$. According to Ahlgren (1977), the K_s value for nitrogen ranged from 1 to 60 $\mu g\,l^{-1}$. From Figure 4, it is evident that more than 80% of the SRP values at station PSN 14 are below 10 $mg\,m^{-3}$ and more than 94% lie below 20 $mg\,m^{-3}$ (extended K_s limit for the OECD study: see Fricker 1980). This may indicate a P limitation.

In contrast, the dissolved nitrogen concentrations were well above twice the K_s value for nitrogen, indicating the relative abundance of dissolved N in the lake water. However, the values shown in Figure 4 represent total dissolved (inorganic + organic) nitrogen and a large percentage of it is contributed by organic fraction. Hence, a possible nitrogen limitation may be masked by the comparatively large dissolved nitrogen fraction.

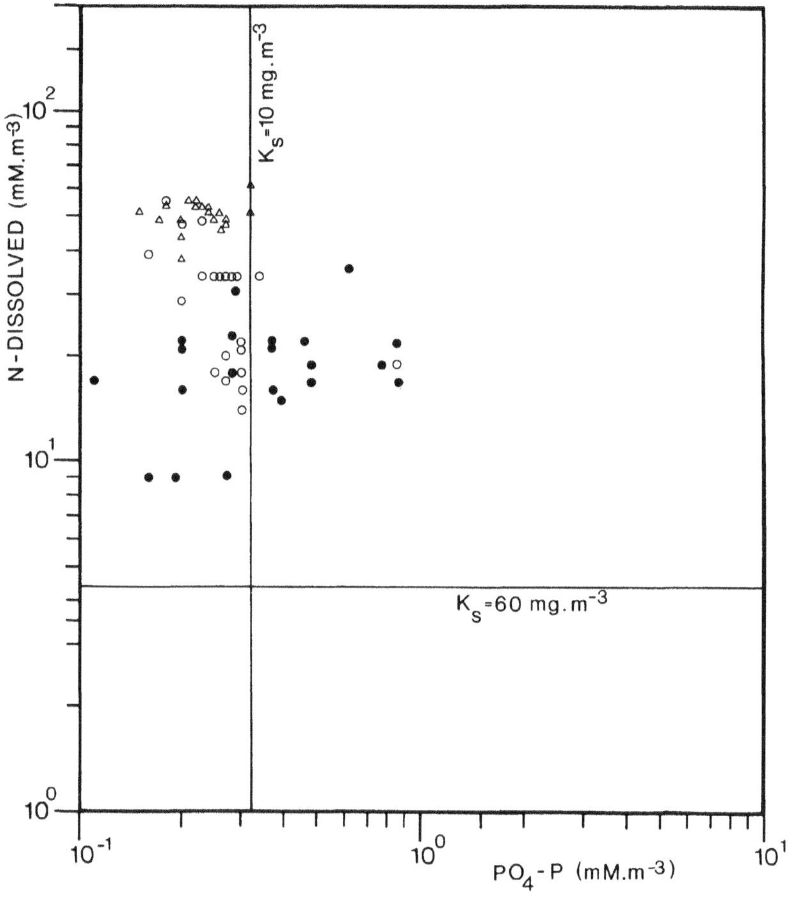

Fig. 4. Dissolved nitrogen to SRP ratio of the diel observations on 8 March 1980 (●), 18 March 1980 (Δ) and 27 March 1980 (○). Half-saturation constants (K_s) for uptake of phosphate (10 $mg\,m^{-3}$) and dissolved nitrogen by algae (see text for explanation) are indicated by the additional co-ordinates.

Physical factors also had an indirect role to play in the vertical distribution of nutrients in the lake. The prevailing calm conditions in the lake, and the high solar radiation characteristic of the tropics, resulted in the development of a strong stratification during the day. This is clearly shown by the distribution of Birgean energy in the water column (Figs. 2 and 3; see also Bauer 1983). The stability established during the day was sufficient to act as a barrier against mixing. This probably resulted in localization of the nutrients and hence may be one of the reasons for the existence of strong nutrient gradients during the afternoon.

It is also probable that vertical movement of the phytoplankton population was restricted by the thermal stratification. Incidently, the highest primary productivity in the water column, chlorophyll (Table 2; see also Dokulil et al. 1983), particulate phosphate, phosphatase activity and increase of SRP concentration were all observed at the same depth (Figs. 1–3). This implies that the nutrient demands of the 'entrapped' phytoplankton population were suplemented by the enhanced enzyme activity during the peak productivity period.

Acknowledgements

Thanks are due to H. Winkler for the long hours he spent at the computer. The criticism of C. Nalewajko and M. Tilzer helped to improve the text. The manuscript benefitted from comments by Harvey Shear, G. Falkner and A. Novak. I am thankful to C. Senaratne for the help in the laboratory and to M. Dokulil and I. de Silva for help in the field.

References

Ahlgren, G., 1977. Growth of *Oscillatoria agardhii* in chemostat culture. 1. nitrogen and phosphorus requirements. Oikos 29: 209–224.

Bauer, K., 1983. Thermal stratification, mixis and advective currents in Parakrama Samudra Reservoir, Sri Lanka. In: Schiemer, F. (ed.) Limnology of Parakrama Samudra – Sri Lanka: a case study of an ancient man-made lake in the tropics. Developments in Hydrology (this volume). Dr W. Junk, The Hague.

Berman, T., 1969. Phosphatase of inorganic phosphorus in Lake Kinneret. Nature 224: 1231–1232.

Berman, T., 1970. Alkaline phosphatases and phosphorus availability in Lake Kinneret. Limnol. Oceanogr. 15: 663–674.

Chu, S. P., 1942. The influence of the mineral composition of the medium on the growth of planktonic algae. I. Methods and culture media. J. Ecol. 30: 284–325.

Delhomme, J. P., 1978. Kriging in hydro sciences. Advances in Water Resources 1: 251–266.

Dillon, P. J. & Rigler, F. H., 1974. The phosphorus chlorophyll relationship in lakes. Limnol. Oceanogr. 19: 767–773.

Dokulil, M., Bauer, K. & Silva, I., 1983. An assessment of phytoplankton biomass and the primary productivity of Parakrama Samudra, a shallow man-made lake in Sri Lanka. In: Schiemer, F. (ed.) Limnology of Parakrama Samudra – Sri Lanka: a case study of an ancient man-made lake in the tropics. Developments in Hydrobiology (this volume). Dr W. Junk, The Hague.

Elgavish, A. & Elgavish, A., 1980. ^{31}P NMR differentiation between intracellular phosphate pools in *Cosmarium* (Chlorophyta). J. Phycol. 16: 368–374.

Fitzgerald, G. P. & Nelson, T. C., 1966. Extractive and enzymatic analysis for limiting or surplus phosphorous in algae. J. Phycol. 2: 32–37.

Fitzgerald, G. P., 1972. Bioassay analysis of nutrient availability. In: Allen, H. E. & Kramer, J. R. (eds.) Nutrients in Natural Waters, pp. 147–169. Wiley Interscience, New York.

Fogg, G. E., 1973. Phosphorus in primary aquatic plants. Water Research 7: 77–91.

Fricker, Hj., 1980. OECD eutrophication program – regional project. Alpine Lakes. (Final report.) Swiss. Federal Board for Environmental Protection, Bern, Switzerland. 234 pp.

Fuhs, G. W., 1969. Phosphorus content and rate of growth in the diatoms *Cyclotella nana* and *Thalassiosira fluviatilis*. J. Phycol. 5: 312–321.

Fuhs, G. W., 1972. Microbial influences on phosphorus cycling. In: Ballantine, R. K. (ed.) The aquatic environment: the microbial transformations and water management implications, pp. 149–169. EPA 430/G-73-008.

Galloway, R. A. & Krauss, R. W., 1963. Utilization of phosphorus sources by *Chlorella*. In: Microalgae and photosynthetic bacteria. Plant Cell Physiol. Spec. Suppl., pp. 569–575.

Golterman, H. L., 1973. Natural phosphate sources in relation to phosphate budgets: a contribution to the understanding of eutrophication. Water Research 7: 3–17.

Golterman, H. L., 1975. Physiological Limnology. Elsevier, Amsterdam. 489 pp.

Gunatilaka, A., 1980. The chemistry of Parakrama Samudra. In: Schiemer, F. (ed.) Parakrama Samudra (Sri Lanka), Limnology Project, Interim Report, pp. 35–53. IIZ, Vienna.

Gunatilaka, A. & Senaratna, C., 1981. Parakrama Samudra (Sri Lanka) Project, a study of a tropical lake ecosystem. II: Chemical environment with special reference to nutrients. Verh. Internat. Verein. Limnol. 21: 1000–1006.

Harvey, H. W., 1953. Note on the absorption of organic phosphorus compounds by *Nitzchia closterium* in the dark. J. Mar. Biol. Assoc. U.K. 31: 475–476.

Healey, F. P. & Hendzel, L. L., 1975. Effect of phosphorus deficiency on two algae growing in chemostats. J. Phycol. 11: 303–309.

Healey, F. P. & Hendzel, L. L., 1976. Physiological changes during the course of blooms of *Aphanizomenon flos-aqaue*. J. Fish. Res. Board. Can. 36: 36–41.

Healy, F. P. & Hendzel, L. L., 1980. Physiological indicators of nutrient deficiency in lake phytoplankton. Can. J. Fish. Aquat. Sci. 37: 442–453.

Hutchinson, G. E., 1967. A treatise on Limnology, vol. 2. Wiley, New York, 1015 pp.

Jones, J. G., 1972. Studies of freshwater bacteria: association with algae and alkaline phosphatase activity. J. Ecol. 60: 59–75.

Jones, J. G., 1973. Studies on freshwater micro-organisms: phosphatase activity in lakes of differing degree of eutrophication. J. Ecol. 60: 777–791.

Ketchum, B. H. & Corwin, N., 1965. The cycle of phosphorus in a plankton boom in the Gulf of Maine. Limnol. Oceanogr. (Suppl. R) 10: 148–161.

Kuenzler, E. J., 1965. Glucose-6-phosphate utilization by marine algae. J. Phycol. 1: 156–164.

Kuenzler, E. J., 1970. Dissolved phosphorus excretion by marine phytoplankton. J. Phycol. 6: 7–13.

Kuenzler, E. J. & Parras, J. P., 1965. Phosphorus of marine algae. Biol. Bull. 128: 271–284.

Kuhl, A., 1962. Inorganic phosphorus uptake and metabolism. In: Lewin, R. A. (ed.) Physiology and Biochemistry of Algae, pp. 211–229. Academic Press, New York.

Kuhl, A., 1974. Phosphorus. In: Stewart, W. D. P. (ed.) Algal physiology and biochemistry, pp. 636–654. Blackwell Scientific, Oxford.

Lean, D. R. S., 1973a. Phosphorus dynamics in lake water. Science 179: 678–680.

Lean, D. R. S., 1973b. Phosphorus movement between its biologically important forms in lake water. J. Fish. Res. Board Can. 30: 1525–1536.

Lean, D. R. S. & Nalewajko, C., 1976. Phosphate exchange and organic phosphorus excretion by freshwater algae. J. Fish. Res. Board Can. 33: 1312–1323.

Lean, D. R. S. & Nalewajko, C., 1979. Phosphorus turnover time and phosphorus demand in large and small lakes. Arch. Hydrobiol. Beih. Ergbn. Limnol. 13: 120–132.

Mackereth, F. Y. H., 1953. Phosphorus utilization by *Asterionella formosa* Haas. J. Exp. Bot. 4: 296–313.

Murphy, J. & Riley, J. P., 1962. A single-solution method for the determination of phosphate in natural waters. Anal. Chim. Acta 27: 31–36.

Oglesby, R. T. & Schaffner, W. R., 1975. The response of lakes to phosphorus. In: Porter, K. S. (ed.) Nitrogen and Phosphorus Food Production, Waste and the Environment, pp. 23–57. Ann Arbor Sci., Ann Arbor.

Overbeck, J., 1962. Untersuchungen zum Phosphathaushalt von Grünalgen: II. Die Verwertung von Pyrophosphat und organisch gebundenen Phosphaten und ihre Beziehung zu den Phosphatasen von *Scenedesmus quadricauda* (Turp.) Brép. Arch. Hydrobiol. 58: 281–308.

Peters, R. H., 1976. Orthophosphate turnover in East African lakes. Oecologia (Ber.) 25: 313–319.

Rautanen, N. & Karkkainen, V., 1951. On the Phosphatase activity of low-phosphorus *Torulopsis utilis*. Acta Chem. Scand. 8: 106–111.

Raveh, A. & Aunimelech, Y., 1979. The total nitrogen analysis in water, soil and plant material with pursulphate oxidation. Water Research 13: 911–912.

Redfield, A. C., Ketchum, B. H. & Richards, F. A., 1963. The influence of organisms on the composition of sea-water. In: Hill, M. N. (ed.) The Sea, pp. 26–77. Wiley Interscience, New York.

Reichardt, W., 1971. Catalytic metabolism of phosphate in lake water and by Cyanophyta. Hydrobiologia 38: 377–394.

Reichardt, W., Overbeck, J. & Stenbing, L., 1967. Free dissolved enzymes in lake water. Nature 216: 1345–1347.

Rhee, G. Y., 1973. A continuous culture study of phosphate uptake, growth rate and polyphosphate in *Scenedesmus* sp. J. Phycol. 9: 495–506.

Rigler, F. H., 1964. The phosphorus fractions and turnover time of phosphorus in different types of lakes. Limnol. Oceanogr. 9: 511–518.

Rigler, F. H., 1966. Radiobiological analysis of inorganic phosphorus in lake water. Verh. Internat. Verein. Limnol. 16: 465–470.

Rigler, F. H., 1968. Further observations inconsistent with the hypothesis that the molybdenum blue method measures inorganic phosphorus in lake water. Limnol. Oceanogr. 13: 7–13.

Rigler, F. H., 1973. A dynamic view of phosphorus in lakes. In: Griffith, E. et al. (eds.) Environmental Phosphorus Handbook, pp. 539–572. John Wiley & Sons, Toronto.

Rother, J. A. & Fay, P., 1979. Some physiological-biochemical characteristics of planktonic blue-green algae during bloom formation in three Salopian meres. Freshwater Biol. 9: 369–379.

Sakamoto, M., 1966. Primary production by phytoplankton community in some Japanese lakes and its dependence on lake depth. Arch. Hydrobiol. 62: 1–25.

Schiemer, F., 1983. The Parakrama Samudra Project – scope and objectives. In: Schiemer, F. (ed.) Limnology of Parakrama Samudra – Sri Lanka: a case study of an ancient man-made lake in the tropics. Developments in Hydrobiology (this volume). Dr W. Junk, The Hague.

Schindler, D. W., 1976. Biogeochemical evolution of phosphorus limitation in nutrient-enriched lakes of the precambrian shield. In: Nriagu, J. O. (ed.) Environmental Biogeochemistry, pp. 647–664. Ann Arbor Sci., Ann Arbor.

Smith, R. E. H. & Kalff, J., 1981. The effect of phosphorus limitation on algal growth rates: evidence from alkaline phosphatase. Can. J. Fish. Aquat. Sci. 38: 1421–1427.

Sproule, J. L. & Kalff, J., 1978. Seasonal cycles in the phytoplankton phosphorus status of a north temperate zone lake (Lake Memphremagog, Que.-Vt.), plus a comparison of techniques. Verh. Int. Ver. Limnol. 20: 2681–2688.

Strickland, J. D. H. & Parsons, T. R., 1968. A practical handbook of seawater analysis. Bull. Fish. Res. Board Can. 167.

Thomas, E. A., 1969. The process of eutrophication in central European lakes. In: Eutrophication: causes, concequences, correctives, pp. 29–49. Nat. Acad. Sci./ Nat. Res. Council Publ. 1700.

Torrani, A., 1960. Influence of inorganic phosphate in the formation of phosphatases by *E. coli*. Biochim. Biophys. Acta 38: 460–469.

Uehlinger, U., 1980. Untersuchungen zur Autoekologie der planktischen Blaualga *Aphanizomenon flos-aquae*. Ph.D. thesis, Swiss Federal Institute of Technology, Zurich.

Viner, A. B., 1973. Responses of a mixed phytoplankton population to nutrient enrichments of ammonia and phosphate, and some associated implications. Proc. R. Soc. Lond. B 183: 351–370.

Vollenweider, R. A., 1968. Scientific fundamentals of the eutrophication of lakes and flowing waters with particular reference to nitrogen and phosphorus as factors in eutrophication. OECD, DAS/CSI/68.27, Paris.

Wetzel, R. G., 1975. Limnology. Saunders, Philadelphia. 743 pp.

Wetzel, R. G., 1981. Long-term dissolved and particulate alkaline phosphatase activity in a hardwater lake in relation to lake stability and phosphorus enrichments. Verh. Internat. Verein. Limnol. 21: 363–395.

Wynne, D., 1977. Alterations in activity of phosphatases during the *Peridinium* bloom in Lake Kinneret. Physiol. Plant. 40: 219–224.

Wynne, D. & Berman, T., 1980. Hot water extractable phosphorus: an indicator of nutritional status of *Peridinium cinctum* (Dinopyceae) from Lake Kinneret? J. Phycol. 16: 40–46.

Wynne, D., 1981a. Phosphorus, phosphatases and the *Peridenium* bloom in Lake Kinneret. Verh. Internat. Verein. Limnol. 21: 523–527.

Wynne, D., 1981b. The role of phosphatases in the metabolism of *Peridenium cinctum*, from Lake Kinneret. Hydrobiologia 83: 93–99.

Author's address:
A. Gunatilaka
Institute of Zoology
University of Vienna
Althanstr. 14
A-1090 Vienna
Austria

5. An assessment of the phytoplankton biomass and primary productivity of Parakrama Samudra, a shallow man-made lake in Sri Lanka

M. Dokulil, K. Bauer & I. Silva

Keywords: irrigation reservoir, phytoplankton, primary productivity, tropics, Sri Lanka

Abstract

Phytoplankton biomass was assessed from microscopic counts and from measurements of chlorophyll-a. Figures ranged from 2 to 35 mg FW l^{-1} (FW, fresh weight) and from 5 to 105 μg Chl-a l^{-1}, respectively. The phytoplankton assemblage is dominated by blue-green algae (*Anabaenopsis raciborskii, Lyngbya circumcreta, Microcystis* spp.) and diatoms (*Melosira granulata*). In addition, various green algal species occur in low numbers. Only *Monoraphidium* exhibits higher numbers.

Primary productivity, assessed by both the ^{14}C technique and the oxygen light and dark bottle method, ranged from 3.8 to 14.7 g O_2 m^{-2} day^{-1}. Most of the productivity can be attributed to the algal fraction smaller than 33 μm. Photosynthetic efficiency was highest in the afternoon as a result of greater phosphorus availability because of increased phosphatase activity.

Community respiration (0.002–0.400 mg O_2 l^{-1} h^{-1}) was stimulated by light and is not a constant fraction of photosynthesis. Specific rates per unit chlorophyll-a were high.

Photosynthesis is, to a great extent, controlled by light and phosphorus availability. Newly formed biomass is exploited by zooplankton (mainly rotifers) and herbivorous fish (mainly *Sarotherodon mossambicus*). Biomass levels in the reservoir are largely affected by the flow-through, which is dependent on irrigation management activity.

1. Introduction

Shallow freshwater impoundments in the tropics are becoming increasingly important as sources of fish protein and various methods including the introduction of new species are frequently tried in the hope of improving the fishery. To judge the success of these changes, it is necessary to know the maximum sustainable fish yield. This, in turn, requires reliable estimates of the primary productivity.

Recently, considerable knowledge of such productivity in tropical freshwater systems has accumulated, especially in Africa (Ganf 1972, 1974a and b, 1975; Lemoalle 1973, 1975, 1979; Melack 1979a and b, 1980; Robarts 1979; Robarts & Southall 1977; Talling 1965, 1966a and b, 1975; Talling et al. 1973) and India (Ganapati & Sreenivasan 1972; Sreenivasan 1964a and b, 1968, 1970). However, relatively little information exists about the factors controlling primary productivity in shallow irrigation reservoirs. These were primarily constructed for irrigation and flood control, which means that hydrological factors, such as water-level fluctuations and flow-through, are potentially important regulating mechanisms of species succession and biomass development in the phytoplankton.

This study provides an estimate of the primary production in the plankton community of Parakrama Samudra, a large shallow irrigation reservoir in the northern dry zone of Sri Lanka. For a general description of the lake, refer to Schiemer (1980, 1981, 1983). In addition, attempts are made

Schiemer, F. (ed.), Limnology of Parakrama Samudra - Sri Lanka
 ISBN 90 6193 763 9

to identify the factors responsible for the control of production on the basis of concurrent hydrological, physico-chemical and plankton data.

2. Materials and methods

Data were collected in Sri Lanka during two research visits in August/September 1979 and March/April 1980. The results reported here mainly refer to station 14 in the northern basin of Parakrama Samudra (PSN), but also included are details of temporal variation of algal biomass at stations PSN 3 and PSN 8. In addition, the investigation of areal heterogeneity was extended to include stations in the middle (PSM) and southern parts (PSS). All stations were sampled at a depth of 0.5 m except PSN 14, where samples were taken from the surface down to 3.5 m at 0.5-m intervals. Samples were collected with a darkened Ruttner-type water sampler of 2 litres capacity and stored in complete darkness to minimize affects due to exposure to high light intensities. All manipulations in the boat were protected from exposure to direct sunlight.

Total incoming radiation (TIR) in the wavelength range from 300 nm to 3 000 nm was continuously recorded from 1 March 1980 until 6 April 1980 by a star pyranometer (Schenk, Austria). Photosynthetic active radiation (PhAR) between 400 nm and 700 nm was quantified both above the surface and underwater with a PhAR quantum radiometer (Lambda Instruments, USA). Spectral characteristics of lake water were obtained using a selenium photocell (Schenk, Austria) and coloured glass filters BG 12, VG 9, and RG 610 (Schott, West Germany). Extinction coefficients for PhAR and the different spectral regions were calculated according to Vollenweider (1955). The total downwelling energy flux (E) was estimated from selenium cell readings, adopting the procedure given by Vollenweider (1961).

The depth of the euphotic zone (z_{eu}) was assumed equivalent to the 1% light level. Secchi depth was taken as the mean of the disappearing and reappearing depth of a white disk of 25-cm diameter (average reflection 85%). In addition, coloured Secchi disks have been tested in a few cases (Elster & Štepánek 1967).

Only the PhAR quantum radiometer and a white Secchi disk were available in 1979.

Chlorophyll-a was estimated from spectrophotometer readings after extraction of the pigment in cold absolute methanol, adopting the procedure of Holm-Hansen & Riemann (1978). Glass-fibre filters (Whatman GF/C) were usually stored overnight in a freezer and extracted on the next day, except for the samples of 8, 15, and 18 March, which had to be stored in the cold until 24 March 1980 because of a methanol shortage. For the same reason, several samples had to be kept in complete darkness at room temperature (approx. 28° C) for up to 30 days during the 1979 visit. Some of the observed irregularities in the phaeopigment concentrations may have been the result of prolonged storage causing degradation of the pigment. However, experiments in the laboratory (Bauer, unpublished) with algal cultures indicated no significant degradation during storage, under simulated Sri Lankan climatic conditions.

Table 1. Average linear dimensions of algal species, geometric shape and mean volumes for single cells, filaments of colonies.

	l (μm)	w (μm)	Geometric shape	Volume (μm^3)
Anabaenopsis raciborskii (filament)	75	3	Cylinder	520
Lyngbya circumcreta (filament)	35	2	Cylinder	135
Lyngbya limnetica (filament)	67	1.5	Cylinder	120
Melosira granulata (filament)	60	10	Cylinder	4 700
Merismopedia punctata (cell)	3.5	2.5	Sphere	17
Merismopedia tenuissima (cell)	2.0	1.5	Sphere	4
Microcystis spp. (colony)	40	35	Ellipsoid	20 000
Monoraphidium irregulare (cell)	14	1	Spindle	18
Mougeotia sp. (filament)	80	4	Cylinder	1 005
Peridinium inconspicuum (cell)	24	18	Ellipsoid	3 500
Scenedesmus spp. (colony)	24	12	Square	860
Synedra acus (cell)	47	3	Square	405

Enumeration of phytoplankton cells was performed using an inverted microscope (Lund et al. 1958). Identification of the algal species follows Rott (1983). Algal biomass was estimated from cell counts and mean algal volume, calculated from 25 to 50 individual measurements, after applying simple geometric formulae (Table 1). Biovolume was transposed to fresh weight assuming a specific gravity of 1.

Photosynthesis was estimated either from changes in oxygen concentration or carbon uptake in light and dark bottles incubated 'in situ'. Dark bottles were wrapped in aluminium foil and kept in light-proof bags.

Oxygen concentration was measured by two different modifications of the Winkler technique. In 1979, the titrations were performed directly in the bottles. After the precipitate had settled, some of the supernatant water was withdrawn with a syringe; the precipitate was then dissolved and titrated with 0.01 n sodium thiosulfate, using soluble starch as the end-point indicator. The precision of ten samples was ± 1.1%. In 1980, the back-titration technique was amperometric end-point detection according to Talling (1973) was used. The reproducibility was better than ±0.5%. From the oxygen changes in the light and dark bottles and the initial O_2 concentration, gross and net photosynthetic rates were calculated.

Table 2. Central key to symbols and units.

Symbol	Unit	Explanation
TIR	$J\ cm^{-2}\ d^{-1}$	Total incoming radiation per day
PhAR	$\mu E\ m^{-2}\ s^{-1}$	Photosynthetic available radiation
I_o	$kJ\ m^{-2}\ h^{-1}$	Subsurface PhAR intensity per hour
ΣI_o	$kJ\ m^{-2}\ d^{-1}$	Subsurface PhAR intensity per day
I_o	$\mu E\ m^{-2}\ s^{-2}$	Subsurface PhAR
I_i	$\mu E\ m^{-2}\ s^{-2}$	Light intensity describing onset of inhibition
I_{opt}	$\mu E\ m^{-2}\ s^{-2}$	Light intensity at depth of optimum photosynthesis
I_K	$\mu E\ m^{-2}\ s^{-2}$	Light intensity describing onset of saturation
I_c	$\mu E\ m^{-2}\ s^{-2}$	Light intensity at compensation depth (A = R)
Iz_{SD}	%	Light intensity at Secchi depth
E	%	Relative light intensity between 400 and 700 nm
ϵ_ν	$ln\ m^{-1}$	Vertical extinction coefficient of PhAR = ϵ_{PhAR}
ϵ_{min}	$ln\ m^{-1}$	Minimum vertical extinction coefficient
ϵ_s	$ln\ m^{-1}$	Self-shading coefficient
z_{SD}	m	Secchi depth
z_{eu}	m	Depth of euphotic zone = depth of 1% light level
z_{opt}	m	Depth of optimum photosynthesis
z_c	m	Compensation depth (photosynthesis = respiration)
z_m	m	Maximum depth
B	$mg\ Chl\text{-}a\ m^{-3}$	Phytoplankton biomass
ΣB	$mg\ Chl\text{-}a\ m^{-2}$	Biomass integrated over euphotic zone
A_{opt}	$mg\ O_2\ l^{-1}\ h^{-1}$	Photosynthetic rate at optimum depth
P_{opt}	$mg\ O_2\ mg\ Chl\text{-}a^{-1}\ h^{-1}$	Specific photosynthetic rate per unit biomass
ΣA	$mg\ O_2\ m^{-2}\ h^{-1}$	Depth integral of photosynthetic rate
$\Sigma\Sigma A$	$g\ O_2\ m^{-2}\ d^{-1}$	Primary productivity per day
R	$mg\ O_2\ l^{-1}\ h^{-1}$	Respiration rate
R_{opt}	$mg\ O_2\ mg\ Chl\text{-}a^{-1}\ h^{-1}$	Specific respiration rate per unit biomass
ΣR	$mg\ O_2\ m^{-2}\ h^{-1}$	Depth integral of respiration rate
$\Sigma\Sigma R$	$g\ O_2\ m^{-2}\ d^{-1}$	Respiration per day (24 h)
q_c	-	Net production per day = $\frac{\Sigma\Sigma A}{\Sigma\Sigma R}$
e	$mg\ O_2\ mg\ Cl\text{-}a^{-1}\ kJ^{-1}\ m^2$	Efficiency
$\overline{NCi}$	$\mu g\ Chl\text{-}a\ l^{-1}$	Net chlorophyll increase per day
$\overline{NP}$	$\mu g\ O_2\ l^{-1}\ d^{-1}$	Average daily net productivity over water column
TBP	$\mu g\ Chl\text{-}a\ l^{-1}$	Theoretical produced biomass during interval
Δt	h	Day length

Photosynthetic carbon fixation was estimated from ^{14}C-uptake experiments according to Steemann-Nielsen (1952). Bottles were inoculated with 1 ml ^{14}C-stock solution (NaH-$^{14}CO_3$, 3 μCi) and exposed for 2–3 h 'in situ'. At the end of the exposure time, the bottles were detached and transferred to the shore in complete darkness. Subsamples (10 ml) were immediately filtered onto 0.2-μm Millipore filters, using a syringe plus a 'Swinnex' filter holder. Filters were transferred to scintillation vials and air-dried. Activity of carbon-14 on the filters was measured by liquid scintillation count-

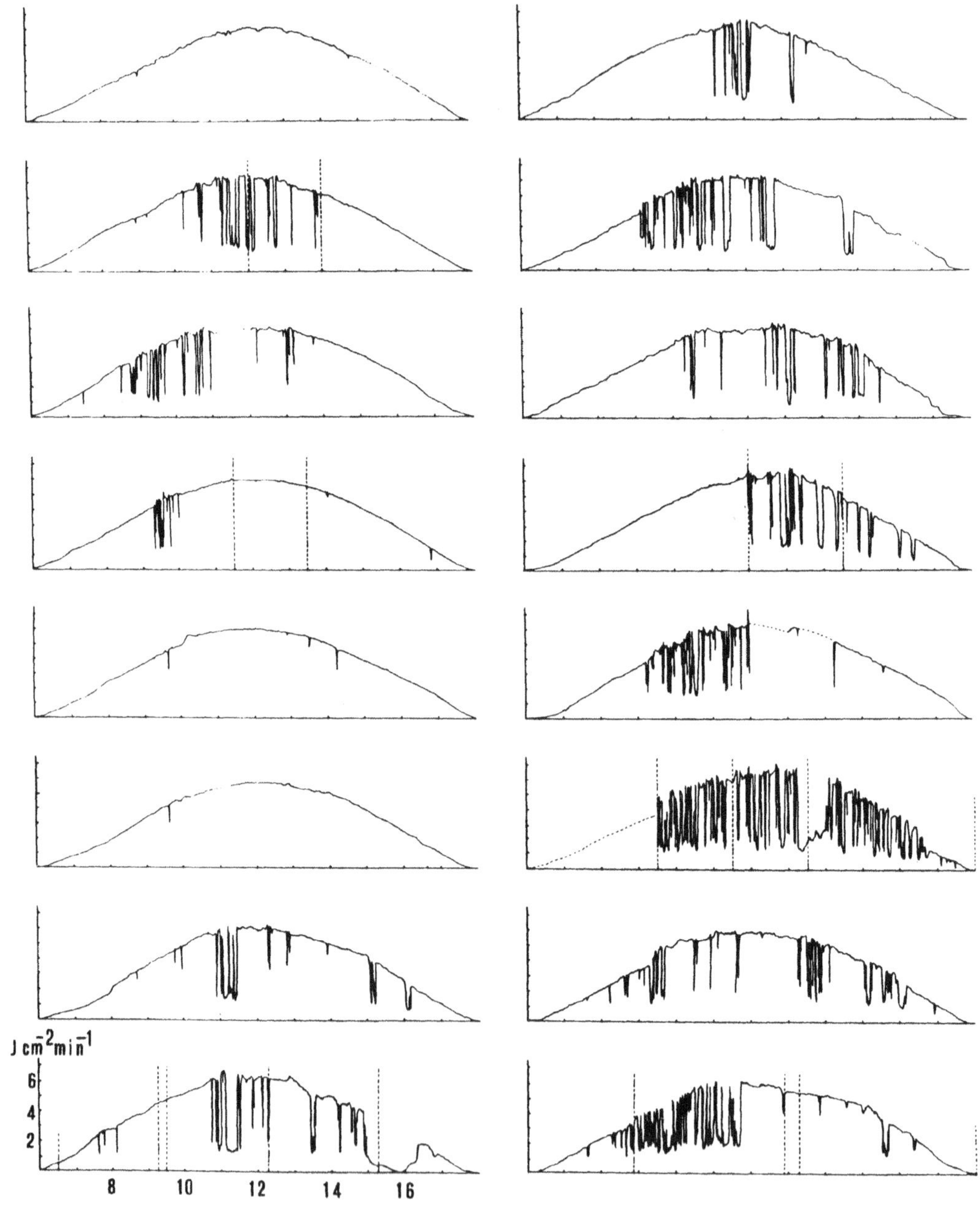

ing. Samples from 1979 were counted by Dr. Dharmawardana, Radio Isotope Centre, University of Colombo, but 1980 samples were brought back to Austria and counted four weeks later.

All counts were corrected for filtration error (e.g. McMahon 1973) by filtering different volumes and extrapolating to 0 ml filtered. Dark ^{14}C uptake was subtracted from the uptake in the light. Radioactive-labelled compounds released during incubation were not estimated.

Total inorganic carbon concentration was calculated from alkalinity titrations by using the tables of Rebsdorf (1972).

Post-incubation size fractionation of carbon-uptake experiments was carried out by passing the water through 33- and 10-μm netting before filtration.

Light dependence of photosynthesis was tested in an experiment on 30 August 1979. A range of seven different light intensities (0, 7.5, 45, 150, 275, 390 and 950 $\mu E\ m^{-2}\ s^{-1}$) were obtained by using glass plates and different layers of thin white paper. Triplicate bottles were exposed in glass jars that were covered with these glass plates and placed in the lake water close to shore. The water in the jars was changed regularly in order to keep the temperature constant.

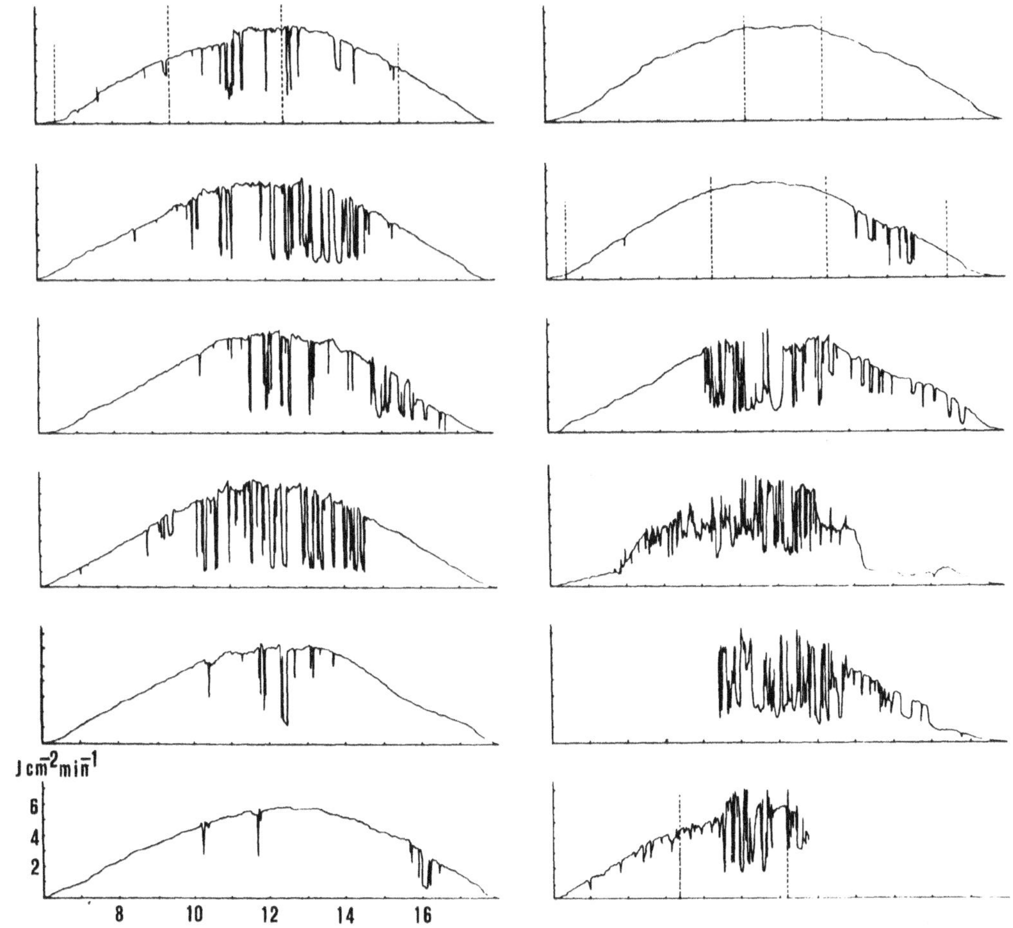

Fig. 1. Daily variation of total incoming radiation (TIR) for the period 1 March 1980 till 6 April 1980. Integrals per day can be identified from Table 3. Broken lines indicate experimental periods. Average hourly radiation over experimental times is given in Table 8.

Symbols and units used in the present study are summarized in Table 2.

The responsibilities are as follows: K. Bauer for the data from the 1979 visit, and M. Dokulil and I. Silva for the material from the year 1980. We acknowledge the help of E. Kumar during one of the mid-day exposures.

3. Results and discussion

3.1. Light conditions

Variation of TIR per day was remarkably small during the period of 1980 (Fig. 1). Based on local noon-time, values were 10.5% ± 8.4% higher for the afternoon period (Table 3). Average radiation amounted to 24.4 ± 1.9 (7.7%) MJ m^{-2} day^{-1}. Photosynthetic active radiation at mid-day was between 2 150 and 2 350 μE m^{-2} s^{-1} in 1979 and ranged from 1 600 to 2 150 μE m^{-2} s^{-1} in 1980. The fraction between 400 and 700 nm (PhAR) of the total daylight radiation varied from 40.9% to 51.1% in 17 comparative measurements on different days at different times (Table 4). The average of 45.3% is close to the figure given by Talling (1957b).

The mean albedo of the water surface, calculated from the same data set, amounts to 5.1%.

An example of underwater PhAR attenuation is given in Figure 2A. Stronger light extinction and hence greater coefficients sometimes occur at greater depths, due to increased light absorption and scattering by denser algal layers. Because of the curved nature of PhAR penetration, a single extinction coefficient is a rough approximation.

However, a coefficient of 9.27 was obtained on a single occasion in August 1979 because of high turbidity due to wind action. In the calm period of September, values ranged from 2.95 to 4.40 (Table 5). During the second visit, extinction coefficients were considerably lower (0.88–1.48) because of calm wind conditions. A similar increase of absorption with depth was observed in different parts of the spectrum (Fig. 2B). The most penetrating component is the red spectral region ($\epsilon_{red} = 0.98$), green being very similar ($\epsilon_{green} = 1.00$). The blue spectral component is most strongly absorbed ($\epsilon_{blue} = 2.46$). Total energy flux (E) calculated from filter readings according to Vollenweider (1961) closely corresponds with PhAR attenuation.

Of the three lake basins, light penetration is distinctly higher in the middle part compared with PSS and PSN (Fig. 2A). Consequently the depth of the euphotic zone was about 5 m in PSM and 3.1–3.9 m in the two other basins in spring 1980. During August and September 1979, the 1% level varied considerably from 0.5 to 2.73 m. These smaller and more variable values were caused by material stirred up from the sediment, an effect of frequent storms, convective mixis and higher chlorophyll-a values.

In consequence, Secchi depth was 30–105 cm in 1979, whereas it ranged from 110 to 195 cm in March and April 1980. Coloured Secchi disks agree with selenium cell measurements of the respective spectral regions. On average the incident light is reduced to 19.4% at Secchi depth (z_{SD}), suggesting a factor of 1.6 relating z_{eu} to the vertical extinction coefficient. In fact the mean value of all observations is 1.72 (Table 8). As shown by Vollenweider (1960), the correlation between Secchi depth and the extinction coefficient is of the general form

$$(\epsilon_\nu)^n \times z_{SD} = \text{const.}$$

Based on 17 pairs of measurements in Parakrama Samudra, $n = 0.84$ and the constant is 1.54 ($r^2 = 0.95$, $P \leq 0.001$; Fig. 3).

The depth of the euphotic zone may be estimated directly from measurement of PhAR attenuation. Calculation of z_{eu} from photocell readings in different spectral regions, however, is very laborious. Talling (1965) was able to show that the euphotic zone may be approximated by b/ϵ_{min}, where b is a constant of 3.7. This constant proved to be applicable in a wide range of African and European lakes (Talling et al. 1973; Bindloss 1974; Ganf 1974a). Jewson (1977) reports a slightly higher figure of 3.93 for Lough Neagh. The mean factor relating z_{eu} to ϵ_{min} in Parakrama Samudra is 4.0 ($n = 12$), which is almost identical with the value mentioned before. Total down-welling radiation will be overestimated, when only the most penetrating spectral component is considered. Most of the difference is eliminated by raising ϵ_{min} to 1.15 ϵ_{min}, a factor somewhat smaller than 1.33 derived by Talling (1957a).

Alternatively the depth of the euphotic zone may be estimated from Secchi-disk readings by a factor

Table 3. Total incoming radiation (TIR) for the morning and the afternoon hours, and per day, recorded at the Freshwater Fisheries Station in Polonnaruwa during the 1980 visit.

Date	0600–1200	1200–1800	Radiation/day
	(J cm^{-2})		(J cm^{-2} day^{-1})
1.3.1980	1202	1391	2593
2.3.1980	1256	1318	2574
3.3.1980	1247	1410	2657
4.3.1980	1275	1353	2628
5.3.1980	1287	1328	2616
6.3.1980	1169	1280	2450
7.3.1980	1101	1284	2385
8.3.1980	1391	930	2321
9.3.1980	1297	1167	2464
10.3.1980	1142	1207	2348
11.3.1980	1243	1230	2472
12.3.1980	1186	1256	2440
13.3.1980	1167	1306	2472
14.3.1980	1044	1112	2156
15.3.1980	No data		
16.3.1980	1212	1244	2456
17.3.1980	1010	1350	2360
18.3.1980	1150	1316	2466
19.3.1980	1305	1239	2544
20.3.1980	1188	1212	2400
21.3.1980	1334	1138	2472
22.3.1980	1285	1296	2581
23.3.1980	No data		
24.3.1980	1195	1345	2541
25.3.1980	1220	1294	2514
26.3.1980	No data		
27.3.1980	1350	1159	2509
28.3–1.4.1980		No data	
2.4.1980	954	1185	2139
3.4.1980	No data		
4.4.1980	1004	747	1751
5.4.1980			~1730
6.4.1980			~2454

empirically derived from systematic combined measurements of both parameters. For Parakrama Samudra, $z_{eu} = 2.75\ z_{SD}$ ($n = 18$) with 95% confidence limits 1.93-3.57.

3.2. Species composition, biomass and light interception

The phytoplankton assemblage of Parakrama Samudra was characterized by blue-green algae and diatoms in August/September 1979. In addition, various green algal species appeared in low numbers. Biomass was dominated by *Melosira granulata, Anabaenopsis raciborskii,* and an uni-

Table 4. Total incoming radiation (TIR) compared with simultaneous measurements of the photosynthetic available radiation (PhAR).

TIR (J cm^{-2} min^{-1})	PhAR μE cm^{-2} min^{-1})	PhAR (%)
5.94	11.7	43.0
6.11	12.0	43.2
5.86	12.0	45.1
6.17	12.0	42.6
5.78	11.4	43.4
5.73	12.1	46.5
4.52	10.5	51.1
5.86	12.9	48.5
1.84	3.4	40.9
2.80	6.0	47.0
5.40	11.6	47.5
1.80	3.8	47.0
5.86	12.0	45.1
3.52	6.6	41.3
5.78	12.0	45.6
4.73	10.2	47.5

dentified *Mougeotia* species (cf. Rott 1983). After the monsoons, between February and April 1980, differences in species composition of the three lake basins were observed. The blue-green alga *Anabaenopsis raciborskii* and the green *Mougeotia* sp. were the most important components in the southern basin (PSS). In the northern basin, these two species were accompanied by *Lyngbya limnetica, L. circumcreta, Merismopedia punctata, M. tenuissima,* and *Monoraphidium irregulare.* In contrast, the deeper basin (PSM), between these two, was dominated by the diatom *Melosira granulata* and the *Mougetia* species, with the cyanophytes having only minor importance. The presumed water bloom of *Mycrocystis* in PSM (Rott 1982) could not be verified in the quantitative samples.

The phytoplankton biomass normally has a homogeneous distribution with depth in windy conditions, but differences may arise in calmer weather (but see Bauer 1983). Two examples of the latter are illustrated in Figure 4. The contribution of the various algal groups is shown as well. On both occasions, phytoplankton biomass exhibited a peak in the lower half of the water column, possibly as a result of light preference, buoyancy regulation or sedimentation during the calmer conditions.

Biomass estimates based on fresh weight (FW) (mg l^{-1}) and chlorophyll-a are in good agreement (Table 6) and the following relation has been calcu-

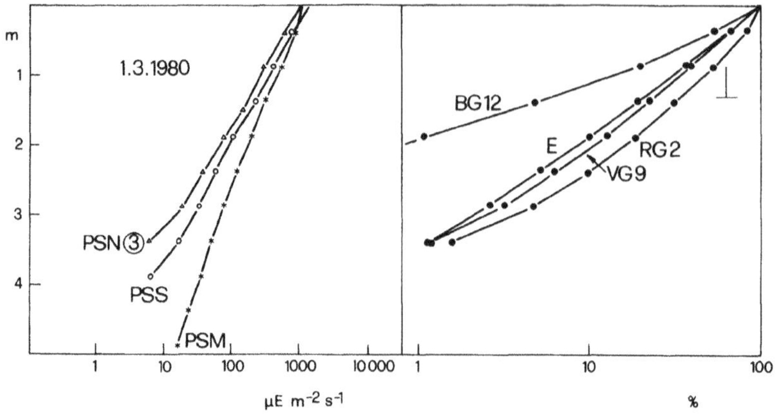

Fig. 2. Semiologarithmic plots of underwater light attenuation. (A) Photosynthetic available radiation (PhAR) in absolute units (μE m^{-2} s^{-1}) in the three basins of Parakrama Samudra. PSS, south basin; PSM, middel basin; PSN 3, northern basin, station 3. (B) Example of relative light intensities in the blue (BG 12), green (VG 9) and red (RG 2) spectral region. The percentage transmission (E) between 400 and 700 nm is calculated from these three spectral regions.

lated: $B_{Chl\text{-}a} = -1.9947 + 0.0025\ B_{FW}$ (n = 24, $r^2 = 0.71$, $P \leq 0.001$). The mean chlorophyll-a content of fresh-weight biomass amounts to 0.25% ± 0.10%, a figure often found in blue-green algal dominated freshwater environments (cf. Table 17.6 in Dokulil 1979).

Algal cells form part of the suspended matter that modifies the underwater attenuation of light. In water bodies of low or moderate productivity, the phytoplankton contributes little to light attenuation with depth. Increasing crops, however, can intercept a substantial proportion of incident light.

One way to describe variations in light penetration due to changes in the algal density is the increment of the vertical extinction coefficient, usually ϵ_{min}, per unit increase in population density (ϵ_s in ln units per mg chlorophyll-a per m^2). In the present study, mean column chlorophyll-a concentration

Table 5. Underwater light characteristics and derived parameters during the 1979 and 1980 research period. For symbols refer to Table 2.

		z_{SD} (m)	ϵ_ν (m^{-1})	ϵ_{min} (m^{-1})	Iz_{SD} (%)	z_{eu} (m)	$\epsilon_\nu \cdot z_{SD}$	z_{eu}/z_{SD}	z_m/z_{eu}
1979									
22.8	PSN 14	0.30	9.21	–	(6.3)	0.50	2.76	1.67	4.0
30.8	PSN 4	0.50	2.95	–	22.8	1.56	1.48	3.12	1.3
7.9	PSN 14	0.40	4.37	–	17.4	1.05	1.75	2.63	1.9
14.9	PSN 3	0.50	4.22	–	12.1	1.09	2.11	2.18	1.8
14.9	PSN 8	0.60	2.79	–	18.8	1.65	1.67	2.75	1.2
1980									
29.2	PSN 14	1.10	1.08	0.91	30.4	4.26	1.19	3.87	0.9
1.3	PSS	1.36	1.32	–	16.6	3.94	1.80	2.57	1.0
	PSM	1.78	0.95	–	18.5	4.85	1.69	2.72	0.8
	PSN 8	1.37	1.22	–	18.8	3.77	1.67	2.75	1.1
4.3	PSN 14	1.25	1.17	1.04	23.2	3.94	1.46	3.15	1.0
7.3	PSM	1.95	0.93	0.81	16.3	4.95	1.81	2.54	0.8
	PSN 8	1.20	1.40	–	18.6	3.29	1.68	2.74	1.2
	PSN 3	1.15	1.28	–	22.9	3.60	1.47	3.13	1.1
17.3	PSM	1.85	0.88	–	19.6	5.23	1.63	2.83	0.8
	PSN 8	1.15	1.47	–	18.4	3.13	1.69	2.72	1.3
	PSN 3	1.10	1.48	1.29	19.6	3.11	1.63	2.83	1.3

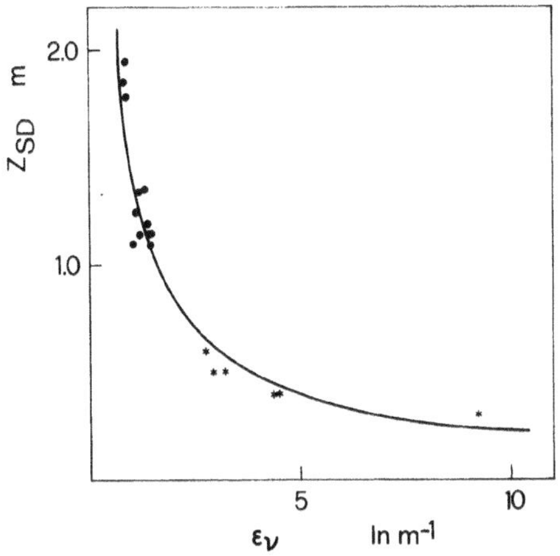

Fig. 3. Relation of the Secchi depth (z_{SD}) to the vertical extinction coefficient: * = 1979, • = 1980; the correlation coefficient is 0.97.

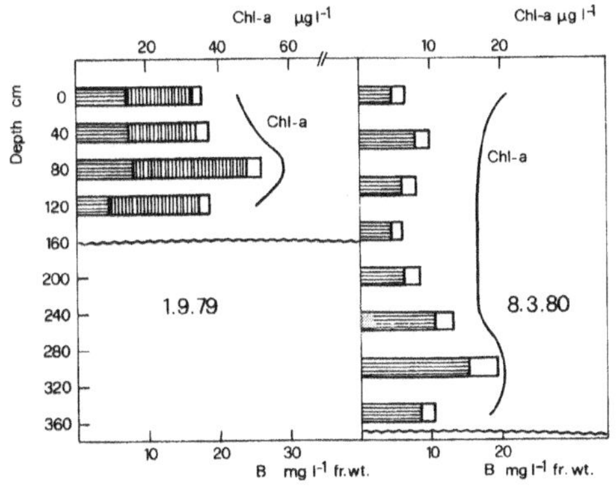

Fig. 4. Examples of vertical distribution of the phytoplankton in 1979 and 1980. Fresh-weight biomass (B) is shown as horizontal bars: [blue-green bar] blue greens, [diatom bar] diatoms, □ all other groups.

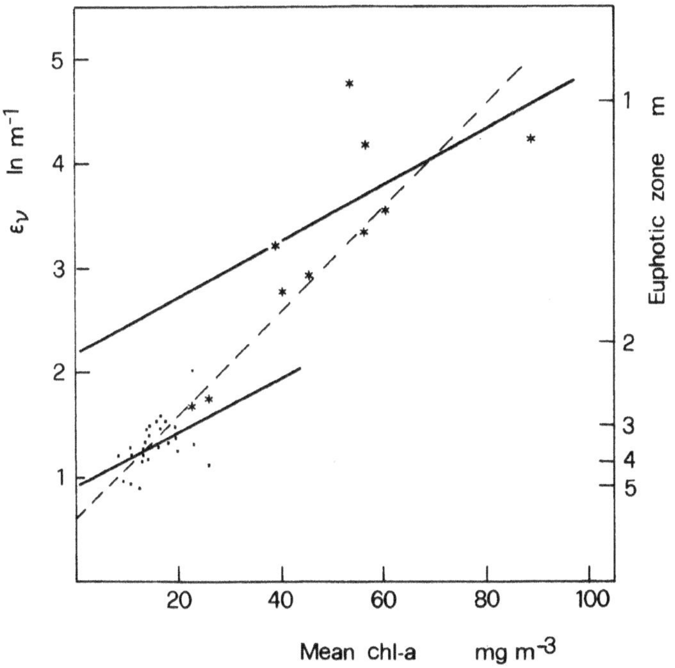

Fig. 5. The relationship between the vertical extinction coefficient (ϵ_ν) for PhAR and the mean chlorophyll-a concentration in the water column: * = 1979, • = 1980. Continuous lines for individual years (not significant); broken line for both years. For further explanation refer to the text.

Table 6. Phytoplankton counts (no. l^{-1}), fresh-weight biomass (FW) chlorophyll-a concentration, and chlorophyll content for selected samples from 1979 and 1980.

1979				1980			
No. l^{-1} $\times 10^6$	FW mg l^{-1}	Chl-a $\mu g\ l^{-1}$	%	No. l^{-1} $\times 10^6$	FW mg l^{-1}	Chl-a $\mu g\ l^{-1}$	%
28.98	23.81	46.2	0.19	29.97	6.45	20.8	0.32
67.65	17.60	45.8	0.26	77.87	10.10	17.4	0.17
57.96	18.7	50.9	0.27	46.79	7.99	16.8	0.21
54.83	25.9	58.1	0.22	30.65	6.13	16.8	0.27
38.06	18.6	50.2	0.27	58.55	8.48	16.6	0.20
34.02	10.9	26.2	0.24	68.75	13.27	16.6	0.14
25.15	6.4	39.8	0.62	129.99	20.53	20.1	0.10
71.79	23.40	60.7	0.26	56.45	10.57	18.5	0.18
55.80	29.50	105.0	0.36	8.42	2.25	6.7	0.30
61.75	20.2	57.1	0.28	8.46	1.99	5.3	0.27
				8.49	2.10	6.7	0.32
				16.91	5.01	13.5	0.27
				23.11	8.45	9.6	0.11
				53.22	8.52	14.0	0.16

Table 7. Light characteristics (PhAR) for the phytoplankton populations of 1979 and 1980. For symbols, refer to Table 2.

Date	Time	I_o ($\mu E\ m^{-2}\ s^{-1}$)	I_i	I_{opt}	I_K	I_o/I_K	I_c	z_c (m)
1979								
30.8	1530	950	550	390	330	2.9	–	–
1.9	0840	1050	690	500	190	5.5	1	1.7
	1125	1200	–	(1200)	215	5.8	13	1.4
5.9	1030	1250	–	(1250)	–	–	–	0.5
7.9	1215	760	–	(760)	233	3.2	–	–
13.9	1425	920	–	(920)	205	4.5	1	1.5
1980								
2.3	1200	1260	435	365	220	5.7	32	2.8
4.3	1145	1800	800	495	320	5.6	68	2.8
8.3	0635	380	–	(380)	250	1.5	8	1.8
	0925	1030	350	225	50	20.6	5	3.5
	1245	1840	300	183	100	18.4	16	3.2
	1600	135	–	(135)	70	1.9	6	2.0
15.3	1200	1200	–	(1200)	150	8.0	34	2.4
18.3	0700	480	–	(480)	120	4.0	75	1.6
	0945	1800	600	542	350	5.1	24	3.6
	1245	1700	1200	940	750	2.3	48	2.9
	1545	900	600	470	235	3.8	88	1.8
25.3	1130	1450	880	800	180	8.1	31	3.1
27.3	0645	450	–	(450)	130	3.5	20	2.4
	1030	1550	–	(1550)	370	4.2	43	2.6
	1345	1100	–	(1100)	390	2.8	7	3.2
	1640	400	–	(400)	230	1.7	5	2.2

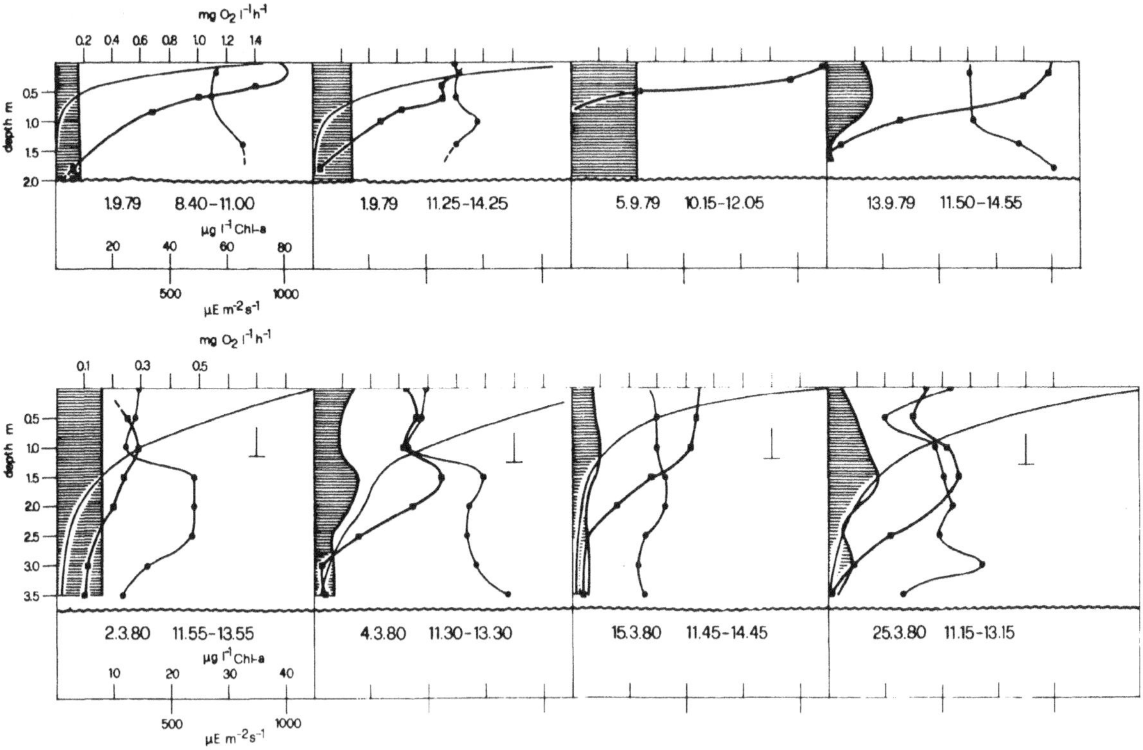

Fig. 6. Depth profiles of gross photosynthesis (■), respiration ▤, chlorophyll-a concentration (•), and light attenuation (continuous line) for the years 1979 and 1980: Secchi depth is indicated in 1980. Note the different scales in 1980.

was plotted against ϵ_{PhAR} (Fig. 5). The ϵ_s value calculated from the slope of the regression for all available data (pre- and post-monsoon periods) is 0.050, which is 0.043 when based on ϵ_{min}, using the relationship $\epsilon_{PhAR} = 1.15\ \epsilon_{min}$ (cf. section 3.1).

This self-shading coefficient is substantially higher than those reported elsewhere, summarized in Harris (1978) and Westlake (1980). The most obvious explanation seems to be that pooling of the data from both visits results in an artificial but highly significant correlation. Separate calculations for 1979 and 1980 gave ϵ_s values of 0.027 and 0.025, respectively, which were not well correlated. Anyhow, individual treatment of the two data sets is feasible on the basis of the observed large difference in background extinction between periods.

During the 1979 period, frequent winds stirred material up from the sediment; whereas under the prevailing conditions in 1980 (deeper water, almost no wind), light attenuation was much less affected by particles other than phytoplankton. The difference in the background extinction between the two visits clearly appears in Figure 5 as the interceptions of the two regression lines with the *y*-axis.

The above-mentioned ϵ_s values were converted to $\epsilon_s$min, giving 0.023 and 0.021, to be comparable with published results. From Harris (1978, Fig. 26) it is evident that the self-shading coefficients of the present study are among the highest so far reported (average cell volumes were respectively 1 500 and 250 μm^3). Owing to the theoretical treatment of light interception by algal cells by Kirk (1976), high ϵ_s values are expected in Parakrama Samudra because of considerable importance of light scattering, especially under the turbid conditions of 1979. The scattering by particles is responsible for variability of the data.

Defining the euphotic zone by the relationship $4.0/\epsilon_{min}$ (cf. section 3.1), the maximum possible quantity of chlorophyll contained in the euphotic zone (ΣB_{max}) will be theoretically equal to 4.0/$\epsilon_s$min (Talling 1965). Thus the maximum chlorophyll-a content within z_{eu} amounts to 174 and 190 mg m^{-2} for 1979 and 1980, respectively. The observed values of ΣB show that these quantities were never reached (Table 8).

3.3. Depth distribution

Photosynthesis and its distribution with depth is perhaps the aspect most thoroughly explored in freshwater systems at present. Selected mid-day profiles in Parakrama Samudra are given in Figure 6 and more information about depth distribution may be obtained from the diurnal studies illustrated in Figure 9A–C.

Suppression of photosynthesis near the surface (light inhibition) occurred in 10 of the 22 experiments in PSN (Table 7). Since photo-inhibition clearly is a light-induced effect, it is essentially a diurnal phenomenon. Early morning and late afternoon profiles generally show little or no light inhibition (Fig. 6 and Table 7), as has been shown frequently by various authors reviewed by Harris (1978). The onset of photo-inhibition (I_i) has been calculated to range from 320 to 1 200 μE m^{-2} s^{-1}

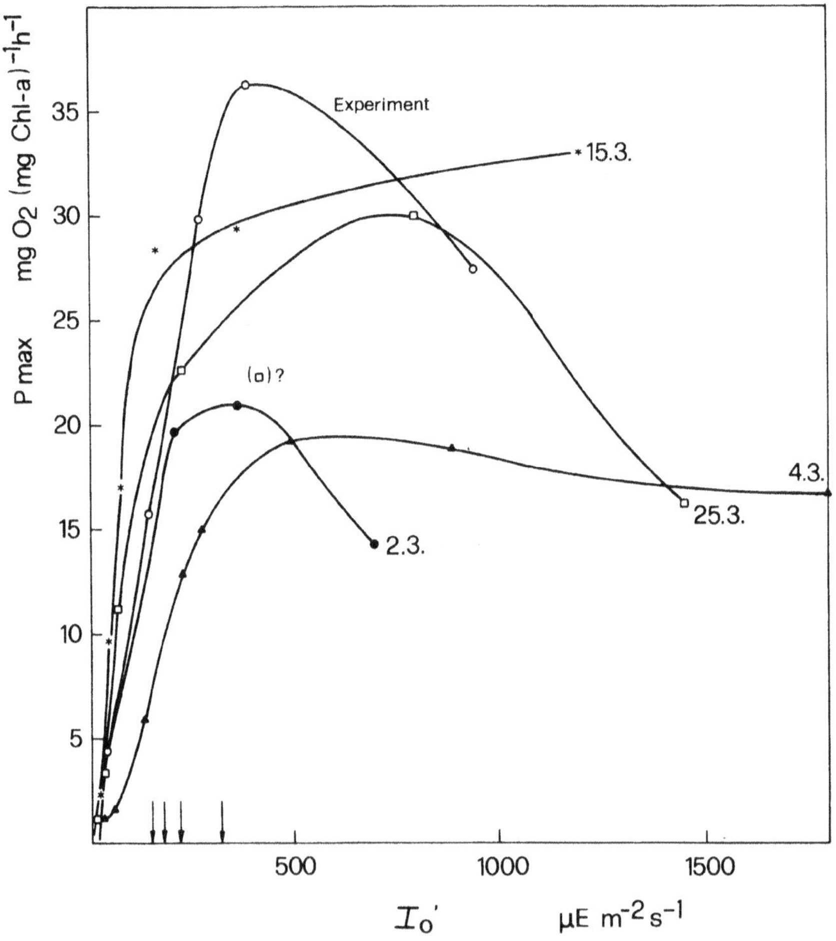

Fig. 7. Relationships between specific photosynthesis and irradiance for mid-day *in situ* exposures and under experimental conditions (cf. section 2). Values of I_K are indicated by arrows on the irradiance axis.

Table 8. Summary of light characteristics, biomass, photosynthesis, respiration and derived parameters for 1979 and 1980. For symbols and units refer to Table 2.

Date	Time	I_o	ΣI_o	ϵ_ν	B	ΣB	A_{opt}	P_{opt}	ΣA	$\Sigma\Sigma A$	R	R_{opt}	ΣR	$\Sigma\Sigma R$	q_c	e
1979																
1.9	0840			4.10	56.2	122.0	1.62	28.8	1 480		0.16	2.8	330			
	1125			3.24	50.9	105.0	1.06	20.8	1 075	12.5[b]	0.27	5.3	500	10.0[b]	1.25	
5.9	1015				53.6	46.0	1.85	34.5	850	8.5[b]	0.47	8.8	490	11.8[b]	0.72	
7.9	1200			4.41	57.7	74.0	1.68[a]	29.1	988[a]	9.9[a]	–	–	–	–	–	
13.9	1150			4.77	50.1	122.0	1.59	31.7	1 470	14.7[b]	0.18	3.6	325	7.8[b]	1.88	
				4.77	50.1	122.0	1.00[a]	20.0	748[a]	7.5[a]	–	–	–	–	–	
19.9	1350				105.0	73.0	1.63	15.6	929	9.3[b]	0.10	0.9	203	4.9[b]	1.90	
26.9	1345				47.7	99.0	1.34	28.0	761	7.6[b]	0.35	7.3	560	13.4[b]	0.57	
1980																
2.3	1155	1 380	11 244	1.32	12.2	64.5	0.29	23.8	850	6.9	0.12	9.8	420	4.8	1.44	0.010
4.3	1130	1 560	11 482	1.17	29.8	91.0	0.45	15.1	915	6.7	0.16	5.4	344	5.4	1.24	0.006
8.3	0630	828		1.15	9.6	46.0	0.37	38.5	430		0.13	13.5	238			0.011
	0915	1 389		1.52	–	–	0.35	–	833		0.16	–	192			0.012
	1230	1 180		1.54	16.8	61.3	0.35	21.0	800		0.23	13.7	313			0.022
	1540	180		1.54	18.5	56.0	0.25	13.5	330		0.25	13.5	182			0.033
	1830	–	10 144	–	–	50.8	–	–	–	7.0	0.07	–	185	4.9	1.42	0.013
15.3	1145			1.49	13.6	50.0	0.45	33.0	830	8.3[b]	0.10	7.4	240	5.8[b]	1.27	
18.3	0630	594		1.16	14.8	49.3	0.31	20.9	400		0.15	7.8	359			0.014
	0930	1 322		1.20	16.0	47.5	0.38	23.8	800		0.11	6.9	211			0.013
	1230	1 314		1.23	12.0	47.5	0.51	42.5	833		0.13	10.8	287			0.013
	1530	360		1.29	9.0	36.9	0.23	25.5	367		0.21	?	272			0.028
	1830	–	10 773	–	–	50.0	–	–	–	7.2	0.07	–	125	4.9	1.47	0.013
25.3	1115	1 548	10 985	1.24	20.3	70.4	0.46	22.7	992	7.1	0.18	8.9	360	5.5	1.29	0.009
27.3	0630	845		1.30	27.0	81.3	0.35	12.9	543		0.11	4.1	261			0.008
	1020	1 581		1.38	16.0	69.4	0.54	33.8	1 100		0.19	11.9	348			0.010
	1330	845		1.58	16.7	58.8	0.63	37.7	833		0.16	9.6	334			0.016
	1630	96		2.02	12.8	80.5	0.22	17.2	375		0.19	14.7	370			0.048
	1830	–	10 963	–	–	48.8	–	–	–	8.6	0.19	–	190	6.1	1.41	0.016
6.4	1010	1 828		1.60	46.7	137.9	0.21[a]	4.5	383[a]	3.8[b]	–	–	–	–	–	–

[a] Converted from ^{14}C uptake by the factor 2.67.
[b] Calculated from standard conversion, $\Sigma\Sigma A \cong 10 . \Sigma A$ and $\Sigma\Sigma R \cong 24 . \Sigma R$.

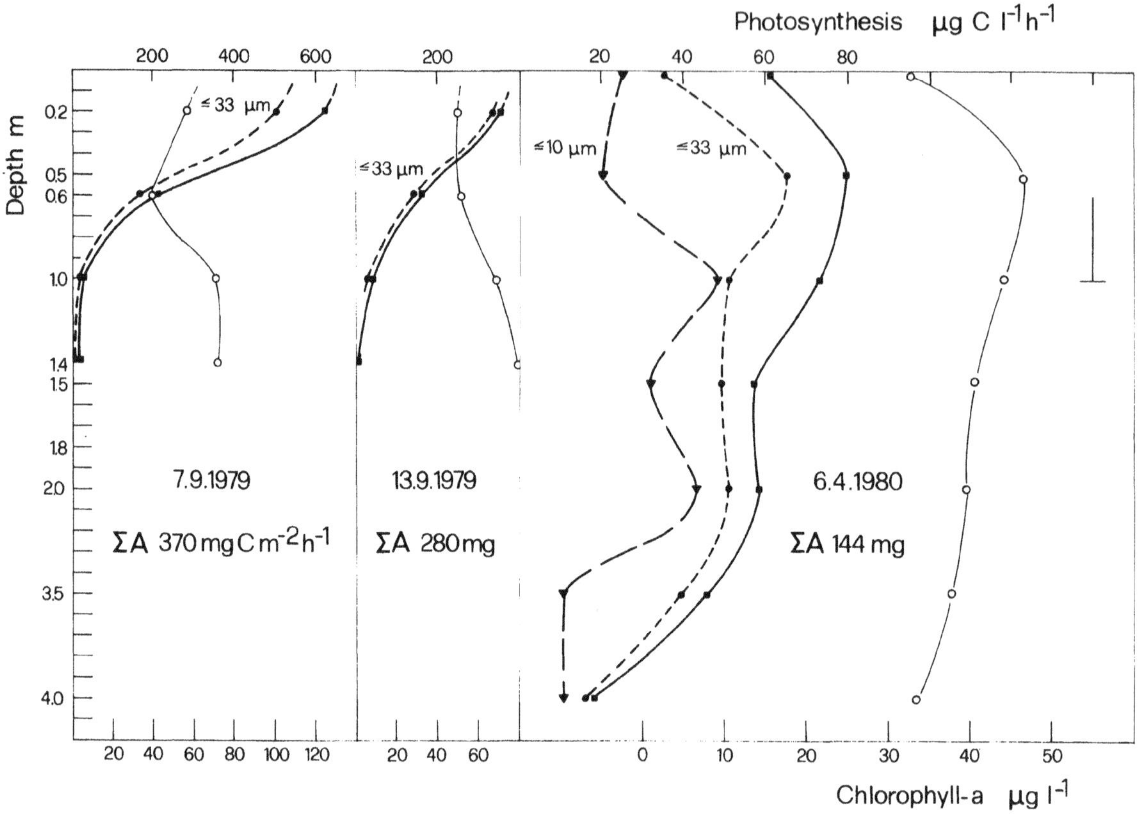

Fig. 8. Size fractionation of carbon-uptake experiments: ■, unfractionated photosynthesis; •, fraction passed through 33-μm netting; V, fraction passed through 10-μm netting; O, chlorophyll-a concentration. Note the different scales for the 1980 experiment.

(corresponding to 16%–70% of I_0). Photosynthetically active subsurface radiation (I_0) must exceed 900 μE m^{-2} s^{-1} to produce inhibition, which results in a depression of A_{opt} of 3.7%–41.1%, depending on light intensity.

On average, light levels of 52% of I_0 appear to be inhibitory. From a broad range of IBP data, Westlake (1980) concluded that light inhibition (I_i) occurs at levels well above 60% I_0.

The photo-inhibition threshold is about 600 μE m^{-2} s^{-1} for Lake Lanao (Lewis 1974), and around 500 in Neusiedlersee (Dokulil 1983). According to the review by Harris (1978), the lower and upper boundaries are 200 and 2 200 μE m^{-2} s^{-1}, respectively. In Neusiedlersee, the reduction of photosynthesis at the surface is, on average, 38.1% of optimum photo-assimilation. The onset of photo-inhibition ranges from 110 to 1 200 μE m^{-2} s^{-1}.

On various occasions, e.g. 5 September 1979, 15 March 1980, and mid-day exposures on 27 March 1980 (Figs. 6 and 9 and Table 7), photo-inhibition was not observed despite subsurface PhAR intensities well above the inhibition threshold. Possible explanations might be the differential adaptation of blue-green algal populations observed by Ward & Wetzel (1980) or rapid adaptation mechanisms by spill-over from photosystem II to PS I discussed by Vincent (1979).

When the highest photo-assimilation rates did not occur in the surface bottle, the depth of optimum photosynthesis (z_{opt}) was 0.2 m in 1979 and ranged from 0.5 to 1.5 m in 1980. The corresponding light intensity at that depth (I_{opt}) varied between 183 and 940 μE m^{-2} s^{-1} (Table 7). There was a tendency for higher values of I_{opt} to accompany higher levels of I_0, which has also been observed by Lewis (1974).

Wherever phytoplankton biomass is homogene-

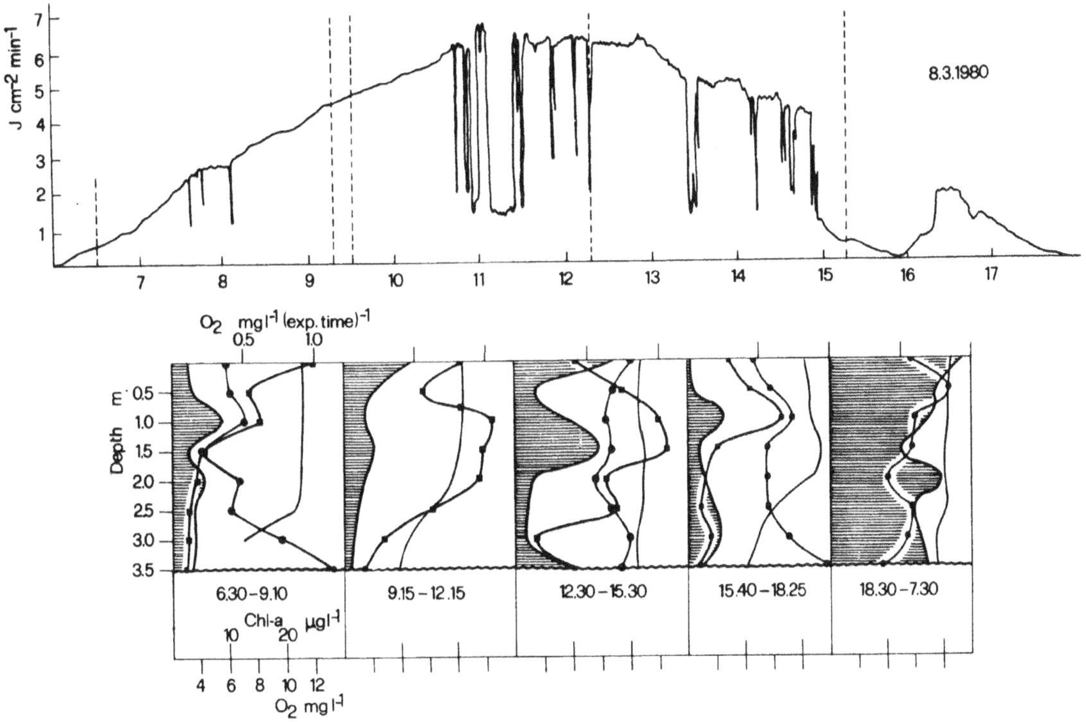
J cm-2 min-1
8.3.1980
O2 mg l-1 (exp. time)-1
Depth m
6.30-9.10
9.15-12.15
12.30-15.30
15.40-18.25
18.30-7.30
Chl-a μg l-1
O2 mg l-1

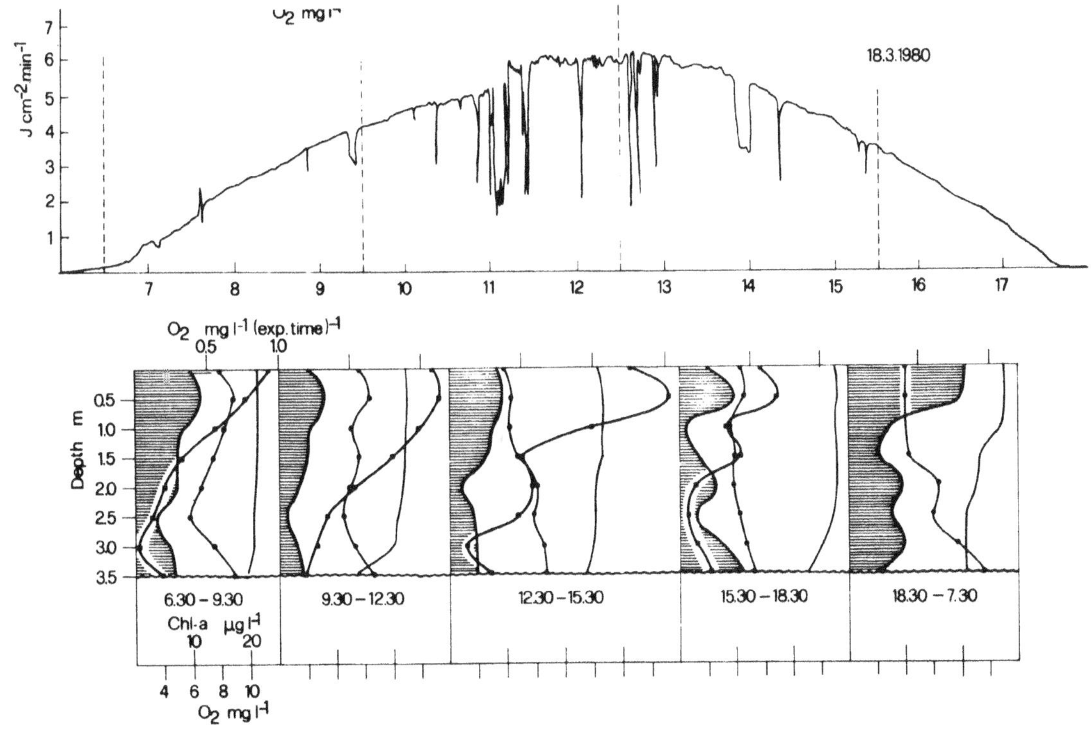
J cm-2 min-1
18.3.1980
O2 mg l-1 (exp. time)-1
Depth m
6.30-9.30
9.30-12.30
12.30-15.30
15.30-18.30
18.30-7.30
Chl-a μg l-1
O2 mg l-1

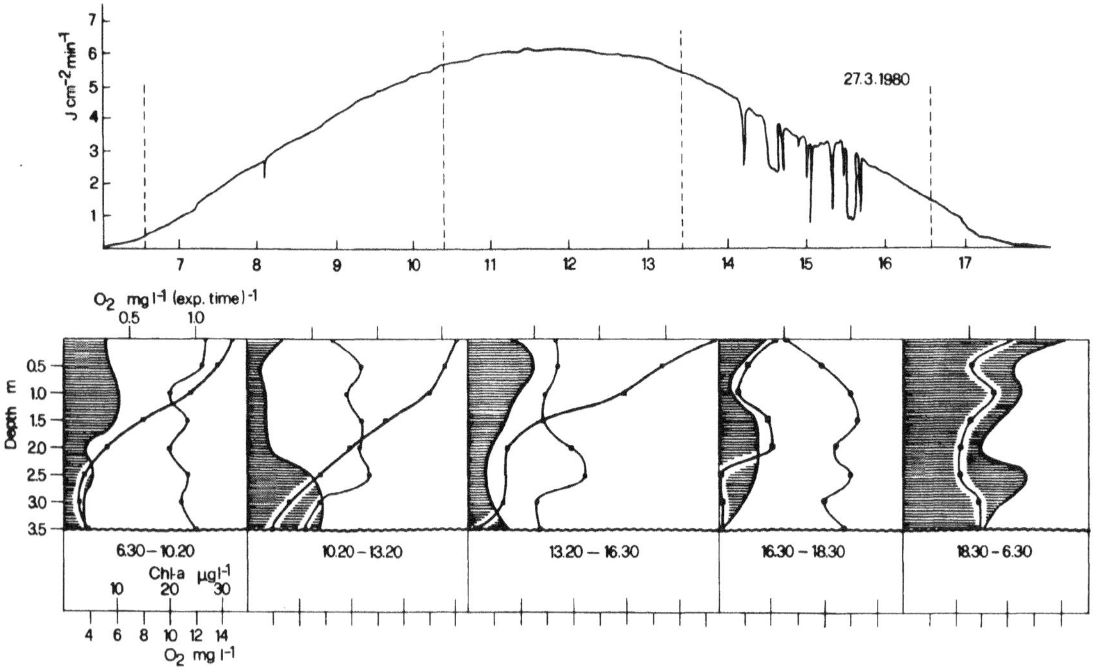

Fig. 9. Diel periodicity on 8 March (A), 18 March (B), and 27 March 1980 (C). (Top half) Variation of total incoming radiation. Dashed lines correspond to the experimental times in the lower half. Integrals for the exposure times and the days can be identified from Table 8. (Bottom half) Depth profiles of photosynthesis (▪), respiration ▒, chlorophyll-a concentration (•) and oxygen concentration (continuous line) during four consecutive in situ exposures during the day and respiration during the night.

ously distributed with depth, the photosynthesis profiles show the characteristic pattern of light-limited rates in the deeper water, rising to a single subsurface maximum at light saturation (Figs. 6, 8 and 9). Other profiles are distinctly dichotomous, corresponding to occasions where there are inhomogeneities in the chlorophyll-a concentration. One consequence of this is that the deeper these biomass maxima occur, the smaller will be their contribution to the total integral photosynthesis, as light attenuation is very rapid.

According to Talling (1957a), the onset of light saturation can be defined by the parameter I_K. In the present study, I_K values were obtained from plots of specific photosynthesis (mg O_2 mg Chl-a^{-1} h^{-1}) against the PhAR intensities at the respective depths (Fig. 7). I_K values were approximated whenever saturation occurred at the lake surface or when irregularities in the profile made definition of the parameter difficult. Figures are in the range of 50–390 μE m^{-2} s^{-1} (Table 7). On average, I_K corresponds to 22.6% of I_o. Rodhe (1965) found I_K ap-

Table 9. Daily gross photosynthetic rates measured by the summation of short-time incubation ($\Sigma\Sigma A_{obs.}$) compared to rates calculated from the 2nd and 3rd short-time period applying five different techniques.

$\Sigma\Sigma A_{obs.}$	$\Sigma\Sigma A_{calc.}$				Talling (1965)		Vollenweider (1965)			
	From $\Sigma I_o/I_o$		From 0.9 Δt		Formula (5)		Formula 8.1.1.		Ref. integral	
	2nd	3rd	2nd	3rd	2nd	3rd	2nd	3rd	2nd	3rd
7.0	6.1	6.9	8.1	8.0	7.9	7.8	7.0	6.6	7.1	7.0
7.2	6.5	6.8	8.0	8.3	7.2	6.7	6.2	5.2	5.4	7.1
8.6	7.6	10.8	11.0	8.3	8.9	9.4	7.3	5.8	6.7	6.8

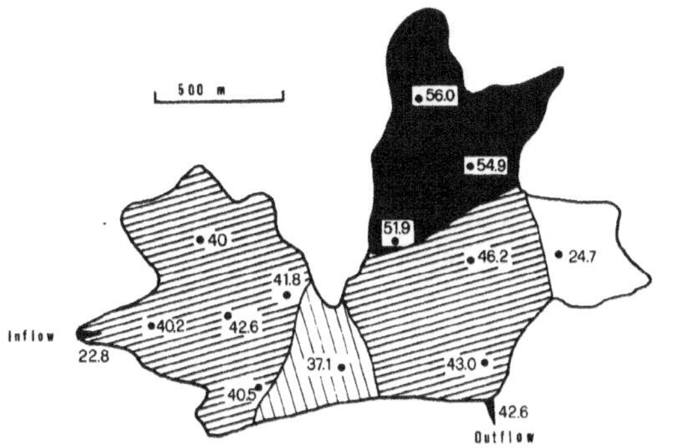

Fig. 10. Horizontal variation of chlorophyll-a concentration in the northern basin (PSN) of Parakrama Samudra.

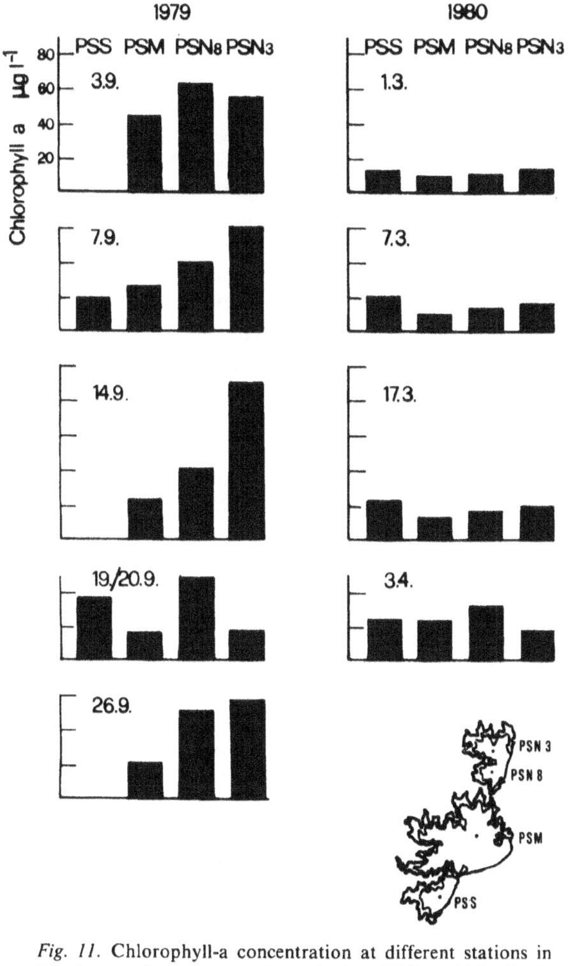

Fig. 11. Chlorophyll-a concentration at different stations in Parakrama Samudra southern basin (PSS), middle basin (PSM), northern basin (PSN), stations 3 and 8.

proximated 16% of th surface radiation, whereas Westlake (1980) concluded, from data gathered in a wide variety of lakes during IBP, that this parameter is about 15% of I_0, but deviations are great.

The relationship of the surface irradiance to the light saturation intensity (I_0/I_K), which might be considered as a measure of the light adaptation of the system, thus amounts to 4.43, with 95% C.L. of 3.27 and 5.98.

The depth at which photo-assimilation balances respiration, and hence net photosynthesis becomes zero, is known as the compensation depth (z_c). This depth varied between 0.5 and 1.7 m during the 1979 period and was in the range of 1.6–3.6 m in 1980 (Table 7). Since respiration here is defined as 'community respiration', the true compensation point is probably underestimated, for reasons discussed in the respiration section (see section 3.7). The mean light intensity at the compensation point (I_c) was calculated to be 1.6% of I_0, with 95% C.L. of 0.9 and 3.1 (Table 7).

Average compensation depth was 1.3 m in 1979 and 3.0 m in 1980, suggesting that only a portion of the biomass is in the euphotic zone at any one instant. The relation of the photosynthetic zone to the non-illuminated part of the column is essential for net column photosynthesis. Depending on whether the euphotic zone is estimated by the light data (z_{eu}) or by photosynthetic measurements (z_c), the ratio for 1979 is 1:2.0 or 1:1.6, respectively. For 1980, the ratios are 1:1.0 and 1:1.3. These ratios and Figure 6 illustrate the lake as shallow both physically and optically in 1980. In 1979, however, the lake is physically shallower but optically deeper. The smaller illuminated zone is a reflection of the background turbidity, already discussed. Greater biomass and more rapid circulation seem to compensate for the 'optical deep' situation in 1979, because daily productivity is similar in both years (Table 8).

The shape of the depth profile might also be characterized by the V/O ratio by Rodhe (1958a), relating optimum photosynthesis per unit volume to column photosynthesis per unit area. This relationship has proved useful for comparing lakes of different depths and trophic states (Rodhe 1958b). For the shallow-water phase of Parakrama Samudra in 1979 and the higher water level in 1980, ratios of 1.48 and 0.57 have been calculated from Table 4. This indicates the increasing importance of the most productive cubic metre within the water column for the total productivity per unit area as lakes become shallower and Secchi depth decreases (Rodhe 1958b).

The importance of algal cell sizes to photosynthetic activity is apparent from comparisons of net plankton and nannoplankton in both marine (e.g. Malone 1971, 1977) and freshwater environments (Pavoni 1963; Nauwerck 1966; Gliwicz 1967; Tilzer et al. 1977). Moreover, nannoplankton plays a major role in energy transfer to the next trophic level as it is more readily grazed upon by zooplankton.

So fractionation experiments were performed on both visits in 1979 and 1980 by passing subsamples from the incubation bottles through 33-μm netting before filtration onto Millipore filters (Fig. 8). In the 1980 experiment, the algal population was further divided into a fraction smaller than 10 μm.

The photosynthetic uptake rates in the fraction smaller than 33 μm were between 57% and 94% of the total ^{14}C uptake and had a similar depth-distribution pattern to the unfractionated samples (Fig. 8). However, the nannoplankton fraction (<10 μm) gave irregular results with depth, varying between 25% and 74% of the total photosynthetic rate. This variability could be attributed to differing quantities of thin and short filaments of blue-green algae (e.g. *Anabaenopsis*) passing through the net (Pavoni 1963). The percentages of chlorophyll-a passing through the net were in close agreement with the percentage share of ^{14}C uptake found in the respective fractions of the phytoplankton population.

The contribution of the <33-μm fraction to integral photo-assimilation is 72%, 85%, and 80% for the three experiments shown in Figure 5, which agrees with results from other lakes (Michael & Anselm 1978; Westlake 1980; Dokulil 1983).

3.4. Diurnal variation

Since photosynthetic experiments are usually carried out for short periods of the day, typically around noon, precautions must be taken if they are to be used for making estimates of daily rates, because any asymmetry that exists will result in substantial errors (cf. Vollenweider 1965; Gächter 1972).

Three time-course incubations were therefore executed dividing each day into four approximately equal time periods. In order to get reliable estimates of the daily net production, a darkened depth profile was also incubated overnight. The details of these experiments are given in Figure 9 and Table 8.

Total daily radiation was only slightly different, being highest on 27 March, with no clouds around mid-day. The lowest radiation was recorded on 8 March because of a rainstorm in the late afternoon.

Photo-inhibition was absent from the morning periods on each of the three days, but gradually developed on 8 and 13 March. No reduction of photosynthesis at the surface was observed on 27 March for reasons that are unclear (cf. section 3.2). Secondary peaks of photosynthesis due to increased chlorophyll-a concentrations occurred only when sufficient light was available. Thus bimodal photosynthetic depth curves were observed in the third period on both 8 and 18 March. The increasing chlorophyll-a concentrations with depth, as in the first and fourth periods on 8 March 1980, were not reflected in the photosynthetic rates since they occurred below the depth of the compensation point, z_c (Fig. 9 and Table 7).

Dark oxygen uptake, representing community respiration, had a similar depth-distribution pattern to the O_2 evolution in the light bottles. Minimal respiration rates occurred during the night. For a more detailed discussion of the respiration, refer to section 3.7.

Integral rates of photosynthesis (ΣB, Table 8) were highest in the second time interval. Gross photosynthetic efficiency (e as mg O_2 mg Chl-a^{-1} kJ^{-1} m^2), however, increased over the day, peaking in late afternoon. Higher rates in the afternoon seem to be due to greater phosphorus availability because of increased phosphatase activity (cf. Gunatilaka & Senaratna 1981) on 18 and 27 March ($r^2 = 0.83$, $P = 0.001$, $n = 8$). Extrapolation of the regression equation e = -0.020 + 0.0049 (PO_4-P) suggests a minimal phosphorus concentration of about 4 μg l^{-1} necessary for light utilization by the specific phytoplankton assemblage.

Table 9 summarizes daily surface integrals, observed (sum of consecutive short exposures) and calculated, from periods 2 and 3 according to five different approaches. The simple assumption of photosynthesis being proportional to light intensity gave good agreement when the instantaneous integrals of period 3 were used for calculation, but failed in the case of period 2. Similar results were obtained using the more sophisticated reference integral method proposed by Vollenweider (1965). All other models were less accurate. Discrepancies between calculated and observed rates were greatest on the third day, possibly because of the strong asymmetry of efficiency on that day.

3.5. Areal heterogeneity

Phytoplankton biomass is unevenly distributed in Parakrama Samudra. One example of areal heterogeneity is depicted in Figure 10 for PSN (below the 174 feet a.s.l. contour line) based on 11 data points from 27 August 1979. Chlorophyll-a concentrations are almost evenly distributed (~40 $\mu g\ l^{-1}$) in the centre of the PSN basin, whereas pigment concentrations are higher or lower in the bays at the northern end of the lake. The uneven distribution is a reflection of local nutrient inputs, currents, and temperature differences. Concentrations at the main outflow through gate 1 were similar to those in the main lake, but those at the inflow were nearly half (Fig. 10).

The differences in chlorophyll-a concentration between the three main lake basins for other occasions during 1979 and 1980 are illustrated in Figure 11. In general, levels of photosynthetic pigment concentration were higher in 1979, ranging from 16 to 90 $\mu g\ l^{-1}$, partly because of the very low water level at that time. In most cases, the middle basin exhibited the lowest pigment content and consequently the deepest underwater light penetration. At the two stations in the north, it was again the northwesterly one (PSN 3) that had the higher chlorophyll-a concentration in most instances. A gradient of increasing pigment concentration from the south to the north was observed on several occasions. There is only limited information available on the levels of chlorophyll-a in the inflow at Ambanganga, but the results obtained so far have been low (2.4-6.1 $\mu g\ l^{-1}$). This inflow is at the southern end of the lake but, as the main outflow is through sluice number 1, near the Rest House in the northern basin, a gradient of pigment concentration may develop from PSS to PSN. This is most likely when outflow rates result in longer retention times. Low phytoplankton biomass levels in the middle basin are a result of lower nutrient concentrations (Gunatilaka & Senaratna 1981) and slow exchange with the southern basin. At the main outflow (gate 1 of PSN) chlorophyll-a concentration was representative of the average concentration in that basin, suggesting some mixing of the water masses.

3.6. Variability in time

Phytoplankton biomass varies in time as well as in space. The data gathered so far are insufficient to show long-term trends, but evidence from the short seasonal periods can be used to outline the possible mechanisms involved in qualitative and quantitative changes of the phytoplankton populations. Alterations in species composition have already been discussed (cf. section 3.2). Quantitative changes are assessed from mean chlorophyll-a concentrations and average daily column net photosynthesis (Fig. 12A and B).

During the 1979 period, chlorophyll concentration in PSN was as low as 10 $\mu g\ l^{-1}$. In the following 12 days, the biomass increased to a maximum of 60 μg chlorophyll-a per litre. Pigment concentration then stabilized at approximately 50 $\mu g\ l^{-1}$ until the end of September (Fig. 12A).

The apparently low biomass on 21 August 1979 is a reflection of the preceding high flow-through. At this time, the daily output rate was three times higher then the input rate (cf. Duncan & Gulati 1981, Fig. 4; Schiemer 1983) because of irrigation management activities. When output levels dropped, chlorophyll-a concentration started to increase at an instantaneous growth rate of +0.31 per day. During the subsequent calm period in September, when input and output levels remained more or less constant, the rate of change of the mean chlorophyll-a content per unit volume of water approached zero or was about -0.02 per day. Figures became more variable between 14 and 26 September.

The development of the biomass during the investigation period in 1980 started from a mean chlorophyll-a concentration of 10.8 $\mu g l^{-1}$ on 29 February (Fig. 12B). It subsequently increased and reached a peak of 25.2 $\mu g\ l^{-1}$ on 4 March. Pigment concentration declined over the following three days and appeared stable at around 14 μg chlorophyll-a until 18 March, except for a higher figure on 17 March. During the remaining period (18 March to 6 April 1980), the phytoplankton biomass steadily increased, reaching a mean chlorophyll-a concentration of 39.4 $\mu g\,l^{-1}$ at the end of the investigation period.

Again variation in the through-put rates due to changes in the amount of water required for irrigation seems to be responsible for the changes in biomass. Due to the higher water level of the lake, the water volume was nearly doubled, resulting in a time lag of about four days between a change in the

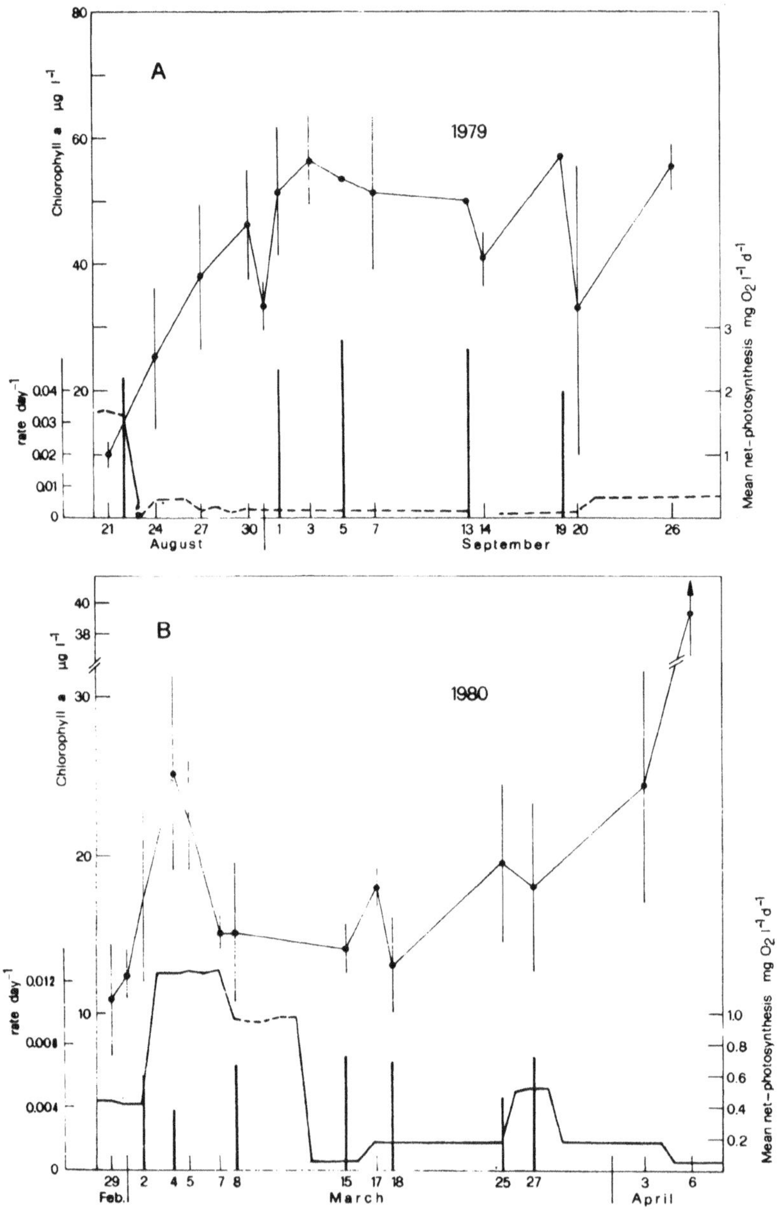

Fig. 12. Development of mean column chlorophyll-a concentration in the northern basin (PSN) of Parakrama Samudra during the two visits in 1979 and 1980. Vertical bars on the chlorophyll curve indicate standard deviation. Vertical columns are mean net photosynthesis in the water column per day. The dashed line is the output rate per day of the reservoir.

output rate and an apparent reaction of the phytoplankton population (fig. 12B). The time delay seems to be the result of lower maximum output rates, which were 0.016 per day compared to 0.035 in the preceding year (Fig. 12A and B).

Fluctuations of the mean net photosynthesis per day were modest, less than two-fold, during both investigations. This approximately corresponded to the variation in radiation reaching the lake surface (cf. section 3.1). Variability of the mean quantity of chlorophyll-a per unit volume of water was 5.7 (min. 4.1, max. 8.2) and 3.6 (min. 2.4, max. 6.1) for the 1979 and 1980 periods, respectively.

3.7. Respiration

Community respiration (R_{com}) ranged from 0.002–0.400 mg O_2 l^{-1} h^{-1}. The shape of the depth profiles is partly dependent on the vertical gradient of phytoplankton concentration (Figs. 3 and 6). Maximum rates of R_{com} often correspond with the depth of optimum photosynthesis. However, there is no significant correlation with either the photosynthetic rate or the algal biomass.

Previous exposure to light apparently stimulates the rate of respiration in the dark (Figs. 6 and 9). During prolonged dark treatment, the rate of respiration decreases markedly, agreeing with observations on blue-green algae both in the field and in the laboratory (Kratz & Mayers 1955; Dokulil 1971; Gibson 1975). Hourly rates during the night (0.002–0.008 mg O_2 l^{-1} h^{-1}) are therefore lower compared with rates during daylight 0.003–0.400 mg O_2 l^{-1} h^{-1}). Dark oxygen uptake of the total water column seems to be related to surface irradiance conditions, but is not significantly correlated.

Specific respiration (R_{opt}) varied between 1.1 and 13.5 mg O_2 (mg Chl-a)$^{-1}$ h^{-1}. Figures are difficult to interpret since they are based on community oxygen uptake comprising zooplankton as well as bacterial respiration. High specific respiration rates may be the result of concentrations from components other than phytoplankton. In addition, part of the oxygen uptake may be attributed to glycolate metabolism or a Mehler reaction, especially at light intensities above I_K (Bidwell 1977; Harris 1978).

Ganf (1974c) reports R_{opt} values in the range of 0.2–4.0 obtained under conditions similar to the present study. The same range of specific community respiration has been found during IBP (Westlake 1980), but rates as high as 10 mg O_2 (mg Chl-a)$^{-1}$ h^{-1} have been measured. Gibson (1975) calculated specific respiration rates as high as 6.3 mg O_2 (mg Chl-a)$^{-1}$ h^{-1} for laboratory experiments with *Oscillatoria redekei* at 20 °C. For mixed natural populations, Jones (1977) estimated R_{opt} at 21 °C to be 2.4. Of the 137 individual measurements presented here, 28 are higher than 8 mg O_2 (mg Chl-a)$^{-1}$ h^{-1}.

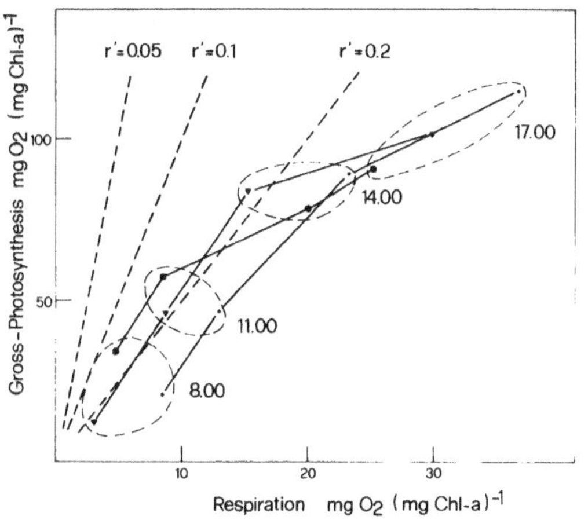

Fig. 13. Solid lines represent the relationship between cumulative daily photosynthesis and cumulative respiration during three separate days. Approximate mid-points of experimental periods are indicated beside the ringed points. Dashed lines were obtained by assuming that respiration (r') is a constant fraction of photosynthesis (1/5, 1/10 or 1/20).

The contribution of zooplankton and bacteria to the community respiration is difficult to assess since no specific experiments have been performed. The oxygen uptake of zooplankton is assumed to be negligible as the population abundance is less and is composed almost entirely by rotifers (Duncan 1983). The bacterial fraction might be somewhere between 10% and 60% of total oxygen uptake, judging from statements by Golterman (1971), Ganf (1974c), Gibson (1975), Jewson (1976) and Jones (1977). Thus, phytoplankton respiration can be in error up to 60%, resulting in improbably high specific rates per unit chlorophyll-a. As a consequence of erroneous respiration figures, gross photosynthesis might be overestimated.

From the diurnal studies of photosynthesis and respiration reported in Figure 9, the cumulative sum of P_{opt} (mg O_2 [mg Chl-a]$^{-1}$) is plotted against the cumulative sum of the corresponding specific community oxygen uptake (Fig. 13). Each curve represents a single day, the results from each time period are ringed, and the approximate mid-point of each period is indicated. From these data, the following conclusions can be deduced: Firstly, respiration increases with increasing photosynthesis asymptotically, approaching a maximum value. Secondly, respiration is not a constant fraction of photosynthesis, but is related by a non-linear function, relative respiration being higher in the afternoon. A similar relationship between P_{opt} and R_{opt} was reported by Ganf (1974c) for the tropical Lake George, Uganda.

Thirdly, a relative respiration (r′) of 0.2 would best fit the measurements from the morning and early afternoon periods. This figure is much larger then the commonly quoted ones of 0.05–0.1 (Talling 1965; Jewson 1976; Harris 1978). Similar high ratios were found by Bindlos (1974) for blue-green and green algal dominated Loch Leven, Scotland, and by Tilzer et al. (1977) for Mikolajskie Lake, Poland, when dinoflagellates formed 90% of the biomass. The latter authors also report a significant increase of P–R ratio in the afternoon. In contrast, Dokulil (1982) found a linear relation of respiration to photosynthesis for the green algal dominated Neusiedlersee, Austria, r′ being between 0.05 and 0.1.

Heighrelative respiration rates or blue-green algae and green algae, 20.5% and 25.4%, respectively, were reported by Bidwell (1977). According to Aruga (1965), the respiration of *Anabaena cylindrica* amounts to 29.4% of optimum photosynthesis at 30 °C.

3.8. Ecological losses

Algal cells are subject to various internal and external influences that affect the size and rate at which phytoplankton populations can grow. Of the internal factors, respiration and release of organic substances are perhaps the most important, ultimately resulting in a reduction of newly formed cell material. Respiration has already been discussed, whereas release had not been considered, because no direct measurements are available, but might partly be responsible for the large discrepancy between the oxygen and the carbon-14 method in the present study (cf. Table 8).

Of the external factors, grazing by zooplankton and herbivorous fish, sedimentation, natural death, and wash-out through the reservoir sluices are considered as the major variables responsible for losses from the population.

To obtain an estimate of the likely total sum of all these processes, theoretical biomass figures were calculated from the mean chlorophyll-a concentration of PSN and mean column net photosynthesis per day presented in Figure 9 and Table 8. Setting the chlorophyll-a content equal to 0.3% of fresh-

Table 10. Theoretical produced biomass (TBP), loss rates per day and turn-over in days calculated from average chlorophyll-a concentration (± standard deviation) and mean column net photosynthesis ($\overline{NP}$). For explanation refer to the text, for symbols to Table 2.

Date *1979*	Chl-a $\mu g\, l^{-1}$	$\overline{NP}$ $\mu g\, O_2\, l^{-1} d^{-1}$	NCI $\mu g\, l^{-1}$	TPB $\mu g\, l^{-1}$	Loss $\%\, d^{-1}$	Turn-over days	Date *1980*	Chl-a $\mu g\, l^{-1}$	$\overline{NP}$ $\mu g\, O_2\, l^{-1} d^{-1}$	NCI $\mu g\, l^{-1}$	TPB $\mu g\, l^{-1}$	Loss $\%\, d^{-1}$	Turn-over days
21.8	10.0 ± 2.2	2 200	25.0				29.2	10.8 ± 3.5					
24.8	25.2 ± 11.2			85.0	23.5	1.0	1.3	12.5 ± 1.5					
27.8	38.0 ± 11.6			100.2	20.7	1.5	2.3	17.5 ± 5.5	600	6.8	19.3	9.3	2.6
30.8	46.3 ± 8.9			113.0	19.7	1.9	4.3	25.2 ± 6.1	371	4.2	31.1	9.5	6.0
31.8	33.1 ± 4.0			71.3	53.6	1.3	5.3	22.5 ± 3.3			29.4	23.5	5.4
1.9	51.6 ± 10.3	2 300	26.1	58.1	11.2	2.0	7.3	15.2 ± 1.1			30.9	25.4	3.6
3.9	56.6 ± 7.0			103.8	22.7	2.2	8.3	15.3 ± 4.4	657	7.5	21.2	27.5	2.0
5.9	53.6	2 800	31.8	108.8	25.4	1.7	15.3	14.2 ± 1.7	714	8.1	67.8	11.3	1.8
7.9	51.3 ± 12.3			117.2	28.1	1.6	17.3	18.0 ± 1.2			30.4	20.4	2.3
13.9	50.1	2 640	30.0	242.1	13.2	1.7	18.3	13.4 ± 3.1	657	7.5	25.8	48.1	1.8
14.9	40.8 ± 4.4			80.1	49.1	1.4	25.3	19.1 ± 5.3	457	5.2	65.9	10.1	3.7
19.9	57.1	1 990	22.6	190.8	14.0	2.5	27.3	17.8 ± 5.3	714	8.1	29.5	19.9	2.1
20.9	32.8 ± 22.8			79.7	58.8	1.5	3.4	24.3 ± 7.2			74.5	9.6	3.0
26.9	55.5 ± 3.7			168.4	11.2	2.5	6.4	39.4 ± 5.2			48.6	6.3	4.8

weight biomass (of section 3.2) and assuming 10% carbon in the fresh weight, net photosynthetic rates ($\overline{NP}$) are converted into 'net chlorophyll' formed per day (NCl). The value thus obtained is then further multiplied with the number of days between observations, assuming photosynthesis to remain constant. Summation with the actual chlorophyll concentration at the start results in an estimation of the amount of mean biomass theoretically produced, if no losses occur (TPB). Comparing this figure with the actual mean chlorophyll-a concentration of that day allows the calculation of the percentage loss per day (columns 4 and 12 of Table 10).

In 1979 loss rates of less than about 25% lead to a significant net increase in biomass and values between 25% and 30% about balance the biomass gains, whereas rates much in excess of 30% per day are sufficient to reduce biomass drastically.

At the higher water level of 1980, loss rates (Table 10) from the phytoplankton population higher than 15% per day result in a decrease of algal concentration (Fig. 12B). Values close to 11% seem to equal net biomass increases. For a significant change in the population to be detected by this technique, total losses must be less than 10% per day.

As has been pointed out (cf. section 3.6), biomass changes are largely governed by the flushing rate of the lake due to regulation of the outflow by irrigation management. Since chlorophyll-a concentration in the outflow approximates the mean concentration of PSN (cf. section 3.5), it seems safe to assume that the greater portion of the calculated losses are due to wash out.

Based on the rotifer concentrations in PSN in 1979 and their feeding rate (Duncan & Gulati 1981; Duncan 1983), the contribution of zooplankton feeding to these losses might be considerable. The same holds good for the material ingested by herbivorous fish. Dokulil (1983) calculated daily food consumption of one juvenile *Sarotherodon mossambicus* to be equivalent to 8.4 mg chlorophyll-a, which is the amount of biomass produced every day by about 1–2 m³ of the lake.

3.9. *Controlling-factors summary*

The preceding sections gave a detailed description of the information currently available on the biomass and the photo-synthetic activity in Parakrama Samudra, as well as some relationships and possible causes. In this section, controlling factors will be summarized and discussed with special emphasis towards explaining the magnitude and variation of net productivity.

Among the factors affecting phytoplankton photosynthesis, the supply of light and nutrients are usually considered as most important. Photosynthesis and respiration are principally temperature dependent, a factor that is less variable under tropical conditions compared to temperate regions because of the smaller variability of radiation and hence water temperature (Straškraba 1980).

Photosynthetic capacity per unit chlorophyll-a (P_{opt}) averaged 25.7 mg O_2 mg Chl-a^{-1} h^{-1} with 95% C.L. of 21.6 and 27.4 (n = 23), respectively (Table 8). This is a value commonly encountered in tropical waters (Talling 1965; Freson 1972; Lemoalle 1973; Ganf 1974a). Some of the capacities in Table 4 and Figure 4 are among the highest so far recorded (Westlake 1980), and might be biased by insufficient chlorophyll extraction.

Algal gross photosynthesis per hour (ΣA in Table 8) was strongly related to incident photosynthetically active radiation (PhAR). The correlation is highly significant ($r^2 = 0.68$, $P \leqslant 0.001$, $n = 15$), suggesting that about 68% of the variation between incubation periods can be explained by the variation in sunlight (Fig. 14). Similarly, net photosynthesis per unit of lake surface (ΣA/ΣR) depended on incoming radiation $r^2 = 0.84$, $P \leqslant 0.001$, $n = 15$). The regression equation $y = 58.12 + 0.476x$ suggests that about 122 kJ m^{-2} h^{-1} are necessary to result in significant net gain by photosynthesis. As has al-

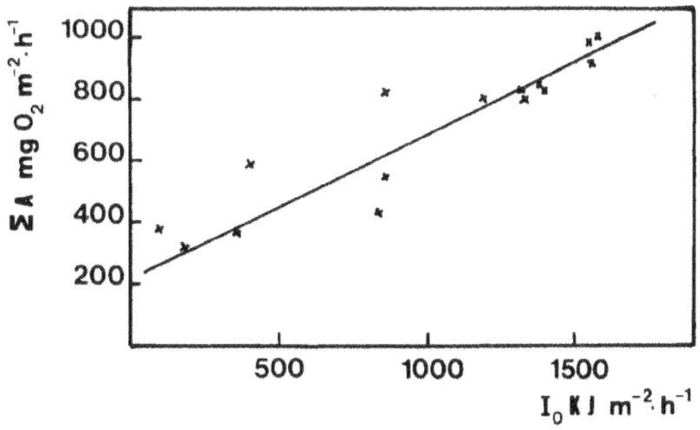

Fig. 14. Relationship between integral photosynthesis (ΣA) and subsurface PhAR intensity I_0; the correlation coefficient is 0.82.

ready been discussed in section 3.4, light utilization (e) depends on phosphorus concentration, 4 μg l^{-1} being the minimum concentration for positive photosynthetic activity. Mean efficiency of light energy utilization (J fixed/J received) is 1.43%, which is comparatively high (Berman & Pollingher 1974; Gächter (1972).

On the basis of these relationships, it seems safe to assume that incoming radiation and phosphorus concentration are the main regulating functions re-

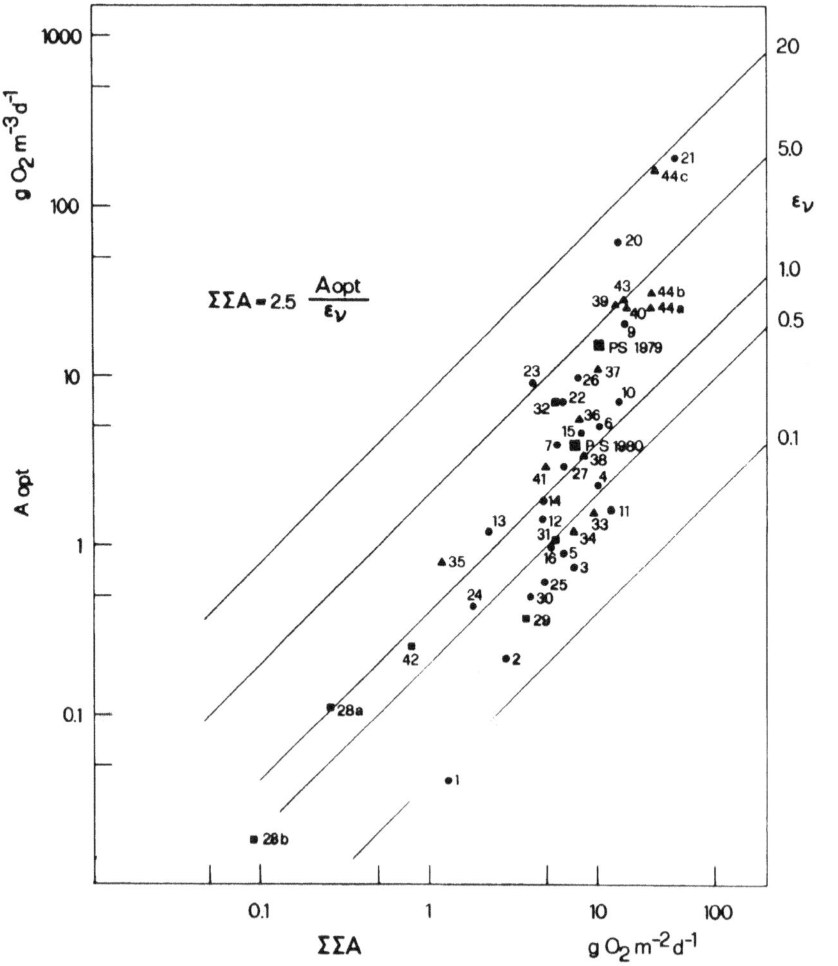

Fig. 15. Relationship between photosynthesis at optimum depth (A_{opt}) and integral photosynthesis per day ($\Sigma\Sigma A$) for various tropical lakes in Africa (•), Asia (▲) and America (■) compared to Parakrama Samudra (PS). Solid lines represent the relationship $\Sigma\Sigma A = 2.5\ A_{opt}/\epsilon_\nu$ for different extinction coefficients. Identification of lakes: 1 – L. Tanganyika; 2 – L. Tanganyika, Kitaza Bay (Melack 1980); 3 – L. Victoria, offshore; 4 – L. Victoria, Pilkington Bay; 5 – L. Victoria, Grant Bay; 6 – L. Victoria, Kovirondo Gulf; 7 – L. Victoria, Winam Gulf (Talling 1965); 8 – L. George (Ganf 1974); 9 – Kazinga Channel; 10 – L. Edwards; 11 – L. Albert; 12 – L. Bunyoni; 13 – L. Mulehe (Talling 1965); 14 – L. Naivasha; 15 – L. Oloidien; 16 – Crescent Island Crater (Melack 1979a); 17 – Naivasha Crater; 18 – Elmentaita; 19 – L. Nakuru (Melack 1979b); 20 – L. Kilotes; 21 – L. Aranguadi (Talling et al. 1973); 22 – L. McIlwaine (Roberts 1979); 23 – L. Chad (Lemoalle 1973); 24 – L. Volta (Biswas 1978); 25 – L. Sibaya (Allanson & Hart 1975); 26 – Gebel Aulia Reservoir; 27 – L. Lubumbashi (Freson 1972); 28a – L. Carioca (W); 28b – L. Carioca (S) (Barbosa & Tundisi 1980); 29 – L. Titicaca (Widmer et al. 1975); 30 – L. Gatun; 31 – L. Madden (Gliwicz 1976); 32 – L. Casthano (Schmidt 1973); 33 – L. Lanao (Lewis 1974); 34 – L. Mainit (Lewis 1973); 35 – Kodaikanal L.; 36 – Yercaud L.; 37 – Ooty L. (Sreenivasan 1964b); 38 – Amaravathi Reservoir (Sreenivasan 1965); 39 – Chingleput Ford Pond (Sreenivasan 1968); 40 – Kulam Pond (Ganapati & Sreenivasan 1972); 41 – Aliyar Reservoir (Sreenivasan 1970); 42 – L. Redondo (Marlier 1967); 43 – Malaysian fish ponds (Prowse 1972); 44 – Malayian fish ponds (Richardson & Lin 1975).

sponsible for the variation of the instantaneous integral of photosynthesis.

Figures from 1979, which were not included in the calculations because of insufficient data on daily radiation, were much more variable (Fig. 14) due to the variability of the euphotic zone because of high turbidity (section 3.1).

Daily rates of gross and net photosynthesis do not show a significant relation to light, for several reasons. Information on day rates based on direct observations by the summation of short-time exposures is rather limited. Rates calculated from midday incubations suffer from crude conversion by the day length in 1979. Extrapolation on short-time experiments in 1980 results in an underestimation of the daily photo-assimilation, because of the asymmetrical nature of light utilization.

Estimates of daily net productivity are largely biased by the degree of precision with which the respiration per day can be evaluated. As contributions from other groups in the plankton (the bacteria and zooplankton) are also measured and because there is a non-linear relation to photosynthesis (section 3.7), algal respiration will be overestimated and therefore net productivity underestimated. Considering these limitations, the ratio of the column photosynthesis to column respiration per day ($q_c = \Sigma A / \Sigma R$) can be calculated. This ratio must be equal to, or greater than, 1 for the population to survive. Examination of the figures in Table 8 reveals that the balance between photosynthesis and respiration was always greater than 1, except for two occasions in 1979, indicating a capacity for growth of the phytoplankton assemblage.

This capacity for growth may ultimately lead to a population increase until self-shading becomes a regulating mechanism, if nutrients are sufficiently available. Since phytoplankton biomass never reached high values in 1980 and because of the high background extinction in 1979, which was even more important for light absorption than phytoplankton, this feedback mechanism is considered unimportant at present.

The newly formed biomass is exploited by zooplankton and herbivorous fish, but losses due to high water renewal in connection with irrigation control are more important for the regulation of phytoplankton biomass levels in the reservoir.

3.10. Comparison with other tropical lakes

Parakrama Samudra is compared to other tropical lakes from Asia, Africa and South America in Figure 15, adopting the V/O diagram by Rodhe (1958a). Comparing the most productive cubic metre for different values of ϵ, the data from Parakrama Samudra are well within the range of photosynthetic activities reported from shallow lakes and ponds of India. Several of the more shallow lakes in America, like Lake Chad or Lake Edward, fall into the same category. Of the lakes from South Africa, only the Lago di Casthano is comparable, whereas all others have much lower photosynthetic rates. It is evident from the diagram that deeper and clearer water bodies may have almost the same integral rate as Parakrama Samudra, but have substantially lower A_{opt} rates because of the more extended euphotic zone. Highly productive systems, in contrast, like Aranguardi in Africa or some of the fish ponds in Malaysia, have volume-based rates that are higher than the surface due to strong self-shading.

Very recently, Hecky & Fee (1981) published detailed data on the primary production of Lake Tanganiyka, indicating a three-fold higher rate of integral primary production compared to the preliminary figure given by Melack (1980), which is presented in Figure 15 (no. 1). Recent information on tropical African soda lakes may be found in Melack (1981).

Acknowledgements

We are deeply indepted to Dr. J. F. Talling and Dr. D. H. Jewson for valuable comments and criticism of the manuscript, and for correcting the English text. We wish to thank K. Moniruzzaman for counting algal samples, R. Reich for drawing some of the diagrams and I. Gradl for typing the manuscript.

References

Allanson, B. R. & Hart, R. C., 1975. The primary production of Lake Sibaya, Kwa Zulu, South Africa. Verh. Internat. Verein. Limnol. 19: 1426-1433.

Aruga, Y., 1965. Ecological studies of photosynthesis and matter production of phytoplankton II. Photosynthesis of algae in relation to light intensity and temperature. Bot. Mag. Tokyo 78: 360–365.

Barbosa, F. A. R. & Tundisi, J. G., 1980. Primary production of phytoplankton and environmental characteristics of shallow quaternary lake at eastern Brazil. Arch. Hydrobiol. 90: 139–161.

Bauer, K., 1983. Thermal stratification, mixis, and advective currents in the Parakrama Samudra Reservoir, Sri Lanka. In: Schiemer, F., (ed.) Limnology of Parakrama Samudra – Sri Lanka: a case study of an ancient man-made lake in the tropics. Developments in Hydrobiology (this volume). Dr W. Junk, The Hague.

Berman, Th. & Pollingher, U., 1974. Annual and seasonal variations of phytoplankton, chlorophyll, and photosynthesis in Lake Kinneret. Limnol. Oceanogr. 19: 31–54.

Bidwell, R. G., 1977. Photosynthesis and light and dark respiration in freshwater algae. Can. J. Bot. 55: 809–818.

Bindloss, M. E., 1974. Primary productivity of phytoplankton in Loch Leven, Kinross. Proc. R. Soc. Endinb. (B) 74: 157–181.

Biswas, S., 1978. Observations on phytoplankton and primary productivity in Volta Lake, Ghana. Verh. Intern. Verein. Limnol. 20: 1672–1676.

Dokulil, M., 1971. Atmung und Anaerobioseresistenz von Süßwasseralgen. Int. Rev. ges. Hydrobiol. 56: 751–768.

Dokulil, M., 1979. Seasonal pattern of phytoplankton. In: Löffler, H. (ed.) Neusiedlersee. The Limnology of a shallow Lake in Central Europe, pp. 203–231. Dr W. Junk, The Hague.

Dokulil, M., 1983 (in preparation). An assessment of the phytoplankton biomass, primary productivity and bacterial production in a turbid shallow lake (Neusiedlersee, Austria).

Dokulil, M., 1983. Aspects of gut passage of algal cells in *Sarotherodon mossambicus*. In: Schiemer, F. (ed.) Limnology of Parakrama Samudra – Sri Lanka: a case study of an ancient man-made lake in the tropics. Developments in Hydrobiology (this volume). Dr W. Junk, The Hague.

Duncan, A., 1983. Composition, density and distribution of the zooplankton in Parakrama Samudra. In: Schiemer, F. (ed.) Limnology of Parakrama Samudra – Sri Lanka: a case study of an ancient man-made lake in the tropics. Developments in Hydrobiology (this volume). Dr W. Junk, The Hague.

Duncan, A. & Gulati, R. D., 1981. Parakrama Samudra (Sri Lanka) Project; a study of a tropical lake ecosystem. III. Composition, density and distribution of the zooplankton in 1979. Verh. Intern. Verein. Limnol. 21: 1001–1008.

Elster, H. J. & Štepánek, M., 1967. Eine neue Modifikation der Secchi-Scheibe. Arch. Hydrobiol. Suppl. 33: 101–106.

Freson, R. E., 1972. Aspect de la limnochimie et de la production primaire au lac de la Lubumbashi. Verh. Internat. Verein. Limnol. 18: 661–665.

Gächter, R., 1972. Die Bestimmung der Tagesraten der planktischen Primärproduktion – Modelle und In-situ-Messungen. Schweiz. Z. Hydrol. 34: 211–244.

Ganapati, S. V. & Sreenivasan, A., 1972. Energy flow in aquatic ecosystems in India. In: Kajak, Z. & Hillbricht-Ilkowska, A. (eds.) Productivity problems of freshwater. Proc. IBP-Unesco Symp. Warszawa, Kraków, pp. 457–475.

Ganf, G. G., 1972. The regulation of net primary production in lake George, Uganda, East Africa. In: Kajak, Z. & Hillbricht-Ilkowska, A. (eds.) Productivity problems of freshwater. Proc. IBP–Unesco Symp. Warszawa, Kraków, pp. 693–708.

Ganf, G. G., 1974a. Incident solar irradiance and underwater light penetration as factors controlling the chlorophyll-a content of a shallow equatorial lake (Lake George, Uganda). J. Ecol. 62: 593–609.

Ganf, G. G., 1974b. Phytoplankton biomass and distribution in a shallow eutrophic lake (Lake George, Uganda). Oecologia 16: 9–29.

Ganf, G. G., 1974c. Rates of oxygen uptake by the planktonic community of a shallow equatorial lake (Lake George, Uganda). Oecologia 15: 17–32.

Ganf, G. G., 1975. Photosynthetic production and irradiance-photosynthesis relationship of the phytoplankton from a shallow equatorial lake (Lake George, Uganda). Oecologia 18: 165–183.

Gibson, C. E., 1975. A field and laboratory study of oxygen uptake by planktonic blue-green algae. J. Ecol. 63: 867–879.

Gliwicz, Z. M., 1967. The contribution of nannoplankton in pelagic primary production in some lakes with varying trophy. Bull. Acad. pol. Sci. Ser. biol. 15: 343–347.

Gliwicz, Z. M., 1976. Plankton photosynthetic activity and its regulation in two neotropical man-made lakes. Pol. Arch. Hydrobiol. 23: 61–93.

Golterman, H. L., 1971. The determination of mineralization losses in correlation with the estimation of net primary production with the oxygen method and chemical inhibitors. Freshwat. Biol. 1: 249–256.

Gunatilaka, A. & Senaratna, C., 1981. Parakrama Samudra (Sri Lanka) Project, a study of tropical ecosystem. II. Chemical environment with special reference to nutrients. Verh. Internat. Verein. Limnol. 21: 994–1000.

Harris, G. P., 1978. Photosynthesis, productivity and growth: the physiological ecology of phytoplankton. Arch. Hydrobiol. Beih. Ergebn. Limnol. 10: 1–171.

Hecky, R. E. & Fee, E. J., 1981. Primary production and rates of algal growth in Lake Tanganyika. Limnol. Oceanogr. 26: 532–547.

Holm-Hansen, O. & Riemann, B., 1978. Chlorophyll-a detemination: improvement in methodology. Oikos 30: 438–447.

Jewson, D. H., 1976. The interaction of components controlling net phytoplankton photosynthesis in a well mixed lake (Lough Neagh, Northern Ireland). Freshwat. Biol. 6: 551–576.

Jewson, D. H., 1977. Light penetration in relation to phytoplankton content of the euphotic zone of Lough Neagh, N. Ireland. Oikos 28: 74–83.

Jones, R. I., 1977. The importance of temperature conditioning to the respiration of natural phytoplankton communities. Br. Phycol. J. 12: 277–285.

Kirk, J. T. O., 1976. A theoretical analysis of the contribution of algal cells to the attenuation of light within natural waters. III. Cylindrical and speroidal cells. New. Phytol. 77: 341–358.

Kratz, W. A. & Mayers, J., 1955. Photosynthesis and respiration of three blue green algae. Plant Physiol. 30: 275–280.

Lemoalle, J., 1973. L'energie lumineuse et l'activité photosynthétique du phytoplankton dans le Lac Tchad. ORSTOM, sér. Hydrobiol. 7: 95–116.

Lemoalle, J., 1975. L'activité photosynthétique du phytoplancton en relation avec le niveau des eaux du Lac Tchad (Afrique). Verh. Internat. Verein. Limnol. 19: 1398–1403.

Lemoalle, J., 1979. Activité photosynthétique du phytoplancton du Lac Tchad. ORSTOM Raport pour la Limnologie Africaine Nairobi, 16–23 décembre 1979.

Lewis, W. M., Jr., 1973. A limnological survey of Lake Mainit, Philippines. Int. Rev. ges. Hydrobiol. 58: 801–818.

Lewis. W. M., Jr., 1974. Primary production in the plankton community of a tropical lake. Ecol. Monogr. 44: 377–409.

Lund, J. W. G., Kipling, C. & Le Cren, E. D., 1958. The inverted microscope method of estimation of algal numbers and the statistical basis of estimation by counting. Hydrobiologia 11: 143–170.

Malone, T. C., 1971. The relative importance of nannoplankton and netplankton as primary producers in tropical oceanic and neritic phytoplankton communities. Limnol. Oceanogr. 16: 633–639.

Malone, T. C., 1977. Light saturated photosynthesis by phytoplankton size fractions in New York bight, USA. Mar. Biol. 42: 281–293.

Marlier, G., 1967. Ecological studies on some lakes of the Amazon Valley. Amazoniana 1: 91–115.

McMahon, J. W., 1973. Membrane filter retention – a source of error in the ^{14}C method of measuring primary production. Limnol. Oceanogr. 18: 319–323.

Melack, J. M., 1979a. Photosynthetic rates in four tropical African freshwaters. Freshwat. Biol. 9: 555–571.

Melack, J. M., 1979b. Temporal variability of phytoplankton in tropical lakes. Oecologia 44: 1–7.

Melack, J. M., 1980. An initial measurement of photosynthetic productivity in Lake Tanganyika. Hydrobiologia 72: 243–248.

Melack, J. M., 1981. Photosynthetic activity of phytoplankton in tropical african soda lakes. Hydrobiologia 81: 71–85.

Michael, R. G. & Anselm, V. M., 1978. Role of nannoplankton in primary productivity studies in tropical ponds. Verh. Internat. Verein. Limnol. 20: 2196–2201.

Nauwerck, A., 1966. Beobachtungen über das Phytoplankton klarer Hochgebirgsseen. Schweiz. Z. Hydrol. 28: 4–28.

Pavoni, M., 1963. Die Bedeutung des Nannoplanktons im Vergleich zum Netzplankton. Schweiz. Z. Hydrol. 25: 219–341.

Prowse, G. A., 1972. Some observations on primary and fish production in experimental fish ponds in Malacca, Malaysia. In: Kajak, Z. & Hillbricht-Ilkowska, A. (eds.) Productivity problems of freshwater. Proc. IBP–Unesco Symp. Warszawa, Kraków, pp. 555–561.

Rebsdorf, A., 1972. The carbon dioxide system in freshwater. A set of tables for easy computation of total carbon dioxide and other components of the carbon dioxide system. Printed Booklet: Freshwater Biological Laboratory, Hillerød, Denmark.

Richardson, J. L. & Jin, L T., 1975. Algal productivity of natural and artificially enriched freshwaters in Malaya. Verh. Internat. Verein. Limnol. 19: 1383–1389.

Robarts, R. D., 1979. Underwater light penetration chlorophyll-a and primary production in a tropical African Lake (Lake McIlwaine, Rhodesia). Arch. Hydrobiol. 86: 423–444.

Robarts, R. D. & Southhall, G. C., 1977. Nutrient limitation of phytoplankton growth in seven tropical man-made lakes, with special reference to Lake McIlwaine, Rhodesia. Arch. Hydrobiol. 79: 1–35.

Rodhe, W., 1958a. Primärproduktion und Seentypen. Verh. Internat. Verein. Limnol. 13: 121–141.

Rodhe, W., 1958b. Primary production in lakes. Some results and restrictions of the ^{14}C-method. Rapp. cons. int. expl. mer. 144: 122–128.

Rodhe, W., 1965. Standard correlations between pelagic photosynthesis and light. Mem. Ist. Ital. Idrobiol. Suppl. 18: 365–381.

Rott, E., 1983. A contribution to the phytoplankton species composition of Parakrama Samudra, an ancient-man-made lake in Sri Lanka. In: Schiemer, F. (ed.) Limnology of Parakrama Samudra – Sri Lanka: a case study of an ancient man-made lake in the tropics. Developments in Hydrobiology (this volume). Dr W. Junk, The Hague.

Schiemer, F. (ed.), 1980. Parakrama Samudra (Sri Lanka) Limnology Project. Interim Report, p. 112, Inst. Internat. Coop. Vienna.

Schiemer, F., 1981. Parakrama Samudra (Sri Lanka) Project, a study of a tropical lake ecosystem. I. An interim review. Verh. Internat. Verein. Limnol. 21: 987–993.

Schiemer, F., 1983. Parakrama Samudra Project – scope and objectives. In: Schiemer, F. (ed.) Limnology of Parakrama Samudra – Sri Lanka: a case study of an ancient man-made lake in the tropics. Developments in Hydrobiology (this volume). Dr W. Junk, The Hague.

Schmidt, G. W., 1973. Primary production of phytoplankton in the three types of Amazonian waters. II. The limnology of a tropical flood-plain lake in central Amazonian (Lago di Castanho). Amazoniana 4: 139–203.

Sreenivasan, A., 1964a. The limnology, primary production, and fish production in a tropical pond. Limnol. Oceanogr. 9: 391–396.

Sreenivasan, A., 1964b. Limnological studies and fish yield in three upland lakes of Madras State, India. Limnol. Oceanogr. 9: 564–575.

Sreenivasan, A., 1965. Limnology of tropical impoundments. III. Limnology and productivity of Amaravathy Reservoir (Madras State), India. Hydrobiologia 26: 501–516.

Sreenivasan, A., 1968. The limnology of and fish production in two ponds in Chinglepat (Madras). Hydrobiologia 32: 131–144.

Sreenivasan, A., 1970. Limnological studies on Parambikulam Aliyar-Project-I, Aliyar Reservoir (Madras State), India. Schweiz. Z. Hydrol. 32: 405–417.

Steemann-Nielsen, E., 1952. The use of radioactive carbon for measuring organic production in the sea. J. Cons. Int. Explor. Mer. 18: 117–140.

Straškraba, M., 1980. The effects of physical variables on freshwater production: analyses based on models. In: Le Cren, E. D. & Lowe-McConnell, R. H. (eds.) The functioning of freshwater ecosystems. Intern. Biol. Progr. 22: 13–84, Cambridge Univ. Press.

Talling, J. F., 1957a. Photosynthetic characteristics of some freshwater plankton diatoms in relation to underwater radiation. New Phytol. 56: 29-50.
Talling, J. F., 1957b. The phytoplankton population as a compound photosynthetic system. New Phytol. 56: 133-149.
Talling, J. F., 1965. The photosynthetic activity of phytoplankton in East African lakes. Int. Rev. ges. Hydrobiol. 50: 1-32.
Talling, J. F., 1966a. Photosynthetic behaviour in stratified and unstratified lake populations of a planktonic diatom. J. Ecol. 54: 99-127.
Talling, J. F., 1966b. The annual cycle of stratification and phytoplankton growth in Lake Victoria (East Africa). Int. Rev. ges. Hydrobiol. 51: 545-621.
Talling, J. F., 1973. The application of some electrochemical methods to the measurement of photosynthesis and respiration in freshwaters. Freshwat. Biol. 3: 335-362.
Talling, J. F., 1975. Primary production of freshwater microphytes. In: Cooper, J. P. (ed.) Photosynthesis and productivity in different environments. Intern. Biol. Progr. 3: 225-247, Cambridge Univ. Press.
Talling, J. F., Wood, R. B., Prosser, M. V. & Baxter, R. M., 1973. The upper limit of photosynthetic productivity by phytoplankton: evidence from Ethiopian soda lakes. Freshwat. Biol. 3: 53-76.
Tilzer, M. M., Hillbricht-Ilkowska, A., Kowalczewski, A., Spodniewski, I. & Turczynska, J., 1977. Diel phytoplankton periodicity in Mikolajskie Lake as determined by different methods in parallel. Int. Rev. ges. Hydrobiol. 62: 279-289.
Vincent, W. F., 1979. Mechanisms of rapid photosynthetic adaptation in natural phytoplankton communities. J. Phycol. 15: 429-433.
Vollenweider, R. A., 1955. Ein Nomogramm zur Bestimmung des Transmissionskoeffizienten sowie einige Bemerkungen zur Methode seiner Berechnung in der Limnologie. Schweiz. Z. Hydrol. 17: 205-216.
Vollenweider, R. A., 1960. Beiträge zur Kenntnis der optischen Eigenschaften der Gewässer und Primärproduktion. Mem. Ist. Ital Idrobiol. 12: 201-244.
Vollenweider, R. A., 1961. Photometric studies in inland waters. Relations existing in the spectral extinction of light in water. Mem. Ist. Ital. Idrobiol. 13: 87-113.
Vollenweider, R. A., 1965. Calculation models of photosynthesis-depth curves and some implications regarding day rate estimates in primary production measurements. Mem. Ist. Ital. Idrobiol. Suppl. 18: 425-457.
Ward, A. K. & Wetzel, R. G., 1980. Photosynthetic response of blue-green algal populations to variable light intensities. Arch. Hydrobiol. 90: 129-138.
Westlake, D. F. (coord.), 1980. Primary production. In: Le Cren, E. D. & Lowe-McConnell, R. H. (eds.) The functioning of freshwater ecosystems. Intern. Biol. Progr. 22: 141-146, Cambridge Univ. Press.
Widmer, C., Kittel, T. & Richerson, P. J., 1975. A survey of biological limnology of Lake Titicaca. Verh. Internat. Verein. Limnol. 19: 1504-1510.

Authors' addresses:
M. Dokulil
Institute of Limnology
Austrian Academy of Sciences
Gaisberg 116
A-5310 Mondsee
Austria

I. Silva
Department of Zoology
Ruhuna University
Matara
Sri Lanka

K. Bauer
Fish Health Service
Animal Health Service Bayern
Senator Gerauer Str. 23
D-8011 Grub bei München
Federal Republic of Germany

6. Some remarks on long-term and seasonal changes in the zooplankton of Parakrama Samudra

C. H. Fernando & R. Rajapaksa

Keywords: long-term seasonal changes, zooplankton, tropical lake

Abstract

A study of zooplankton samples taken from Parakrama Samudra revealed long-term changes in species composition. However, these changes were probably due to low water levels following a catastrophic cyclone in 1978. Species composition seems to have returned to a stable condition after restoration of high water levels. Seasonal changes in composition and densities of plankton were not marked over a three-years sampling period.

1. Introduction

The zooplankton of Parakrama Samudra is typical in composition of that found in similar shallow tropical lakes. The number of species is smaller than that found in temperate regions, especially in the Cladocera and Copepoda, and the Rotifera are dominated by the genus *Brachionus*. The present study is based on collections of zooplankton made at irregular intervals between 1957 and 1972 and regular samples taken weekly from February 1978 to December 1980. Zooplankton composition in different water bodies in Sri Lanka has been studied intensively by Fernando (1980a). The species composition in Parakrama Samudra can therefore be compared with that in different types of water bodies. The species composition of the zooplankton has varied during the period of study, earlier resembling the fauna in a large lake, later that of a pond. This is probably due to the low water levels maintained in the reservoir following the disastrous cyclone in November 1978. This low water level was maintained until mid-1980. When typical high water levels were restored, the zooplankton composition reverted to that of a large lake.

2. Methods and materials

Ten zooplankton samples were collected during the period 1957–1972. Weekly samples were collected beginning in February 1978 and continued until the end of 1980 using a no. 25 mesh (64 mμ) and 25-cm-diameter net vertically from a depth of 1.5–4.0 m. All these samples were taken at a fixed point in the northern basin (PSN 2; see Schiemer 1983). During the period of sampling, the depth at this sampling site varied from 1.5 to 4.0 m. The earlier samples were taken with no. 10 and no. 25 mesh nets. All zooplankton samples were preserved in 5%–10% formalin and stored for study.

3. Results

The zooplankton identified in all the samples we examined are given in Table 1. The seasonal variation of total zooplankton for the period 1978–1980 is given in Figure 1. These data were obtained by counting the zooplankton from two of the four samples taken each month (i.e. every two weeks). A minimum of 100 individuals of each sample were counted.

Schiemer, F. (ed.), Limnology of Parakrama Samudra – Sri Lanka
 ISBN 90 6193 763 9

A total of 32 species of Rotifera, 19 species of Cladocera, three species of Cyclopoida, two species of Calanoida and one species of Harpacticoida was recorded. In addition, four species of Protozoa, two species of Ectoprocta and members of the groups Ostracoda, Nematoda, Decapoda, and Insecta were found (Table 1).

Among the species recorded throughout the period, many typically limnetic species found in Sri Lanka reservoirs (Fernando 1980a) were common. These are *Brachionus caudatus, B. falcatus, B. calyciflorus, Asplanchnella brightwelli, Keratella tropica* and *Lecane bulla* (Rotifera); *Diaphanosoma excisum, Ceriodaphnia cornuta* and *Moina micrura* (Cladocera); *Mesocyclops leuckarti, Thermocyclops crassus* and *Phyllodiaptomus annae* (Copepoda). Fernando (1980a, 1980b) has shown that these species are among a relatively small number of species occurring in the limnetic zooplankton throughout South-East Asia.

Some of these species mentioned occurred in most of the samples examined. These are *Brachionus caudatus, B. falcatus, Keratella tropica, Lecane bulla* and *Mesocyclops leuckarti.* No Cladocera occurred regularly in our samples though the common species were often found.

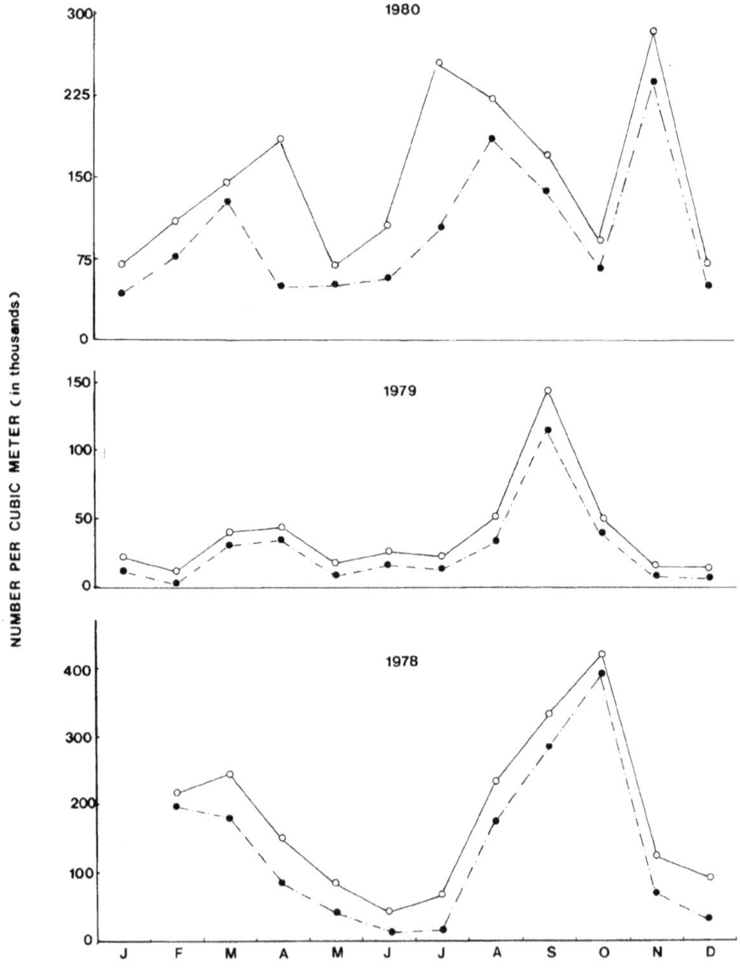

Fig. 1. Seasonal variation in density of total zooplankton and Rotifera in Parakrama Samudra. Numbers are per cubic meter (total zooplankton - Rotifera, Copepoda, Cladocera, chironomids, protozoans and ostracods): O—O total zooplankton, ●—● Rotifera.

Table 1. The occurrence of zooplankters in Parakrama Samudra during 1957–1980. All samples were collected in Topawewa. Only single samples were taken between 1957 and 1972 except in 1965 when two samples were taken.

Species	27.x.57	3.i.65	2.viii.68	7.iii.69	28.xii.70	6.i.72	Weekly samples 1978	1979	1980
Rotifera									
Anuraeopsis coelata De Beauchamp				+			+	+	+
A. fissa (Gosse)					+	+	+	+	
Asplanchnella brighwelli Gosse			+				+	+	+
A. sieboldi (Leydig)	+		+		+	+			+
Brachionus angularis Gosse	+	+	+		+	+			+
B. calyciflorus Pallas	+	+		+	+	+		+	+
B. caudatus Hauer	+			+	+	+	+	+	+
B. falcatus Zacharias	+	+		+	+	+	+	+	+
B. forficula Wierzeiski			+	+	+	+	+		
B. quadridentatus Hermann		+	+			+	+		+
B. rubens Ehrenberg	+								
B. urceus (L.)									+
B. urceolaris O. F. Müller				+		+	+		
Collotheca ornata (Ehrenberg)			+		+		+		
Conochilus sp.	+								
Conochiloides dossuarius (Hudson)		+	+	+					
Euchlanis dilatata Ehrenberg		+	+				+		+
Filinia longiseta (Ehrenberg)	+			+					
Floscularia sp.			+						
Hexathra intermedia Wisniewski			+	+					+
Keratella tropica Apstein	+		+	+	+	+	+	+	+
Lecane bulla (Gosse)		+		+		+	+	+	+
L. luna (O. F. Müller)		+	+			+	+	+	+
L. lunaris (Ehrenberg)				+					+
L. obtusa (Murray)					+	+			
Mytilina ventralis Ehrenberg		+					+	+	+
Pompholyx complanata Gosse		+	+						
Polyarthra vulgaris Carlin				+					
Trichocerca similis (Wierzejski)				+	+	+	+	+	+
T. stylata (Gosse)			+						
Testudinella patina (Hermann)	+	+						+	+
Trichotria tetractis Ehrenberg									+
Cladocera									
Sididae									
Diaphanosoma excisum Sars	+		+	+	+	+			+
D. sarsi Richard	+	+				+			+
D. modigliani Richard									+
Pseudosida bidentata Herrick				+					
Daphniidae									
Ceriodaphnia cornuta Sars	+	+	+	+	+	+			+
Daphnia lumholtzi Sars	+					+			
Simocephalus vetulus (O. F. Müller)					+		+	+	
Macrothridicidae									
Ilyocryptus spinifer Herrick							+		+
Macrothrix spinosa King		+	+				+	+	+
M. triserialis Brady	+		+		+			+	+
Bosminidae									
Bosminopsisdietersi									+

Table 1. (Continued).

Species	27.x.57	3.i.65	2.viii.68	7.iii.69	28.xii.70	6.i.72	Weekly samples 1978	1979	1980
Chydoridae									
Alona harpularia Sars							+	+	+
A. karua King		+				+		+	+
A. verrucosa Sars	+	+		+			+	+	+
Chydorus barroisi Richard		+		+		+	+	+	+
Chydorus eurynotus Sars		+				+			
Leydigia acanthocercoides (Fischer)							+		+
Copepoda									
Ectocyclops phaleratus (Koch)	+				+				
Mesocyclops leuckarti (Claus)[a]	+	+	+	+	+	+	+	+	+
Thermocyclops crassus (Fischer)				+	+		+		+
Harpacticoida									
Elaphiodella sp.	+					+		+	+
Calanoida									
Phyllodiaptomus annae (Apstein)	+	+	+	+	+	+			+
Tropodiaptomus australis Kiefer				+			+	+	+
Protozoa									
Arcella sp.	+	+			+		+	+	+
Centropyxis sp.									+
Difflugia sp.		+	+	+				+	+
Vorticella sp.							+		+
Ectoprocta									
Lophobodella carteri (Hyatt)					+				+
Plumatella repens L.		+	+			+	+	+	+
Nematoda		+	+		+		+		
Ostracoda	+		+				+	+	+
Decapoda *(Caridina)*		+	+	+					
Insecta									
Ephemeroptera	+					+			
Odonata							+	+	
Diptera									
Culicidae		+		+	+				+
Chaoborus sp.	+			+					
Chironomidae	+	+			+	+	+	+	+

[a] According to Kiefer (1981), this species is confined to Europe. However, this familiar name is retained here.

Fernando (1983) lists all the species of zooplankton found in samples collected from 1957 to 1972. These samples contained many common limnetic crustaceans. In 1978 and 1979, these limnetic species were rare (Table 1). Schiemer (1980) also found no limnetic Cladocera and only a few *Phyllodiaptomus annae* in samples collected in August–September 1979. During 1978 and 1979, following the disastrous cyclone and consequent low (shallow) water, the zooplankton was dominated by Rotifera and littoral species. Also during this period, many benthic species were collected with the zooplankton, presumably due to their suspension in wind-stirred shallow water. In 1980, following a rise in water level, the zooplankton composition reverted to the condition present before the cyclone and more closely resembles a typical limnetic situation in Sri Lanka (Table 1 and Fig. 1).

Testacean Protozoa occurred regularly in the samples taken in 1978–1980. These protozoans are common in tropical zooplankton. Fernando (1980a) found Testacea in small and large lake zooplankton in Sri Lanka. Egborge (1979) found testacean protozoans in Nigerian lakes and impoundments.

Ostracoda occurred in most of the samples taken in 1978–1980. This group of crustaceans is common in zooplankton in the tropics (though not temperate regions). They have been recorded in Sri Lanka (Apstein 1907, 1910), Africa (Green 1965, 1973; Egborge 1981), India (Mandal 1979), and South America (Brehm 1939; Deevey et al. 1980).

Among the interesting findings is *Diaphanosoma modigliani* in 1980 (Table 1). This species, typically limnetic, described from lake Toba, Indonesia (Richard 1893), was only recently recorded in Sri Lanka (Rajapaksa & Fernando 1982).

Figure 2 shows the seasonal changes in numbers of total zooplankton and Rotifera in Parakrama Samudra for 1978–1980. Based on the 1978–1980 results, no regular seasonality of any groups of zooplankton is evident.

This can be seen in Figure 2, where the seasonal

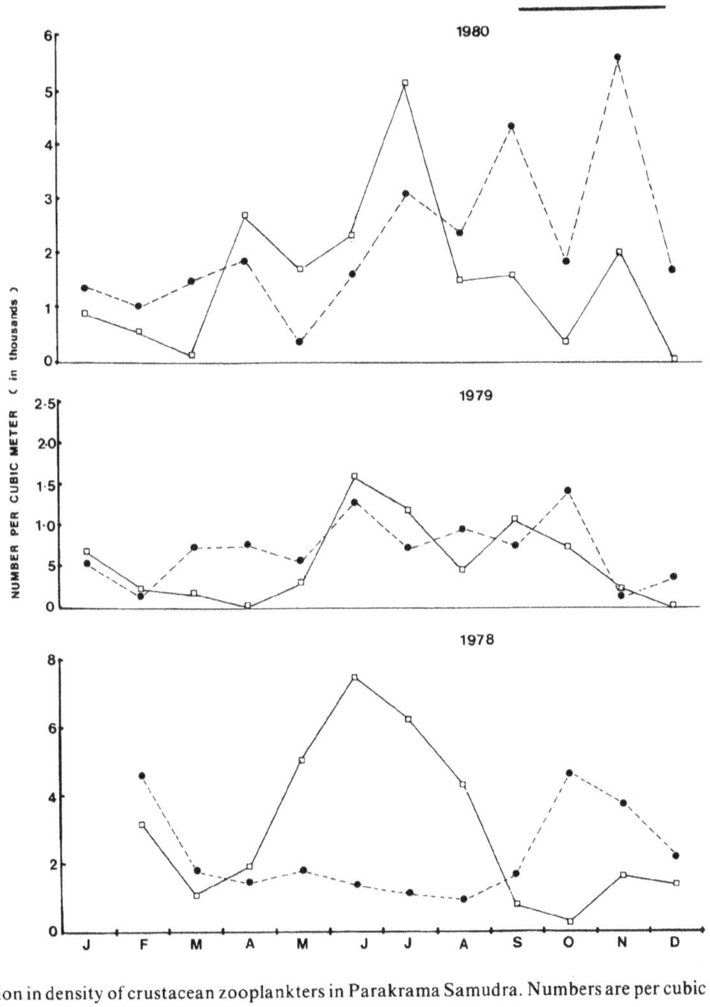

Fig. 2. Seasonal variation in density of crustacean zooplankters in Parakrama Samudra. Numbers are per cubic meter: ●—● Copepoda, □—□ *Cladocera*. —— limnetic crustaceans present.

Table 2. Relative occurrence of phytoplankters in Parakrama Samudra during the period of February 1978–December 1980.

Genera/species	1978											1979												1980											
	F	F	A	M	J	J	A	S	O	N	D	J	F	M	A	M	J	J	A	S	O	N	D	J	F	M	A	M	J	J	A	S	O	N	D
Cyanophyceae (blue-green algae)																																			
Anabaena sp.	+		+		+	+		+		+	+			+	+	+	+			+	+			+	+			+	+	+		+		+	+
Chroococcus sp.	+	+				+	+	+	+		+		+	+	+		+		+		+			+	+	+		+	+	+		+			+
Coelosphaerium sp.			+	+			+	+					+			+	+	+	+	+			+			+	+						+		
Gomphosphaeria sp.		+					+					+			+		+		+				+					+	+						
Merismopedia sp.	+	+				+			+	+	+		+					+	+	+					+	+	+	+		+	+	+	+	+	+
Lyngbya cornuta	+			+	+	+					+	+		+		+			+	+	+	+		+				+	+	+			+		+
Lyngbya sp.	*	*	*	*	+	+	+	*	*	+	+	+	+	+	+	*	+	+		+	*	+	+	+	+	+		+	+	+	+	+		+	+
Microcystis aeruginosa	*	*	+	+	+	*	*	*	*	*	*	+	+	+	*	+	+	+	*	*	+	+	+	+	+	*	*	*	+	*	*	+	+	+	+
Microcystis sp.	*	*	*	*	+	+	+	+	*	*	+	*	*	+	+	+	*	*	*	+	*	*	+	*	*	*	+	+	+	+	+	*	*	+	+
Oscillatoria sp.	+	+		*	+			+	+	+	+				+	+	+		+	+	+			+		+	+		+		+	+	+		
Raphidiopsis sp.	*	*	+	+				+				+				+			+	+	+			*	*		+				+	*	*	+	+
Chlorophyceae (green algae)																																			
Coelastrum sp.	+				+	+			+				+		+			+				+	+					+					+		+
Cosmarium sp. (Zygnemaphyceae)			+				+	+			+	+		+	+		+	+		+				+		+	+		+	+		+	+	+	+
Dictyospherium sp.		+		+	+			+	+		+			+			+	+	+		+				+	+	+				+	+	+		+
Mougeotia sp.	+		+							+		+	+			+	+		+	+	+				+			+	+				+		
Pediastrum biradatum		+	+	+	+			+			+	+			+	+		+	+	+	+	+	+		+	+			+	+	+			+	+
P. simplex	+	+	*	+	+	+	+		+	+	+		+	+	+	+		+	+		+	+	+	+			+	+		+		+	+	+	+
P. tetras		+			+		+						+				+		+	+	+					+			+					+	
Scenedesmus sp.	+	+			+	+	+	+	+	+	+	+	+	+	+	+	+	+	+	+	+	+	+	+	+	+	+	+	+	+	+	+	+	+	+
Spirogyra sp.		+						+	+								+				+			+							+				
Staurastrum sp.	+	+		+		+	+	+	+	+		+	+				+		+	+		+		+	+	+			+		+	+	+	+	+
Tetraedron sp.	+	+	+	+			+	+	+			+	+	+	+					+	+	+	+	+		+	+	+				+		+	+
Bacillariophyceae (diatoms)																																			
Gomphonema sp.				+		+																	+		+		+			+		+			
Melosira granulata	*	*	+	+	+	+	*	*	+	*	+	+	+	*	+		+	+	*	*	+	+	+	+	+	+	+	+	+	+	*	*	+	+	+
Navicula sp.	+	+		+		+		+				+		+	+	+		+			+			+		+				+	+	+			+
Pinnularia sp.			+	+				+									+				+									+	+	+			
Stephanodiscus sp.			+	+		+	+			+			+	+	+		+	+	+	+	+			+	+			+	+		+	+			+
Synedra sp.	+	+	+			+		+	+	+	+	+			+	+	+	+				+		+	+	+	+			+	+		+		
Dinophyceae (dinoflagellates)																																			
Peridinium sp.	+	+		+	+		+		+		+	+		+	+	+			+	+	+		+		+	+	+	+	+		+	*	+	+	+
Euglenophyceae (euglenoids)																																			
Trachelomonas sp.			+		+	+			+	+	+	+	+	+		+	+	+		+	+	+		+	+	+		+	+	+		+		+	+

* Abundant.
\+ Present.

changes in density of the Cladocera and Copepoda are shown. During the period of October–December 1980, the typically limnetic cladocerans, namely *D. modigliani, D. sarsi* and *Ceriodaphnia cornuta,* were found but in small numbers. However, the period under study is atypical because of the low water levels. The Rotifera dominated the zooplankton numerically throughout most of this period under study. This was specially so during the visits of the Austrian–Sri Lankan team in 1979 and 1980 (Duncan 1982; Duncan & Gulati 1981; Schiemer 1980).

A very cursory study of the phytoplankton shows that the blue-gree algae (Cyanophyceae) dominated very strongly (Table 2). This is normal in shallow tropical reservoirs, especially in those which like Parakrama Samudra are eutrophic. The density of algae was also very high especially in 1978–1980. In earlier samples, algal densities were lower.

4. Discussion

The composition of zooplankton in Parakrama Samudra underwent marked changes during the period 1957–1980. Throughout this period, there was a very high density of fish as shown by fish catches. The fish yield ranged from 150 to 500 kg/ha/year^{-1} according to De Silva and Fernando (1980). It is generally accepted that high fish densities reduced the proportion of larger zooplankton (Brooks & Dodson 1965; Hrbáček 1962). It appears that this effect is much more pronounced when the water level is low, as it was during the period 1978–1979. During this period, Rotifera dominated and so did littoral crustaceans. It appears that the low water level (i.e. littoral conditions) was unfavourable for the 'limnetic' species which comprised the usual crustacean component. Fish predation alone cannot account for the virtual absence of crustacean zooplankton in 1978–1979. In the presence of similar fish densities as in Parakrama Samudra in 1980 and even perhaps higher densities of fish in Beira (Colombo) lake, where Mendis (1964) recorded catches of 2 000 kg/ha/annum, the crustaceans in the zooplankton were fairly numerous (Costa & De Silva 1978; unpublished data of authors).

In shallow ponds where high densities of fish occur, the plankton may be almost exclusively Rotifera. Dr. M. D. Dickmen (personal communication 1981) found this to be so in fish ponds with high densities of tilapia in Costa Rica. A similar result was reported by Moitra & Mukherjee (1972) in Calcutta fish ponds. Hurlbert & Mulla (1981) have discussed the effects of fish predation on zooplankton composition in small ponds. Rotifera dominated in high fish densities usually. Egborge (1981) claims that crustacean zooplankton maxima in a Nigerian reservoir were associated with high turbidity (or the rainy season) and the rotiferan maxima corresponded to periods of high transparency of the water (i.e. in the dry season). Apstein (1907, 1910) recorded an abundance of rotifers in both dry and rainy seasons in Lake Gregory and in Colombo Lake, Sri Lanka. In studies in some tropical lakes in Africa, Robinson and Robinson (1971) and Burgis (1974) recorded high densities of zooplankton only in the dry season.

In Parakrama Samudra, high densities of copepods, rotifers and cladocerans were observed in February 1978 (dry season), and also the latter part of 1980 (July–September and November–rainy season). The low density of zooplankton which occurred during most of the study period may be due to the atypical changes which occurred after the cyclone. The study by Hurlbert & Mulla (1981) shows clearly the multiple effects of fish predation on zooplankton composition. In small ponds where spatial heterogeneity of zooplankton is less pronounced and fish are more evenly disturbed, zooplankton cannot escape heavy predation pressure even in limited portions of the habitat. In Parakrama Samudra, at high water levels, perhaps fish predation on zooplankton is most severe only in the shallowest regions. At low water levels, predation is perhaps even and high, giving the typical rotifer-dominated pattern. Also worth noting is the uniformly high densities of algae recorded throughout the period 1978–1980. High algal densities in ponds is associated with low crustacean densities.

Acknowledgements

We wish to thank Mr. W. W. John Fernando, Marawila, Sri Lanka, for collecting our samples from 1978 to 1980. Dr. Bryce Kendrick, Department of Biology, University of Waterloo, made many useful comments which improved our presentation. Professor D. G. Frey, Indiana University, Bloomington, Indiana, provided the 1965 samples.

References

Apstein, C., 1907. Das Plankton im Colombo See auf Ceylon. Zool. Jb. (Abt. Syst.) 25: 201–244.

Apstein, C., 1910. Das Plankton des Gregory Sees auf Ceylon. Zool. Jb. (Abt. Syst.) 29: 661–680.

Brehm, V., 1939. La Fauna microscopica del Lago Pétén, Guatemala. AN. Esc. Nac. Cienc. Biol. Mex. 1: 173–202.

Brooks, J. L. & Dodson, S. I., 1965. Predation, body size and composition of plankton. Science 150: 28–35.

Burgis, M., 1974. Revised estimates for the biomass and production in Lake George, Uganda. Freshwat. Biol. 4: 535–541.

Costa, H. H. & De Silva, S. S., 1978. The hydrobiology of Colombo (Beira) Lake. III. Seasonal fluctuations of plankton. Spolia Zeylan. 32: 35–53.

De Silva, S. S. & Fernando, C. H., 1980. Recent trends in the fishery of Parakrama Samudra, an ancient man-made lake in Sri Lanka. In: Tropical ecology and development. Proc. Fifth international Symposium of Tropical Ecology, J. I. Furtado (ed.), Kuala Lumpur, pp. 927–937.

Deevey, E. S., Jr., Deevey, G. B. & Brenner, M., 1980. Structure of zooplankton communities in the Peten Lake District, Guatemala. In: Kerfoot, W. C. (ed.) Evolution and ecology of zooplankton communities, pp. 669–678, University Press of New England, Hanover, New Hampshire.

Duncan, A., 1983. The composition, density and distribution of the zooplankton of Parakrama Samudra. In: Schiemer, F. (ed.) Limnology of Parakrama Samudra – Sri Lanka: a case study of an ancient man-made lake in the tropics. Developments in Hydrobiology (this volume). Dr W. Junk, The Hague.

Duncan, A. & Gulati, R. D., 1981. Parakrama Samudra (Sri Lanka) Project, a study of a tropical lake ecosystem. III. Composition, density and distribution of the zooplankton in 1979. Verh. Internat. Verein. Limnol. 21: 1007–1014.

Egborge, A. B. M., 1979. Rhizopoda (Protozoa) in a Nigerian impoundment. Nigerian Field 44: 14–20.

Egborge, A. B. M., 1981. The composition, seasonal variation, and distribution of zooplankton in Lake Asejire, Nigeria. Rev. Zool. afr. 95: 137–180.

Fernando, C. H., 1980a. The freshwater zooplankton of Sri Lanka with a discussion of tropical freshwater zooplankton composition. Int. Revue ges. Hydrobiol. 65: 85–125.

Fernando, C. H., 1980b. The species and size composition of tropical freshwater zooplankton with special reference to the oriental region (South East Asia). Int. Revue ges. Hydrobiol. 65: 411–426.

Fernando, C. H., 1983 (in press). Lakes and reservoirs of South East Asia (oriental region). In: Taub, B. (ed.) Lake and reservoir ecosystems. Elsevier/North-Holland, Amsterdam.

Green, J., 1965. Zooplankton of Lakes Mutanda, Bunyonyi, and Mulehe. Proc. Zool. Soc. Lond. 144: 385–402.

Green, J., 1973. A new species of *Oncocypris* (Ostracoda) from a crater lake in West Cameroon. J. Zool. Lond. 171: 251–256.

Hrbáček, J., 1962. Species composition and the amount of zooplankton in relation to fish stock. Rospr. Cest. Akad. Ved. Rada Prir. Ved. 72: 1–116.

Hurlbert, S. H. & Mulla, M. S., 1981. Impacts of mosquitofish *(Gambusia affinis)* predation on plankton communities. Hydrobiologia 83: 125–151.

Kiefer, F., 1981. Beitrag zur kenntnis und geographischer verbreitung von Mesocyclops leuckarti auctorum. Arch. Hydrobiol. Suppl., 62: 148–190.

Mandal, B. K., 1979. Limnological studies of a freshwater fish pond at Burdwan, West Bengal, India. Jap. J. Limnol. 40: 10–18.

Mendis, A. S., 1964. A contribution to the limnology of Colombo Lake. Bull. Fish. Res. Stn. Ceylon 17: 213–220.

Moitra, S. K. A. & Mukhejee, S. K., 1972. Studies on the freshwater plankton of a fish pond at Kalyani, West Bengal. Vest. Čs. Spol. Zool. 36: 24–28.

Rajapaksa, R. & Fernando, C. H., 1982. Cladocera of Sri Lanka with remarks on some species. Hydrobiologia 94: 49–69.

Richard, J., 1893. Entomostraces recueillis par M. E. Modigliani dans de Lac Toba (Sumatra). Ann. Mus. civ. Stor. Nat. Genova 11: 565–578.

Robinson, A. H. & Robinson, P. K., 1971. Seasonal distribution of zooplankton in the northern basin of Lake Chad. J. Zool. Lond. 163: 25–61.

Schiemer, F., 1980. An interim review of the Parakrama Samudra Project. In: Schiemer, F. (ed.) Parakrama Samudra Sri Lanka Limnology Project, pp. 9–33. Publ. Inst. Internat. Cooperation, Vienna.

Schiemer, F., 1983. Parakrama Samudra Project – scope and objectives. In: Schiemer, F. (ed.) Limnology of Parakrama Samudra – Sri Lanka: a case study of an ancient man-made lake in the tropics. Developments in Hydrobiology (this volume). Dr W. Junk, The Hague.

Author's address:
C. H. Fernando & R. Rajapaksa
Department of Biology
University of Waterloo
Waterloo, Ontario
Canada, N2L 3G1

7. The composition, density and distribution of the zooplankton in Parakrama Samudra

A. Duncan

Keywords: tropical, rotifers, size, through-flow, food, predation

Abstract

The composition, density and distribution of the zooplankton of Parakrama Samudra (Sri Lanka), an irrigation reservoir, were studied in March and April 1980. Twenty-four samples from six sites and on four occasions revealed the presence of a zooplankton consisting of rotifers and protozoans, mainly ciliates, and the virtual absence of crustacean zooplankton (apart from *Diaphanosoma excisum* and *Phyllodiaptomus annae* in a *Ceratophyllum* bed). Compared with a previous visit in August and September 1979, the rotifers consisted of similar but fewer species, attained lower densities (about one-fifth, with an overall density of 664 individuals per litre) and were uniformly distributed in time and space. The protozoans achieved higher densities than this, ranging from 958 to 4 443 individuals per litre, with *Lionotus* sp. contributing most. The character of the zooplankton in 1980 and its differences from that in 1979 are discussed in relation to flushing rates, dilution, food availability and size-selective predation.

1. Introduction

Duncan & Gulati (1981) recorded a zooplankton in Parakrama Samudra during August and September 1979 which contained 14 species of widely recognised tropical rotifers, some protozoans and virtually no planktonic crustaceans. They also noted that the rotifers present seemed small-bodied, ranging from 20 to 150 μm in length. Earlier plankton samples from the reservoir examined by Fernando (1980) recorded the presence of 13 species of cladocerans and 5 species of copepods as well as 32 species of rotifers over the years 1952–1972, but some of these were benthic forms. A second visit to Parakrama Samudra during March and April 1980 provided an opportunity to examine the causes for the presence of such a distinctive zooplankton in 1979–80.

This research forms part of a collaborative study on the limnology of Parakrama Samudra, which is an ancient irrigation 'tank' situated in the dry zone of Sri Lanka (7°55′N; 81°E; 58.5 m a.s.l.). The reservoir consists of three separate basins (PSS, PSM, PSN: see map in Schiemer 1983) connected by narrow channels and receives water into PSS from the River Ambanganga. Water for irrigation exits via three sluices under the control of the Irrigation Engineer at Polonnaruwa and, since the main outflow is in the PSN basin, the water flows from south to north. The area and volume of the whole reservoir varies throughout the year, depending upon the needs of the two rice crops and the advent of the monsoonal rains, and was 18.2 km^2 and 65.1 $\times$ 10^6 m^3 in February/March 1980 compared with 11.7 km^2 and 27.9 $\times$ 10^6 m^3 in August/September 1979. Because of the topographical differences in the three basins, the ratios of their volumes (PSN:PSM:PSS) changes at different water levels and was 1:7.5:0.8 in 1979 and 1:4.37:0.77 in 1980. These differences affect the magnitude of the relative flow-through rates of water (rates as a percentage of the basin volume) which Duncan & Gulati (1981)

Schiemer, F. (ed.), Limnology of Parakrama Samudra – Sri Lanka
 ISBN 90 6193 763 9

suggest could be one of the factors influencing the densities of the rotifer populations of PSN by physical removal or dilution.

During the 1980 visit, which took place after the monsoonal rains of November–December, the volume of water contained in PSN was three times greater than during the earlier visit and its depth doubled to about 4 m. Although the absolute rates of inflow and outflow were equally high in February 1980 and August 1979, the greater reservoir volume in February resulted in reduced relative flushing rates to about one-third of those of August. The absolute flow rates were similar because these were the penultimate months prior to harvesting the two rice crops whereas irrigational needs were minimal in March 1980 and September 1979 when the rice crops dried out. Thus, during the latter two months, flushing rates were similarly very low. Since rotifer life cycles last for days or hours at tropical temperatures, it is the *daily* relative flow rates which are likely to affect their densities and the time course of these in PSN are illustrated in Figure 1 throughout the two periods of study.

Both periods of study are presented in Figure 1 for comparative purposes. The pattern of daily water flow rates was very different during the two visits. During the 1980 one, only occasionally were values as high as 10%–12% d^{-1} attained: there was a period in late February–early March when more water was removed than put in and on two dates inputs were as high as 10% d^{-1}. During the rest of the time, there was a balance, more or less, between water passing in and out of the basin, at low daily rates of 1%–4% d^{-1}. In 1979, the PSN basin was subject to rates as high as 30% d^{-1} during August, which was also a very windy month, followed by a calm September with almost no input of water and low outputs, but the latter increased to 10% d^{-1} towards the end of the month.

Some attempt will be made to evaluate the significance of these differences during the two periods to the ability of the rotifers to develop populations in PSN along with other factors affecting rotifer density such as their capacity for increase at the prevailing temperatures, the food available to support this and the existence of predation.

2. Methods

The same methods were adopted for routinely monitoring the zooplankton as are described by Duncan & Gulati (1981). Twenty-four quantitative samples were collected on 1, 17, and 31 March and 3 April 1980 with a 2-litre perspex Ruttner sampler at 0.5-m depth and at the same fixed routine stations

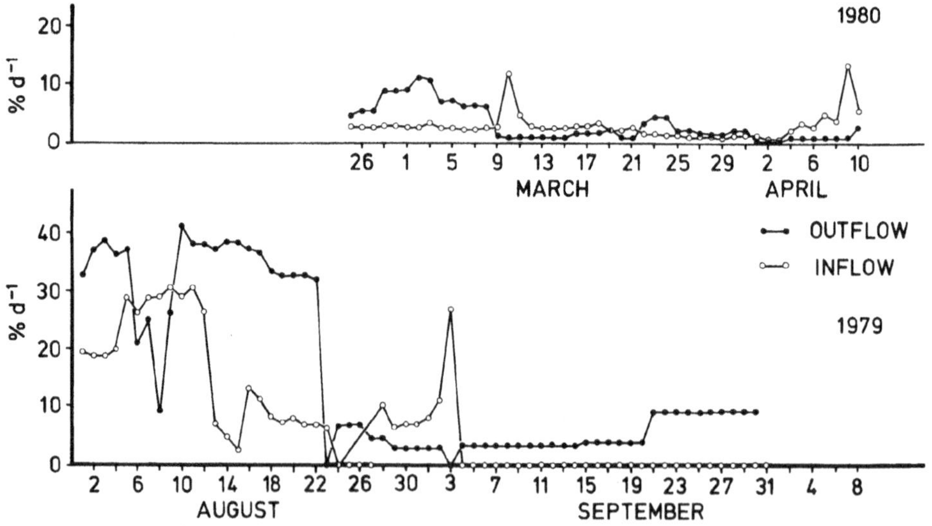

Fig. 1. The daily water inflow and outflow rates of Parakrama Samudra as a percentage of the volume of the northern basin (PSN).

(PSS, PSM, PSN 8 and PSN 3; see Fig. 2 in Duncan & Gulati 1981). One litre was seived through a 33-μm mesh and the rotifers retained on the mesh preserved in formalin. Prior to counting, an undisturbed sample was reduced in volume by pipetting and the rotifers in a drop subsample (usually about 1/200th of the sample) under a cover-slip were counted using a Leitz Orthoplan; up to eight drops per sample were counted. Because of the low densities present, the coefficients of variation were high (33%–58% of the mean). During the counting procedure, the rotifer bodies and eggs were drawn to scale under high power with the aid of a drawing attachment and subsequently measured for the dimensions.

3. Results

Table 1 lists the species of zooplankton present in the open water of the three basins of Parakrama Samudra (PSS, PSM and PSN) during March and April 1980, together with their densities, calculated here as the mean number per litre for the four sampling occasions at the routine stations; the original densities for each date are given in Table 5. As during the previous visit in August and September 1979, planktonic crustacean species were absent from the open-water samples, but a variety of crustaceans were collected by net from amongst the *Ceratophyllum* plants growing in 1–2 m of water near station 1. Amongst these were the planktonic species *Diaphanosoma excisum* and *Phyllodiaptomus annae*. The numerically most abundant taxonomic group of the open-water plankton was the Protozoa, mainly ciliates. Rotifers were much less abundant, with an overall average of 664 ind l^{-1}, compared with 3 261 l^{-1} during the previous visit in 1979.

The same protozoan genera were recorded in the

Table 1. Species in the zooplankton of Parakrama Samudra during March and April 1980 and a comparison of their relative abundance in Parakrama Samudra North with that in August and September 1979.

	Mean no. l^{-1} 1980				Mean no. l^{-1} 1980		PSN (3 + 8) 1979	
	PSS	PSM	PSN 8	PSN 3	*n*	%	*n*	%
Rotifer species								
Brachionus urceolaris							*	
Brachionus falcatus							*	
Brachionus angularis		45		170	85	14	662	20
Brachionus caudatus			76	178	127	21	189	6
Brachionus forficula				69	35	6	150	5
Anuraeopsis navicula	52	277	87	69	78	13	91	3
Trichocerca similis							248	8
Trichocerca sp.			28		14	2	79	2
Gastropus sp.							*	
Polyarthra sp. (large)			70		35	6	46	1
Polyarthra sp. (small)			28		14	2	599	18
Asplanchnella brightwelli							112	3
Ptygura sp.		54					121	4
Filinia longiseta							320	10
Collotheca sp.	49	62	87	102	95	16	172	5
Lecane sp.	68			69	34	6		
Unidentified spp.	562	337	1683		84	14	473	15
Mean density of rotifers	731	775	544	657	601	100	3 262	100
±CV%	45	53	50	43				
Protozoan species								
Vorticella sp.	531	329	37	55				
Lionotus sp.	3 827	413	973	1 207				
Larger ciliates	85	77	47	0				
Difflugia sp.	0	139	485	300				
Mean density of Protozoa	4 443	958	1 542	1562				

Table 2. The size frequency distribution of planktonic rotifers of Parakrama Samudra during March and April 1980 and a comparison of the relative abundance of the size classes in Parakrama Samudra North with that in 1979.

	Mean no. l^{-1} 1980				Mean no. l^{-1} for PSN (3 + 8) 1980		1979	
Body length (μm)	PSS	PSM	PSN 8	PSN 3	n	%	n	%
20– 29							12	*
30– 39	68		65		33	5	87	3
40– 49	101	58					76	2
50– 59	102	130	75		38	6	285	9
60– 69	99	167	107	69	88	15	667	20
70– 79	120	200	78	486	281	47	735	24
80– 89	68	93	77		39	7	229	7
90– 99	68	42	70	102	86	14	205	6
100–109		85	72		36	6	261	8
110–119							109	3
120–129								
130–139							51	2
140–149	105						222	7
150–							78	2
Not measured							244	7
Mean density of rotifers (nos. per litre)	731	775	544	657	601	100	3 261	100
Mean body length (μm)	75.8	78.9	73.5	77.1	75.4		83.2	

Table 3. Mean body length and mean volume of attached eggs of the rotifer species from Parakrama Samudra North in 1979 and 1980, together with the size frequency distribution of the volumes of loose rotifer eggs. Values are presented as means plus or minus the coefficient of variation.

	1979				1980			
Rotifer species	*n*	Body length μm %	*n*	Egg volume $10^3\ \mu m^3$ %	*n*	Body volume μm %		Egg $10^3\ \mu m^3$ %
Brachionus falcatus					1	117.0	1	130.7
Brachionus angularis	8	70.0 ± 15	2	37.4 ± 6	2	85.5 ± 11		
Brachionus caudatus	8	127.1 ± 18	3	148.0 ± 9	7	70.6 ± 12	2	64.2 ± 13
Brachionus forficula	18	86.3	6	64.9 ± 35				
Anuraeopsis navicula	8	82.2 ± 11			5	75.2 ± 10		
Trichocerca similis	7	153.0 ± 10						
Trichocerca sp.	30	78.1 ± 16			1	109		
Gastropus sp.	2	48.5						
Polyarthra sp. (large)	7	111.5 ± 20			2	92.0 ± 2		
Polyarthra sp. (small)	49	62.6 ± 15	1	26.8	1	31.0		
Asplanchnella brightwelli	7	185.3 ± 17						
Ptygura sp.	136	75.9 ± 17	21	13.8 ± 28				
Filinia longiseta	169	84.2 ± 15	8	31.6 ± 28				
Collotheca sp.	39	64.9 ± 36			5	67.8 ± 34	1	20.8 (net)
Species no. 17	59	69.6 ± 19	1	94.0				
Loose eggs							3	6.8 ± 31
			17	16.1 ± 10			7	14.9 ± 21
			54	22.3 ± 5			2	24.9 ± 10
			26	37.6 ± 16				
			33	86.6 ± 52			3	60.1 ± 12
			66	110.6 ± 1				
			4	141.1 ± 7				

1979 samples as are listed in Table 1 and the species composition of the rotifers living in PSN was also similar to that found earlier, but *Ptygura* sp. and larger species such as *Trichocerca similis, Asplanchnella brightwelli* and *Filinia longiseta* were apparently absent or present at very low densities. The comparison shown in Table 1 of the percentage composition of the rotifer fauna of PSN during the two periods of study reveals that the brachionid species increased their contribution by 10%–40% of the individuals in 1980, but that there were relatively fewer *B. angularis* and more *B. caudatus;* there was also an increased percentage occurrence of *Anuraeopsis navicula* and *Collotheca* sp., but decreased presence of *Polyarthra* spp. In 1980, *Anuraeopsis navicula* was the most abundant species in PSM.

The size frequency distribution of rotifers from all three basins for the four sampling occasions and irrespective of species identity is given in Table 2 and shows that they consisted of very small individuals, falling mainly between the lengths of 30–100 μm compared with the 1979 range of 20–150 μm. This is confirmed by the lower mean body size of 75 μm in 1980 compared with 83 μm for 1979. For PSN, this was due to both the absence of the larger species mentioned above as well as to the absence or scarcity of the larger individuals of *Brachionus caudatus* which attained a large mean size in 1979 (Table 3). It has already been noted that this species grows to a large size in neighbouring waters (120–250 μm) in which the larger species *B. calyciflorus* (210–350 μm) is also present (Duncan & Gulati 1981). The very limited size range of rotifers living in PSN during March and April 1980 (Table 2) is very striking. Egg size was calculated as the average volume from measured maximal and minimal dimensions applied to both the oblate and prolate spheroid formulae (Hale 1958). From Table 3, which presents the mean volume of the attached eggs of various species, it appears that the smaller *B. caudatus* of 1980 produced eggs of smaller volume, about 43% of those of 1979, and that the size range amongst the loose eggs belonging to rotifers shows an absence of larger ones.

Only three species of rotifers were found carrying eggs in the routine samples of March and April 1980, *Brachionus caudatus* plus two unidentified forms, although there were a few *B. falcatus, B. forficula* and *Collotheca* sp. with eggs in live net samples. There were, however, loose eggs identifiable as belonging to rotifers and the numerical density of these as well as of the attached eggs in the three basins is given in Table 4. The densities of the attached plus loose eggs were of the same order of magnitude as the rotifers themselves and the estimate of fecundity which can be calculated ranged between 0.500 and 1.008 eggs per animal (Table 4). A high value was found in PSN 3 and this was the site of highest brachionid fecundity in 1979 (for 27 August 1979, 0.17 and 0.44 eggs per animal; for 7–14 September 1979, 0.22 and 0.76 eggs per anim-

Table 4. The fecundity, egg densities and egg sizes of rotifers in Parakrama Samudra during March and April 1980 and on 1 September 1979.

	1980				1979
	PSS	PSM	PSN 8	PSN 3	PSN 14
A) Mean number of eggs per litre					
Attached eggs	0	47	70	106	274
Loose eggs	369	734	202	418	1288
Total	369	781	272	524	1562
B) Fecundity (total egg density divided by total rotifer density)					
Eggs per animal	0.505	1.008	0.500	0.798	0.402
C) Size frequency distribution of attached plus loose eggs (numbers per litre)					
Volume classes (10^3 μm^3)					
0– 9.9		392	116	41	
10– 19.9	236	291	123	155	38
20– 29.9	133			177	55
30– 39.9		46		45	36
40– 49.9				61	
50– 59.9		52	33		
60– 69.9				45	6
70– 79.9					
80– 89.9					33
90– 99.9					1
100–109.9					
110–119.9					66
120–129.9					
130–139.9					
140–149.9					7
Total	369	781	272	524	242
D) Mean egg volume (10^3 μm^3) using formulae for oblate and prolate spheroids					
Mean	20.07	15.06	18.33	24.47	60.90
± Coeff. var. (%)	28	90	95	77	

Table 5. The densities of the planktonic rotifers in the different basins of Parakrama Samudra at various dates in March and April 1980 and a comparison of their mean densities in Parakrama Samudra North with those for the whole reservoir.

Date	Numbers per litre at 0.5-m depth					Mean numbers per litre at 0.5-m depth	
	PSS Inflow	PSS	PSM	PSN 8	PSN 3	PSN (3 & 8)	PS
March 1	–	796	786	540	882	711	751
March 17	–	1 020	1 160	394	837	616	853
March 24	–	–	–	894	180	537	–
				620[a]	546[a]		
				462[b]	618[b]		
March 31	–	–	209	705	726	716	–
April 3	366	376	945	185	660	423	541
Mean density		731	775	544	657	600	664
± Coeff. var. (%)		45	53	50	43	37	45

[a] Depth of sample 2.5 m.
[b] Depth of sample 3.5 m.
– No sample.
Mean ± CV (%) for PSN 8 on 24 March = 695 ± 33%; for PSN 3 = 449 ± 52%.

al, for PSN 8 and PSN 3, respectively: Duncan & Gulati 1981). In general, however, rotifer fecundity was lower in 1979 than in 1980; the best 1979 population estimate was obtained during the diurnal study of 1 September 1979 at PSN 14 and the value of 0.404 eggs per animal has been calculated for the depths 40–80 cm comparable to the routine samples (0.5 m). From Table 4, it can be seen that the size frequency distribution of egg sizes from the three basins shows the same reduced range commented upon earlier and this is confirmed by the small mean egg volumes of 15–20 × 10^3 μm^3 in all the basins compared with the three times greater mean of 60.9 × 10^3 μm^3 for 1 September 1979.

In contrast to the situation in 1979, the rotifer populations remained at densities of less than 1 000 ind. l^{-1} in all three basins of the reservoir throughout March and at the beginning of April 1980. Any apparent tendency for densities in PSS and PSM to be higher than those in PSN is probably insignificant, bearing in mind the high coefficients of variation associated with the 1980 counting procedure. There was also uniformity of rotifer number with depth, as is shown in the counts for several depths at two stations in PSN collected on 24 March 1980. Table 1 showed that the protozoans were more numerous than the rotifers in all three basins, but particularly in PSS, and, in all cases, it was the ciliate *Lionotus* sp. which contributed most to these high numbers. Without *Lionotus* sp., the densities and distribution of these two taxonomic groups within the zooplankton are very similar. Unfortunately, the relatively high abundance of the planktonic protozoans was not appreciated until after the completion of the 1979 counting.

Table 6. Densities of larval planktonic fish (4–15 mm length) in the surface waters of Parakrama Samudra during August and September 1979.

Date	Numbers per m³			
	PSM	PSN 5–8	PSN 6–4	PSN 3–5
August 18	–	0	–	–
August 24	–	0	–	0
September 2	–	0	–	64.7
September 13	–	–	–	103.1
September 21	–	20.9	–	140.2
September 25	–	31.3	49.5	151.6
September 27	0.06	162.5	–	28.0

Duncan & Gulati (1981) reported the presence of planktonic fish larvae during the 1979 visit which were subsequently identified provisionally as belonging to *Hyporhamphus gaimardi* and *Ehirava fluviatilis*. They were caught by a Clarke–Bumpus sampler hauled quantitatively behind a boat through the surface waters down to about 1-m depth and between two stations spanning either the southern section of PSN (stations 5–8) or its northern part (stations 5–3). The number of fish caught was divided by the metred volume of water filtered and the larval densities obtained are given in Table 6 as numbers of individuals per m^3. This table shows that the larvae first appeared on 2 September

1979 in the northern part of PSN and these measured 4–10 mm in length. Numbers increased by about 4.96 larvae m^{-3} day^{-1} until 25 September 1979, by which time the largest planktonic fish measured about 15 mm, although the very small sizes were also present. The fish larvae appeared later in the southern part of PSN and at much lower densities until 27 September 1979, when the bulk of the larval population seems to have moved from the north to the south. Although the appearance of these larvae coincided with a large input of water from PSM, densities were very low there and the evidence points to breeding within PSN rather than to immigration. The final concentration is rather high and the finding of recognisable *Trichocerca* and other rotifer loricae in the guts of a few fish larvae cleared in polyvinyl lactophenol suggests that these fish may represent a serious predation upon members of the rotifer fauna.

In 1980, there was no Clarke–Bumpus sampler available to check whether these two fish species were present as planktonic larvae. In 1980, shoals of young *Sarotherodon mossambicus* were patrolling the inshore areas of PSN which were also inhabited by young fish of other species. These young fish were collected semi-quantitatively by means of a lift net of about 1 m^2 area in order to determine whether any were feeding upon the zooplankton. The net was hauled vertically through the water and caught young fish because the water was turbid. Such vertical hauls were operated in 0.5–1.0 m of water above the bare sand near the Rest House in PSN and in various depths in or besides the *Ceratophyllum* weed bed near the northern shore of PSN. Hauls were replicated and Table 7 lists the fish species caught at the different sites together with their size range and an approximate estimate of their densities. The two main species caught in this way were *Rasbora daniconius* and *Ehirava fluviatilis*. A large number of very small *S. mossambicus* were collected by dip net in 10 cm of water at the shoreward edge of the weed bed. The size of the *E. fluviatilis* ranged from 18 to 31 mm in total length, that is a little larger than the biggest planktonic larvae described earlier; whereas *R. daniconius* showed a wider size range, from 10 to 33 mm, with a few larger at 42 mm. Numbers were much greater within the *Ceratophyllum* bed and particularly of *R. daniconius*. *Ehirava fluviatilis* was the main species inhabiting the bare sandy area near the Rest

Table 7. The young fish species, their size range and approximate numbers from the littoral of Parakrama Samudra North, collected during March and April 1980 with a lift net of 1 m^2.

	Size range (mm)	Numbers caught	Numbers per haul
A) Rest House in 0.5–1.0 m of water over sand (21 hauls), 19 March			
Ehirava fluviatilis	19–31	173	8.2
Rasbora daniconius	30–60	19	0.9
B) *Ceratophyllum* bed near station PSN 1, 9 April			
1) Inshore of weed bed in 10 cm water (dip net)			
Sarotherodon mossambicus	8–27	192	–
2) Inside *Ceratophyllum* bed in 1.5–2.0 m water (2 hauls)			
Ehirava fluviatilis	20–27	31	15.5
Rasbora daniconius	10–17	56	
Rasbora daniconius	21–33	37	
Rasbora daniconius	42	2	47.5
Puntius filamentosus	14–35	16	8.0
Puntius dorsalis	20–22	2	1.0
3) Beyond *Ceratophyllum* bed in 2.0 m open water (6 hauls)			
Ehirava fluviatilis	18–23	22	3.7

House, but at about half the density of the population of the weed area. Only *Ehirava* was caught in open water about 2 m in depth some distance beyond the *Ceratophyllum* bed; the 22 fish caught in one of the six hauls attempted showed a very narrow size range (18–23 mm), just larger than the length of the larger planktonic individuals. These semi-quantitative catches suggest that quite larger numbers of young fish, belonging mainly to two species, were present in 1980 in addition to the very large numbers of shoaling young *S. mossam. cus* in the shallow littoral areas of PSN.

The virtual absence of crustacean zooplankton in 1980 as in 1979 favours the argument that the dense young fish populations are effectively eliminating the planktonic crustaceans in this shallow basin which offers little in the way of macrophytic refuges. It is therefore of interest that in the one small region where a bed of *Ceratophyllum* managed to grow and to provide such a refuge, there existed a crustacean fauna with two of the five planktonic species recorded by Fernando (1980). It was from this region in 1980 that the very small *S. mossambicus* (about 10 mm) and *E. fluviatilis* (23-27 mm) were found to contain crustacean remains in their guts. In the case of *Ehirava*, remains of *Phyllodiaptomus annae* were identifiable amongst the larger benthic prey such as chironomid and ceratopogonid larvae, nematodes, chydorid cladocerans and ostracods; *E. fluviatilis* of the same size range collected from the Rest House littoral contained only benthic forms (chironomid larvae, mayfly nymphs, nematodes, ostracods and benthic cyclopoid copepods).

Although it is generally recognised that young fish are effective predators of the crustacean zooplankton, there is less evidence of fish larvae catching and eating planktonic rotifers effectively enough to change the nature and abundance of the rotifer fauna. This is a role normally played by invertebrate planktonic predators such as *Chaoborus*. However, rotifers were found in large numbers in the stomachs of the smaller *E. fluviatilis* (20–24 mm) caught in the *Ceratophyllum* bed in 1980. These were mainly brachionids and *Trichocerca*, but *Difflugia* also appeared numerously. The brachionids, which proved to be *B. caudatus* when identifiable, ranged in length from 54 to 129 μm, with a mean of 80.3 μm $\pm$ 26% SD ($n = 11$) where whereas, at this time, the mean size of rotifers at the nearest routine sampling station PSN 3 was 77 μm (Table 2) and there were no large individuals. Some brachionids had attached eggs, but these were small (12.8×10^3 $\mu m^3 \pm 41\%$; $n = 3$) compared with the larger rotifer eggs loose in the gut contents (48.8×10^3 $\mu m^3 \pm 38\%$; $n = 5$). The young fish appeared to be selecting the larger individuals which tend to produce large eggs.

4. Discussion

Duncan & Gulati (1981) have suggested that the periods of decline in 1979 to low rotifer densities in PSN were associated with either high flushing rates (inflow and/or outflow) or dilution with water containing few rotifers whereas, during periods of low through-put of water, the rotifer populations were able to exploit their capacity for increase at the high temperatures prevailing to the limit of their food supply and attained quite high densities of more than 5 000 ind l^{-1}. The time course of daily inflows and outflows relative to the basin volume of PSN, which is given in Figure 1, indicates when these periods occurred in 1979, and Figure 3 in Duncan & Gulati (1981) can be consulted for the corresponding changes in time of rotifer densities. In contrast, the overall density of rotifers in 1980 was less than 1 000 ind l^{-1} and did not change significantly throughout the two months of study; there also appeared to be a uniform distribution with depth. In general, the relative water flow rates in 1980 were low and more or less in balance, although on several occasions inputs or outputs reached 10% d^{-1} (Fig. 1). Under such conditions of generally long retention times, one might except the presence of exponentially increasing populations of rotifers attaining quite high densities together with a crustacean plankton, but neither of these was observed.

To what extent physical removal and/or dilution were responsible for the observed fluctuations in 1979 or the absence of them in 1980 can be assessed by constructing a balance equation of the main daily gains and losses of the rotifer populations of PSN equated to the observed daily changes in the standing crop:

$$B \cdot N_n + Q_i \cdot N_m - Q_o \cdot N_n - D \cdot N_n = \Delta N_n / \Delta t \qquad (1)$$

where, N_n and N_m represent rotifer densities (ind m^{-2}) for PSN and PSM, respectively; Q_i and Q_o, the daily inflows and outflows as a fraction of the PSN basin volume; B and D, the daily finite birth and death rates of the rotifer populations; and $\Delta N_n / \Delta t$, the daily change in the rotifer population density of PSN (ind m^{-2} d^{-1}).

An attempt can be made to quantify some of these as daily rates per m^2, for those periods when the lake is well mixed horizontally and vertically: that is, during August, late September and in March. The major inaccuracies are due to infrequent sampling which imposes an assumption of linear change between samples and infrequent estimates of fecundity, from which the daily finite birth rates (births ind^{-1} day^{-1}) was calculated, using the longer egg development duration of 11 h from the PEG regression (Bottrell et al. 1976) rather than the shorter one determined for the PSN *B. caudatus* (Duncan 1983). No attempt was made to measure the finite death rates of rotifers (deaths ind^{-1} day^{-1}), which is probably mainly loss by predation, and this was assessed by difference:

$$D \cdot N_n = \Delta N_n / \Delta t - (B \cdot N_n + Q_i \cdot N_m - Q_o \cdot N_n) \quad \text{ind } m^{-2} d^{-1} \qquad (2)$$

Table 8. Daily finite birth and death rates in the rotifer populations of Parakrama Samudra North during 1979 and 1980, calculated as described in the text, and the rotifer deaths per fish per day as an estimate of predation pressure.

		PSN 8	PSN 3
August 21–31, 1979	Births ind^{-1} day^{-1}	0.37	0.96
	Deaths ind^{-1} day^{-1}	0.05–0.18	0.42–1.05
	Deaths per fish larva per day	0	0
September 14–25, 1979	Births ind^{-1} day^{-1}	0.48	1.66
	Deaths ind^{-1} day^{-1}	0.29–0.42	1.44–1.70
	Deaths per fish larva per day	54.3×10^3	53.5×10^3
March 1980	Births ind^{-1} day^{-1}	0.92	2.92
	Deaths ind^{-1} day^{-1}	0.82	2.82
	Deaths per *Ehirava* per day	30×10^3–120×10^3	110×10^3–420×10^3

There are uncertainties in these calculations in addition to the inaccuracies mentioned above: the magnitude and lag effects of rain drainage; whether the input to PSN corresponds in time and amount to that of PSS; how well the reservoir is mixed during the periods of the calculation. However, there is one interesting result for 1979: when the calculated values for daily deaths (ind m^{-3} d^{-1}) is divided by the larval and juvenile fish densities (ind m^{-3}), the resulting estimate of rotifers eaten per fish per day comes to $54 \times 10^3 \pm 17\%$ ($n = 6$) or 2 250 rotifers per fish per hour. This calculated average omitted the samples for 27 September because there is evidence of a major change in the fish distribution. Direct measurements of feeding rates would confirm how realistic are these estimated rates of daily predation losses.

The other factors which could account for the continuing low densities of rotifers and virtual absence of crustaceans in the open-water zooplankton are low food supply and size-selective predation.

Only rough estimates of the concentration of particles of a size available as food to planktonic rotifers (and crustaceans) have been obtained for PSN. There were measured as part of the feeding studies which are reported in detail in Duncan & Gulati (1983). Water from off the Rest House shore was filtered through a mesh of 33-μm hole size in order to provide a food medium (the filtrate) containing particles of edible size. Table 1 in Duncan & Gulati shows that the food concentration in 1979 was 3–4 times higher (11 mgC l^{-1}) than in 1980 (3 mgC l^{-1}) and this introduces the possibility of food limitation as a factor controlling the 1980 rotifer densities. The level of the 1980 rotifer fecundities, cited in Table 4, and of the daily finite birth rates, cited in Table 8, does not support this suggestion, since both are higher than those for 1979. The question of whether 3 mgC l^{-1} of edible food particles represents a limiting food level for either rotifers or crustaceans under tropical conditions is discussed in the paper on feeding (Duncan & Gulati 1983).

There is a considerable amount of indirect evidence that the young fish predation is the most important factor affecting the nature of the zooplankton of Parakrama Samudra as it was found in March and April 1980, and maybe also in August and September 1979. Young fish predation would account for the absence of crustaceans in the zooplankton apart from those living in the refuge area of *Ceratophyllum* bed, the lowness of the rotifer densities despite their high fecundities and the smallness of the rotifer body sizes, all features of the 1980 visit. The reduction of body size appears to be rather extreme: in 1979, the range of sizes in different species was small compared with published lengths; there was an even further loss of larger individuals of some species and an absence of larger species in 1980; and this was accompanied by lower mean volumes of eggs. This contrasts with the presence of forms like *Trichocerca similis* and larger *Brachionis caudatus* in the stomachs of small *Ehirava fluviatilis* as well as *Phyllodiaptomus annae* in fish a few millimetres longer. There is only fragmentary evidence on the densities in 1980 of the young of this species in the littoral and open water (and none for its planktonic larva), but assuming a uniform distribution, the postulated densities were 4, 8 and 16 fish per m^2 (Table 7). By adopting for 1980 the same procedure outlined earlier for 1979 in order to calculate rough estimates of the daily finite death rate for the rotifers (based upon the daily finite birth rates presented in Table 8), it is possible to obtain the daily loss per individual rotifer. These are very high: 0.82 and 2.82 individuals dying per rotifer per day for PSN 8 and PSN 3, respectively. If one assumes that these deaths were due solely to predation by *Ehirava* between 19 and 31 mm long and taking a rotifer population density of 600×10^3 ind m^{-3}, the numbers of rotifers eaten per fish per day ranges from 0.11×10^6 to 0.42×10^6 (or 4 400–17 600 rotifers per fish per hour) for fish densities ranging from 4 to 16 per m^2. These are values calculated for PSN 3, where the finite death rate was 2.82 (Table 8). For death rates of 0.82 (Table 8) found in PSN 8, the numbers of rotifers eaten per fish per day at the three levels of fish densities ranged from 0.03×10^6 to 0.12×10^6 rotifers per fish per day (or 1 281–5 125 rotifers per fish per hour). As a percentage of the rotifer areal standing crop (2.4×10^6 m^{-2}) in March, these predation rates represent 1.3%–17.5% d^{-1}; as a percentage of the rotifer daily birth per m^2 column, these predation rates represent 1.3%–6.0% d^{-1}. Again, direct measurement of feeding rates plus a real attempt at estimating fish densities would enable us to judge whether these are realistic estimates of the importance of the predatory pressure from young fish upon the planktonic animals in Parakrama Samudra during the monsoonal months when most of the indigenous fish begin to breed (H. H. Costa, personal communication) or whether there are other causes of rotifer loss.

Acknowledgements

I acknowledge with great thanks the financial assistance I received from the British Council in order to carry out this work. I thank the Irrigation Engineer at Polonnaruwa for permission to use his data in Figure 1 and Prof. H. H. Costa for his identification of larval and juvenile *Hyporhamphus gaimardi.*

References

Bottrell, H. H., Duncan, A., Gliwicz, Z. M., Grygierek, E., Herzig, A., Hillbricht-Ilkowska, A., Kurasawa, H., Larsson, P. & Weglenska, T., 1976. A review of some problems in zooplankton production studies. Norw. J. Zool. 24: 419–456.

Duncan, A., 1983. The influence of temperature upon the duration of embryonic development of tropical *Brachionus* species (Rotifera). In: Schiemer, F. (ed.) Limnology of Parakrama Samudra – Sri Lanka: a case study of an ancient man-made lake in the tropics. Developments in Hydrobiology (this volume). Dr W. Junk, The Hague.

Duncan, A. & Gulati, R. D., 1981. Parakrama Samudra (Sri Lanka) Project – a study of a tropical lake ecosystem. III. Composition, density and distribution of the zooplankton in 1979. Verh. Internat. Verein. Limnol. 21: 1001–1006.

Duncan, A. & Gulati, R. D., 1983. Feeding studies with natural food particles on tropical species of planctonic rotifers. In: Schiemer, F. (ed.) Limnology of Parakrama Samudra – Sri Lanka: a case study of an ancient man-made lake in the tropics. Developments in Hydrobiology (this volume). Dr W. Junk, The Hague.

Fernando, G. H., 1980. The freshwater zooplankton of Sri Lanka, with a discussion of tropical freshwater zooplankton composition. Int. Rev. ges. Hydrobiol. 65: 85–125.

Hale, L. J., 1965. Biological laboratory data. Science Paperbacks & Methuen. 147 pp.

Schiemer, F., 1983. The Parakrama Samudra Project – Scope and objectives. In: Schiemer, F. (ed.) Limonology of Parakrama Samudra – Sri Lanka: a case study of an ancient man-made lake in the tropics. Developments in Hydrobiology (this volume). Dr W. Junk, The Hague.

Author's address:
Annie Duncan
Department of Zoology
Royal Holloway College
Englefield Green
Surrey TW20 9TY
United Kingdom

8. A diurnal study of the planktonic rotifer populations in Parakrama Samudra Reservoir, Sri Lanka

A. Duncan & R. D. Gulati

Keywords: diurnal, depth distribution, recruitment, mortality, rotifers

Abstract

One station in Parakrama Samudra was sampled at four depths in a 1.6-m water column on seven occasions on 1 September 1979. Sixteen species of rotifers were present, of which ten were carrying eggs. Counts revealed changes in the percentage composition of rotifer species as well as in their distribution with depth throughout the day. The abundance of rotifers ranged from 3.78×10^6 Ind m^{-2} to 10.04×10^6 Ind m^{-2}, with the lowest values occurring during the early afternoon.

Hourly recruitment rates were somewhat constant throughout the day, with a mean value of 0.54×10^6 m^{-2} h^{-1} $\pm$ 21% and with a daily rate of 14.11×10^6 m^{-2} d^{-1}. The greatest losses of rotifers coincided with periods of the day when environmental variables such as temperature and dissolved oxygen concentrations were most extreme but also when light penetrated most deeply. This may have provided good conditions for visual predation by planktonic fish larvae and their feeding may account for more than half of the daily losses of rotifers. Coinciding with this, there are reductions in the percentage of brachionids, *Trichocerca* spp. and *Filinia longiseta* as well as a reduced presence of larger-sized animals.

1. Introduction

Previous work on tropical lakes has revealed the importance of the diurnal cycle in the economy of a water body (Talling 1957; Ganf & Blazka 1974; Ganf 1974; Ganf & Horne 1975; Moriarty et al. 1973; Lewis 1979) and this has been shown to be fundamentally important for the thermal regime, nutrient chemistry, primary production and fish feeding and behaviour in Parakrama Samudra (Bauer 1982; Dokulil et al. 1983; Dokulil 1983; Gunatilaka & Senaratne 1981; Schiemer & Hofer 1983; Hofer & Schiemer 1983). As the other papers on the zooplankton of this reservoir are based upon samples taken at one depth and at one time of day, attention is focussed in the present chapter on the depth distribution of the zooplankton and its fecundity and how these change within the period of 24 h. Other variables were measured simultaneously with the zooplankton sampling and, although they have been reported upon elsewhere in this volume, they are also considered here where relevant for the interpretation of the zooplanktonic situation. The zooplankton consisted largely of rotifers during the period of the diurnal study and this chapter concentrates upon this group of animals. Two important aims were: (a) to derive a value for the daily recruitment rate of the rotifer community on this day in September 1979, and (b) to determine the time course during the day of recruitment and mortality.

2. Methods

The study was based on station 14 in Parakrama Samudra North (see Fig. 3 in Schiemer 1983). It started at 0600 on 1 September 1979 and continued

Schiemer, F. (ed.), Limnology of Parakrama Samudra – Sri Lanka
 ISBN 90 6193 763 9

at 2-h intervals until the following morning for some measurements, but the last successful zooplankton samples were taken at 1745, the ones at 2300 having failed. A manned boat anchored on station facilitated frequent measurements of solar irradiance, the underwater light attenuation and the water temperature conditions. Most biological samples were taken during the daylight hours at 2-h intervals for phytoplankton, zooplankton and chlorophyll-a. Dark and light bottles were suspended from a buoy for 2-h exposure periods throughout the day. The details of the methodology for all measurements other than those on zooplankton can be obtained by consulting Bauer (1983) and Dokulil et al. (1983).

2.1 *Rotifer counts*

Twenty-eight quantitative samples were collected from all depths in a 1.6-m water column at intervals of 0—40 cm, 40—80 cm, 80—120 cm and 120—160 cm, using a 2-litre plastic Ruttner sampler. One litre of the sample from each depth was seived through a 33-μm mesh and preserved in formalin for counting. The samples were counted in Europe, using either a Leitz Orthoplan with a drawing attachment or a Wild stereomicroscope.

The drop method described in Duncan & Gulati (1981) and in Duncan (1983) was employed for counting animals. Prior to counting, the undisturbed and concentrated sample was reduced in volume by pipetting and the rotifers in a drop subsample (usually about 1/200th of the sample) were spread out under a cover-slip and counted totally. Usually 2—3 drops per sample were counted, with a level of standard deviation given in Table 3. The numbers of eggs attached to rotifer bodies were counted as well as loose eggs recognised as belonging to rotifers. All the organisms and eggs encountered whilst counting were drawn to scale at 500 × magnification, which permitted the calculation of linear dimensions. Egg volume was calculated by substituting the minimum and maximum radii (a, b) into the following formulae: $V = 4/3a^2b$ and $V = 4/3ab^2$ (Hale 1952) and then taking the average. All volumes are expressed as $10^6\ \mu m^3$.

2.2 *Estimation of rotifer fecundity*

Fecundity was estimated from egg densities divided by rotifer densities. Two types of estimates were derived: (a) the fecundity of species with attached eggs, assuming that none of the loose eggs belonged to these, and (b) the fecundity of the rotifer community by dividing the attached plus loose egg density by the density of the total rotifer populations, including those species never seen with an attached egg. Although the sizes of loose eggs are known as well as the range of sizes of eggs attached to particular species, there is a considerable overlap and loose eggs could not be assigned to particular species with certainty.

2.3 *Hatching rates of the rotifer community*

Hatching rates for each time and depth sample were calculated from the second type of fecundity mentioned above. It was necessary to consider each time and depth sample separately because of the wide range of temperature which developed in the water column during the day and because the duration of embryonic development is affected by temperature. The regression relating duration of embryonic development to temperature which was used to calculate hatching rates was one obtained for *Brachionus caudatus* from Parakrama Samudra and is as follows:

$$Y = 116.71\ e^{-0.108X}$$

where Y is the duration in hours and X the temperature in degrees Celsius (Duncan 1983b). This regression gives shorter durations for the same temperature than the relationship obtained for the bigger species, *B. calyciflorus*, which came from the Milk Factory Tank. The durations for this species were similar for the same temperature to those calculated from the generalised formula published by PEG (Bottrell et al. 1976). It was considered more realistic to adopt the locally derived regression as the brachionid eggs measured during the diurnal study were small in size and Duncan (1983b) suggested that the shorter durations of PSN *B. caudatus* were due to the smallness of their eggs. Less justifiable is the application of brachionid durations to other species of rotifers whose embryonic development has not been studied. Hatching rates were calculated as number per m^3 for each time and depth occasion and integrated, using a spline-fit function, with respect to depth and time of day.

2.4 Mortality rates of the rotifer community

Total loss of rotifers (D) in the water column was obtained from the difference of hourly recruitment rate (B) and hourly change of standing crop ($\Delta N/\Delta t$):

$$D = B - \Delta N/\Delta t \quad (\text{numbers m}^{-2}\ \text{h}^{-1}).$$

The hourly instantaneous rate of numerical increase (r) was calculated for the inter-sample period and the change in standing crop was obtained from $N(e^r - 1)$, where N is the column density of rotifers.

3. Results

When the routine station 14 of Parakrama Samudra North was sampled at different times of the day on 1 September 1979, it was found that the zooplankton consisted of rotifers and protozoans with no crustaceans (apart from nauplii of benthic cyclopoid copepods).

Table 1 lists the species of rotifers present in the order of their percentage frequency, calculated as an average of the 28 samples collected. Of the 16 species present, *Filinia longiseta, Ptygura* sp. and an unidentified species which may be a brachionid each contributed more than 10% to the rotifer community whereas *Trichocerca similis*, the small *Polyarthra* sp., the summed brachionid species, *Collotheca* sp. and a smaller *Trichocerca* sp. each provided 5%–10% of the total numbers.

Table 1. Percentage occurrence of species of rotifers present at station 14 in Parakrama Samudra North on 1 September 1979 ($n = 28$ samples).

Species		%
Filinia longiseta		22.56
Ptygura sp.		18.65
Unidentified species		12.45
Trichocerca similis		9.35
Polyarthra sp. (small)		8.76
Brachionus spp.		8.24
B. angularis	4.06%	
B. caudatus	1.14	
B. forficula	2.92	
B. urceolaris	0.12	
Collotheca sp.		6.66
Trichocerca sp.		5.67
Anuraeopsis navicula		4.43
Polyarthra sp. (large)		1.65
Asplanchnella brightwelli		1.38
Gastropus sp.		0.20
Keratella tropica		0.01

100% = 7.614×10^6 individuals per m^2.

Table 2 lists the mean length of most species as well as their mean egg volume, calculated according to the formulae mentioned in methods. The largest species present were *Asplanchnella brightwelli, Trichocerca similis, Brachionus caudatus*, the large *Polyarthra* sp. and *Brachionus urceolaris*; *Gastropus* sp. was the smallest species and the rest fell between 62 μm and 86 μm in length. The species carrying the largest eggs were *B. caudatus* and *B. forficula*, followed by *B. angularis, F. longiseta* and the small *Polyarthra* sp., whereas the smallest eggs were borne by *Ptygura* sp. There were considerable numbers of loose rotifer eggs present in the samples and it proved possible to group these into six size classes according to their linear dimensions and the mean volume (Table 2). An attempt was made to identify the origin of these loose eggs by comparison with eggs of known species, assuming that they had been detached during the sieving procedure. It may be that group (1) eggs belong to *Ptygura* sp., group (3) to *B. angularis* and group (6) to *B. caudatus*. There is too much overlap in size to attempt to assign the other groups of loose eggs to species. Ten of the 16 species of rotifer were breeding, judging by the presence of attached eggs. Seven of these are listed in Table 2 and the other three in Table 5.

3.1 Environmental variables

1 September 1979 was the first calm day after two weeks of strong winds (Fig. 3 in Duncan & Gulati 1981) and the flow-through regime imposed upon the reservoir was also not a severe one as both the daily inflows and outflows were less than 10% of the reservoir's volume (PSN).

The rotifers were subject to considerable changes in environmental variables during the course of the day. The time course of water temperature at different depths throughout the day and the appearance of a thermocline (29 °–30 °C between 0 and 1 m) from 1000 onwards are shown in Figure 1. This thermocline became more intense and deeper during the day and, at the same time, the upper water layers developed further temperature differences of 1 °C, within 1-m depth, so that there were strong

Table 2. The mean sizes of rotifer species and their eggs in Parakrama Samudra on 1 September 1979.

Species	Animals		Attached eggs	
	n	Length ± SD µm	*n*	Volume ± SD 10^6 µm^3
Brachionus urceolaris	2	96.0 ± 34%	2	0.0374 ± 6%
Brachionus angularis	8	70.0 ± 15%	2	0.0374 ± 6%
Brachionus caudatus	8	127.1 ± 18%	3	0.1480 ± 9%
Brachionus forficula	18	86.3 ± 12%	6	0.0649 ± 35%
Anuraeopsis navicula	8	82.2 ± 11%		
Trichocerca similis	71	153.0 ± 10%		
Trichocerca sp.	30	78.1 ± 12%		
Gastropus sp.	2	48.5		
Polyarthra sp. (large)	7	111.5 ± 20%		
Polyarthra sp. (small)	49	62.6 ± 15%	1	0.0268
Asplanchnella brightwelli[a]	7	185.3 ± 17%		
Ptygura sp.[a]	136	75.9 ± 17%	21	0.0138 ± 28%
Filinia longiseta	169	84.2 ± 15%	8	0.0316 ± 28%
Collotheca sp.	39	64.9 ± 36%		
Species B	59	69.6 ± 19%	1	0.0940
Loose eggs		linear dimensions (µm)		
		(1) 36.4 × 26.7	17	0.0161 ± 10%
		(2) 42.9 × 27.9	54	0.0223 ± 5%
		(3) 53.8 × 34.2	26	0.0376 ± 16%
		(4) 63.9 × 46.7	33	0.0866 ± 52%
		(5) 72.9 × 47.9	66	0.1106 ± 1%
		(6) 80.8 × 50.8	4	0.1411 ± 7%

[a] Species which contract upon preservation in formalin.

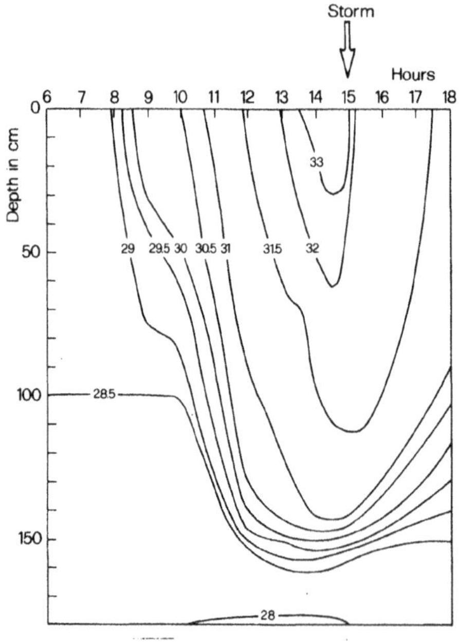

Fig. 1. The thermal stratification at station 14 in Parakrama Samudra North on 1 September 1979.

density gradients within the water column between 1300 and 1500. The advent of a strong but brief storm then mixed the thermally stratified water layers above 1.4-m depth, but was not strong enough to break down the lower thermocline, which had a temperature range between 29 °C and 31 °C within the lowest 40-cm-depth range and a density difference of 61×10^{-5} g cm^{-3}. If rotifers can move vertically against such a density gradient or if they are shifted by the thermally derived density currents proposed by Bauer (1982), the range of density difference which they would encounter would be greatest during the day (1 000–1 600) and least overnight and in the early morning. This is shown in Figure 2 together with several other environmental variables to which rotifers may respond; light penetrates most deeply during the period from 1 000 and 1 400/1 600 and both dissolved oxygen concentration and pH reveal their greatest vertical range then.

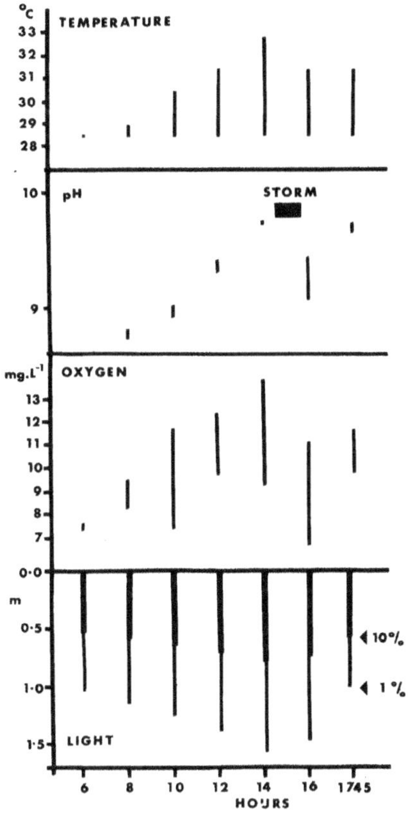

Fig. 2. The ranges of temperature, pH, concentration of dissolved oxygen and depth penetration of light at station 14 in Parakrama Samudra North on 1 September 1979.

3.2 Changes in the rotifer community during the day

The percentage occurrence of rotifer spp. in Table 1 were derived from the basic data presented in Table 3 and obtained from samples collected at four depth ranges (0–40, 40–80, 80–120 and 120–160 cm) at seven times of day (0600, 0800, 1000, 1200, 1400, 1600 and 1745). The areal abundance of each species on each sampling was obtained by integrating their concentrations with depth and the mean density per unit volume by dividing the areal value by the depth (1.6 m). Both of these values are presented in Table 4. Over the period sampled, there were between 3.784×10^6 and 10.036×10^6 rotifers per m^2 column and a mean density of between 2.365×10^6 and 6.272×10^6 rotifers per m^3. The changes in the density of the rotifer community were large during the period studied (Fig. 3). Numbers were high during the early morning (0600–1000), but declined by about 30% per 2 h from 1000 to 1400 and then increased again by about 50%–70% every 2 h from 1400 to 1745; the value at 1745 was 125% of the early morning one. Both the decline and the increase were large enough to be real and not to be a counting deviation. *Filinia longiseta* was the main species of rotifer to show this pattern of decline and increase during the day (Fig. 3). This species contributed over 30% of the total numbers in the early morning, declined to only 4% by 1400 and increased again in the afternoon to a value of 27% by the

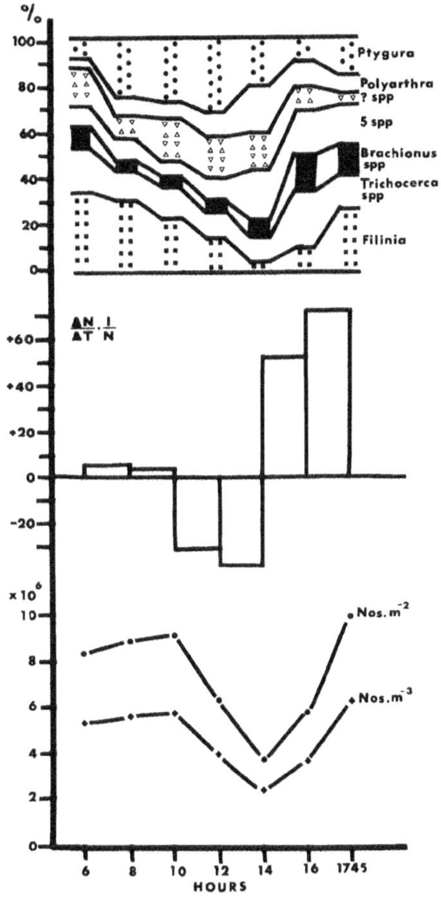

Fig. 3. The abundance of the rotifer community, change in standing crop and the percentage composition of rotifer species at station 14 in Parakrama Samudra on 1 September 1979.

Table 3. The depth distribution of planktonic rotifers and their eggs throughout the day on 1 September 1979 at station 14 in Parakrama Samudra North. Densities given as numbers per litre.

A. Densities of rotifers

Time of day	0–40 cm	40–80 cm	80–120 cm	120–160 cm
0600	4437 ± 34%	3520 ± 30%	6760 ± 19%	7168 ± 1%
0800	3731 ± 11%	4953 ± 12%	5540 ± 14%	7952 ± 1%
1000	3504 ± 1%	4982 ± 3%	5490 ± 28%	9000 ± 1%
1200	2808 ± 48%	3549 ± 33%	4158 ± 10%	4288 ± 1%
1400	961 ± 46%	1802 ± 37%	4365 ± 29%	2320 ± 28%
1600	3480 ± 10%	2272 ± 25%	3855 ± 20%	4881 ± 12%
1745	6012 ± 12%	6090 ± 7%	6204 ± 24%	6785 ± 26%

B. Densities of rotifer eggs

Time of day	0–40 cm	40–80 cm	80–120 cm	120–160 cm
0600 (a)[a]	512	264	416	640
(b)	1792	2024	1976	1408
0800 (a)	182	0	277	710
(b)	1365	1557	2078	710
1000 (a)	110	846	366	225
(b)	986	1316	1464	2925
1200 (a)	104	136	445	0
(b)	624	2457	1040	1200
1400 (a)	0	169	121	116
(b)	266	281	1210	928
1600 (a)	115	0	0	503
(b)	517	568	1338	2020
1745 (a)	350	500	282	115
(b)	1320	810	1410	1215

[a] (a) attached eggs; (b) loose eggs.

evening. The two *Trichocerca* spp., the *Brachionus* spp. and five of the sparser species showed similar changes with time. Not all species changed in this way; *Ptygura* sp., the small *Polyarthra* sp. and some of the unidentified species were proportionately more abundant during the mid-day period and less so both early in the morning and later in the afternoon.

Table 4. The diurnal changes in the abundance of the total rotifer community at station 14 in Parakrama Samudra North on 1 September 1979.

Time of day	Numbers $m^{-2} 10^6$	Numbers $m^{-3} 10^6$
0600	8.381	5.278
0800	8.832	5.514
1000	9.179	5.737
1200	6.233	3.896
1400	3.784	2.365
1600	5.792	3.622
1745	10.036	6.272

The areal abundance was obtained by integration with depth and the mean density per m^3 by dividing the areal value by 1.6 m.

3.3 Depth distribution of rotifer species and rotifer sizes

An analysis of the percentage occurrence of species at different depths and times (Fig. 4) reveals that these changes in abundance are accompanied by quite large differences in the position of rotifers in the water column at various times of day. *Filinia longiseta* was the most important species at all depths in the 0600 sample but its proportion gradually declined, at first in the upper layers of water but later by 1200 throughout most of the water column. *Ptygura* sp. showed a reverse pattern of increasing its percentage contribution from low values at 0600 to quite high ones at 1200, particularly in the upper water layers.

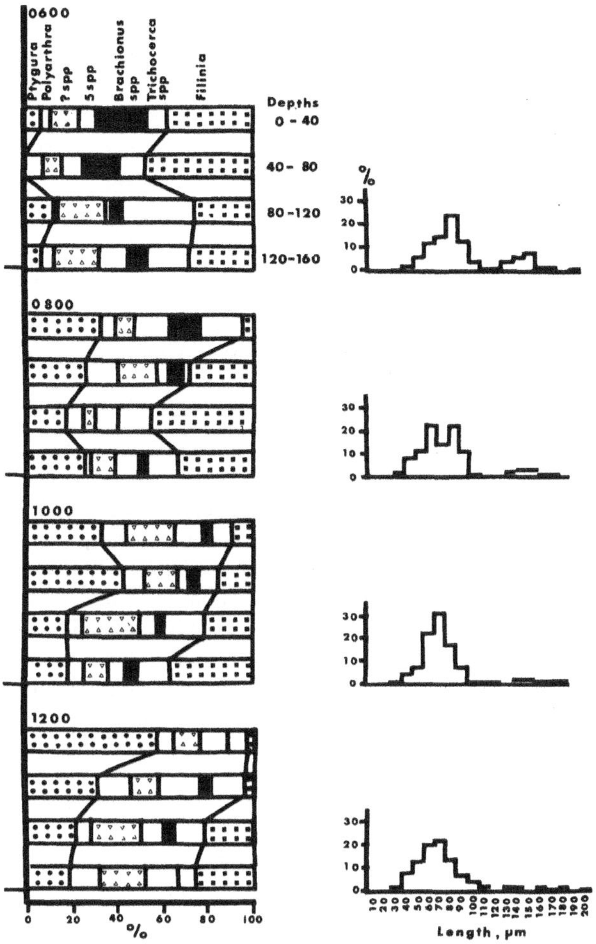

Fig. 4. The depth distribution of rotifer species and the size frequency distribution between 0600 and 1200 on 1 September 1979 at station 14 in Parakrama Samudra.

Accompanying this change in species' vertical distribution, there is also a shifting downwards of the size frequency distribution of rotifer sizes. This is illustrated in Figure 4 and shows both a lowering of the main modal size class from 80 to 90 μm and a loss of the sparse larger size classes from 0600 to 1200 (the only samples for which size analysis is available).

3.4 Estimates of fecundity

These refer only to species with attached eggs (Table 5). There were more species carrying eggs early in the morning and late in the evening than during the mid-day period. The former probably indicates egg laying overnight and was shown by the brachionid species, *Filinia longiseta*, and several of the unidentified species. *Ptygura* sp., on the other hand, was almost continuously present with attached eggs.

These fecundities may be underestimates if the sieving procedure detached eggs, resulting in a considerable increase in the numbers of loose eggs present in the samples.

The densities of loose eggs were much greater than those of attached eggs and also there were

Table 5. The species of planktonic rotifer carrying eggs at different times of day on 1 September 1979 and estimates of their fecundity.

	Ne/Nt: attached egg density divided by the density of the species, using depth-integrated values						
Species	0600	0800	1000	1200	1400	1600	1745
Brachionus angularis	0.384	0	0	0	0.391	0.271	0
Brachionus caudatus	1.000	0	0	0	0	0	0
Brachionus forficula	1.000	0.500	0	0	0	0.138	0.526
Polyarthra sp. (small)	0	0	0.200	0	0	0	0
Ptygura sp.	0.125	0.143	0.226	0.169	0.061	0	0.092
Filinia longiseta	0.080	0.081	0	0	0	0	0.031
Collotheca sp.	0	0	0	0	0	0	0.115
Species A	0.499	0	0.536	0	0	0	0
Species B	0.167	0.167	0	0	0.082	0	0
Species C	0	0	0.200	0	0	0	0
No. spp. with eggs	7	4	4	1	3	2	4

fewer loose eggs in the upper water layers from 1200 to 1745 (Table 3). When attached and loose eggs are summed and the total egg density is divided by the total number of rotifer individuals, the estimated community fecundity is 3–14 times greater than that calculated from attached eggs only. Whether this summation procedure is valid depends upon the viability of detached loose eggs and whether they hatch at the same embryonic development rate as attached eggs. The depth distribution of the rotifer community fecundity derived from all the eggs is plotted in Figure 5 for the various sampling periods. There is a marked change-over during the day from high fecundities in the top metre of water early in the morning to high fecundities in deeper water later in the day (up to the storm at 1530).

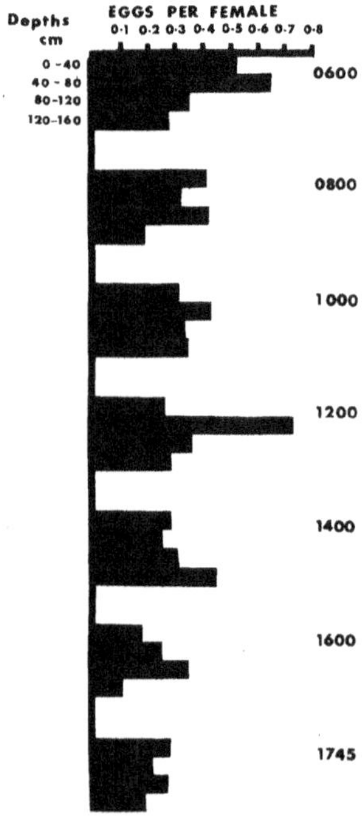

Fig. 5. The depth distribution of the fecundity of the rotifer community at station 14 in Parakrama Samudra on 1 September 1979.

3.5 Hatching and recruitment rates in the rotifer community

The temperatures within the water column on 1 September 1979 ranged from 28 °C to 33 °C; the rotifer hatching rates for each depth-time sample were calculated separately. The temperatures were substituted in the equation relating temperature and embryonic development duration of *Brachionus caudatus* from Parakrama Samudra North (Duncan 1983b):

$$Y = 116.71\ e^{-0.108X}$$

where Y is the duration in hours and X the temperature in degrees Celsius. Table 6E gives the results of these substitutions for each depth-time and Table 6C presents the inverse, i.e. the hourly turn-over rates (1/De). Table 6E contains the hourly hatching rates per female, obtained by multiplying the fecundities in Table 6D by the turn-over rates in

Table 6. The calculation of daily recruitment rates per m² of the rotifer community in Parakrama Samudra North on 1 September 1979.

	0600	0800	1000	1200	1400	1600	1645
A. Temperature							
0– 40 cm	28.5	29.0	30.5	31.5	33.0	31.75	31.25
40– 80 cm	28.5	28.5	29.5	31.25	32.0	31.75	31.25
80–120 cm	28.5	28.5	29.0	31.0	31.5	31.50	30.50
120–160 cm	28.0	28.0	28.0	30.0	31.0	30.25	29.25
B. Duration of embryonic development in hours							
0– 40 cm	5.4	5.1	4.3	3.9	3.3	3.8	4.0
40– 80 cm	5.4	5.4	4.8	4.0	3.7	3.8	4.0
80–120 cm	5.4	5.4	5.1	4.1	3.9	3.9	4.3
120–160 cm	5.7	5.7	5.7	4.6	4.1	4.5	5.0
C. Hourly turn-over rate (1/De)							
0– 40 cm	0.185	0.196	0.233	0.256	0.303	0.263	0.250
40– 80 cm	0.185	0.185	0.208	0.250	0.270	0.263	0.250
80–120 cm	0.185	0.185	0.196	0.244	0.256	0.256	0.233
120–160 cm	0.175	0.175	0.175	0.217	0.244	0.222	0.200
D. Fecundity (total eggs divided by total rotifers)							
0– 40 cm	0.519	0.415	0.313	0.259	0.278	0.183	0.278
40– 80 cm	0.650	0.314	0.434	0.731	0.250	0.250	0.215
80–120 cm	0.354	0.425	0.333	0.357	0.310	0.347	0.273
120–160 cm	0.286	0.179	0.350	0.280	0.450	0.517	0.196
E. Hourly hatching rates (eggs female^{-1} h^{-1})							
0– 40 cm	0.096	0.081	0.073	0.066	0.084	0.048	0.070
40– 80 cm	0.120	0.058	0.090	0.183	0.068	0.066	0.054
80–120 cm	0.066	0.079	0.065	0.087	0.079	0.089	0.064
120–160 cm	0.050	0.031	0.061	0.061	0.110	0.115	0.039
F. Hourly recruitment rates (numbers litre^{-1} h^{-1})							
0– 40 cm	426	302	256	185	81	167	421
40– 80 cm	422	287	448	649	123	150	329
80–120 cm	446	438	357	362	345	343	397
120–160 cm	358	247	549	262	255	561	265
G. Depth-integrated hourly recruitment rates (numbers m^{-2} h^{-1})							
$\times 10^6$	0.662	0.511	0.644	0.588	0.323	0.486	0.565
H. Time-integrated daily recruitment rates (numbers m^{-2} day^{-1})							
			14.107×10^6				

Table 6C. Multiplication of the hourly hatching rates per female (Table 6E) by the number of females per litre (Table 3) provides an estimate of the hourly recruitment rate per litre for each depth and time (Table 6F). The hourly column rate was obtained by integration with depth for each sampling occasion (Table 6G). Finally, the hourly column rates were integrated with respect to time, making the assumptions that the 0600 rate extended back to 0000 and that the 1745 rate extended forwards to 2400.

The hourly recruitment rates were somewhat constant throughout the day, apart from the low value of 0.32×10^6 m^{-2} h^{-1} at 1400, and fell within the limits of the mean ± coefficient of variation (0.54×10^6 m^{-2} h^{-1} ± 21%). The daily recruitment rate of 14.11×10^6 m^{-2} d^{-1} represents about 14% of the maximal density at 1745 (10.036×10^6 m^{-2}), which is a time of day when visual predation is least effective.

3.6 Losses from the rotifer community

The order of magnitude of losses can be estimated by subtracting from the hourly recruitment the hourly changes in the standing crop (Table 7). The

Table 7. Hourly losses of rotifers estimated as the difference between recruitment rates and changes in the standing crop in Parakrama Samudra North on 1 September 1979.

Hours	N $\times 10^6$	e^r	$\overset{*}{N}$ $\times 10^6$	$\Delta N/\Delta t$ $\times 10^6$	B $\times 10^6$	D $\times 10^6$
0600–0700	8.38	1.026	8.60	+0.22	0.66	0.44
0700–0800		1.026	8.82	+0.22		0.44
0800–0900	8.83	1.019	8.99	+0.17	0.51	0.34
0900–1000		1.019	9.16	+0.17		0.34
1000–1100	9.18	0.824	7.55	–1.61	0.64	2.25
1100–1200		0.824	6.22	–1.33		1.97
1200–1300	6.23	0.779	4.85	–1.38	0.59	1.97
1300–1400		0.779	3.78	–1.07		1.66
1400–1500	3.78	1.237	4.68	–0.90	0.32	1.22
1500–1600		1.237	5.78	+1.10		–0.78
1600–1700	5.79	1.317	7.63	+1.84	0.49	–1.35
1700–1800		1.317	10.04	+2.43		–1.94
1800–1900	10.04	0.986	9.90	–0.14	0.57	0.71
0500–0600		0.986	8.38	–0.11		0.06
0600–0700	(8.38)					

Key: N (numbers m^{-2}) is the observed standing crop of rotifers; $\overset{*}{N}$ (numbers m^{-2}) is the predicted standing crop from Ne^r; $\Delta N/\Delta t$ (numbers $m^{-2} h^{-1}$) is the hourly change in standing crop; B (numbers $m^{-2} h^{-1}$) is the hourly recruitment rate from Table 6G; and D (numbers $m^{-2} h^{-1}$) is the hourly loss of rotifers from ($B - \Delta N/\Delta t$).

procedure was to calculate the hourly instantaneous rate of numerical increase (r) for the inter-sample period and to apply e^r to the concentration of rotifers at the beginning of the hour in order to obtain the concentration at the end of the hour and the hourly change in standing crop. The loss was estimated by subtraction of this hourly change in numbers from the numbers recruited during that hour. Losses should be positive: any negative value implies either a gain in numbers due to births or an underestimation of the recruitment rate (e.g. by too low fecundities).

The time course of losses of rotifers (Table 7) follows the same pattern of numbers already illustrated in Figure 3. The losses are relatively low (4%–5% of the standing crop per hour) from 0600 to 1000, but were very high (25%–32% h^{-1}) between 1000 and 1500 when light penetrated most deeply into the water and when most environmental variables showed their greatest range (Fig. 2). From 1500 to 1800, the loss rates became negative, suggesting that some input of rotifers occurred from a source other than reproduction and this may be associated with the advent of the storm after 1530. The overnight loss rate is low, but is not reliable since it is based upon an assumed concentration of rotifers for the 0600 sample on the next day.

4. Discussion

Previous papers (Duncan & Gulati 1981; Duncan 1983a) on the spatial distribution of the rotifer populations of Parakrama Samudra were based largely upon samples from 0.5-m depth and collected during the morning. The present results show that there are changes throughout the day at one site in both the relative proportions of species and the abundance of the total numbers of individuals. These differences are striking: for example, the proportion of *Filinia longiseta* sp. could vary from 30% to 4% and *Ptygura* sp. from 7% to 31%. The numerical range throughout the day was from 3.78×10^6 to 10.04×10^6 ind m^{-2}, a factor of times 2.7. In addition, there are changes in the depth distribution of the predominant species and a shift to smaller rotifers throughout the day.

Some of the changes may be associated with the very large variations in environmental conditions to which the rotifers were subjected. Total numbers, depth distribution and percentage composition of species showed their largest change during the period from 1000 to 1400, when light penetrated most deeply and when the depth range of temperature, dissolved oxygen concentrations and pH were greatest. There is the possibility that most species were avoiding or succumbing to the warm (more than 30 °C), well-lit surface layers of water with very high concentrations of oxygen (about 180% saturated at its maximal value). The strong thermal density gradients within the water column during the lit part of the day may inhibit vertical movement of some animals and account for the generally greater densities of animals below 80-cm depth (Table 3). The largest changes in the rotifer community occurred after the storm and, although the deeper thermocline was not broken down, the violent water mixing may account for the input into the water column at station 14 of large numbers of rotifers which contributed to the negative loss rates reported in Table 7. Elsewhere in the reservoir, higher densities of rotifers existed, as was shown by Duncan & Gulati (1981) for station 3 in the northern bay of Parakrama Samudra North on 27 August 1979.

Another possible cause of loss of rotifers is predation. The most likely predators of rotifers were the planktonic fish larvae of *Hyporhamphus gairmardi* and *Ehirava fluviatilis*, of which there were 64.7 ind m^{-3} present on 2 September 1979 near station 14 (Duncan 1983a). A tentative value for the daily consumption of rotifers by the fish larvae of 54.3×10^3 per larva was derived from rotifer mortality rates and not from measured fish feeding rates. This estimates a potential daily consumption of 3.51×10^6 rotifers $m^{-3} d^{-1}$. The planktonic fish larvae possess large eyes and may be visual predators, feeding during the daytime in the lit parts of the water column. It is during the period from 1000 to 1500, when the light penetrates most deeply (Fig. 2), that the greatest loss rates of rotifers are recorded (Table 7). The summed losses for this 5-h period come to 9.07×10^6 rotifers m^{-2} or 5.67×10^6 rotifers m^{-3}. Assuming that this is the main feeding period for the fish larvae, it appears that this density of fish larvae could account for 62% of the rotifer losses from 1000 to 1500. The rotifer species which show the greatest decline in the surface waters during this period (Fig. 4) are the brachionids, *Trichocerca* spp. and *Filinia longiseta*, and there is a loss of the larger rotifer size classes.

Acknowledgements

We acknowledge with thanks the financial assistance received from The British Council and the Netherlands Foundation for the Advancement of Tropical Research to carry out this work.

References

Bauer, K., 1983. Thermal stratification, mixis and advective currents in Parakrama Samudra Reservoir, Sri Lanka. In: Schiemer, F. (ed.) Limnology of Parakrama Samudra - Sri Lanka; a case study of an ancient man-made lake in the tropics. Developments in Hydrobiology (this volume). Dr W. Junk, The Hague.

Bottrell, H. H., Duncan, A. Gliwicz, Z. M., Grygierek, E., Herzig, A., Hillbricht-Ilkowska, A., Kurasawa, H., Larsson, P. & Weglenska, T., 1976. A review of some problems in zooplankton production studies. Norw. J. Zool. 24: 419-456.

Dokulil, M., 1983. Aspects of gut passage of algal cells in *Sarotherodon mossambica* (Pisces, Cichlidae). In: Schiemer, F. (ed.) Limnology of Parakrama Samudra - Sri Lanka; a case study of an ancient man-made lake in the tropics. Developments in Hydrobiology (this volume). Dr W. Junk, The Hague.

Dokulil, M., Bauer, K. & Silva, I., 1983. An assesment of the phytoplankton biomass and the primary productivity of Parakrama Samudra, a shallow man-made lake in Sri Lanka. In: Schiemer, F. (ed.) Limnology of Parakrama Samudra - Sri Lanka; a case study of an ancient man-made lake in the tropics. Developments in Hydrobiology (this volume). Dr W. Junk, The Hague.

Duncan, A., 1983a. The composition, density and distribution of the zooplankton in Parakrama Samudra. In: Schiemer, F. (ed.) Limnology of Parakrama Samudra - Sri Lanka; a case study of an ancient man-made lake in the tropics. Developments in Hydrobiology (this volume). Dr W. Junk, The Hague.

Duncan, A., 1983b. The influence of temperature upon the duration of embryonic development of *Brachionus* species (Rotifera). In: Schiemer, F. (ed.) Limnology of Parakrama Samudra - Sri Lanka; a case study of an ancient man-made lake in the tropics. Developments in Hydrobiology (this volume). Dr W. Junk, The Hague.

Duncan, A. & Gulati, R. D., 1981. Parakrama Samudra (Sri Lanka) Project - a study of a tropical lake ecosystem. III. Composition, density and distribution of the zooplankton in 1979. Verh. Internat. Verein. Limnol. 21: 1001-1008.

Ganf, G. G., 1974. Diurnal mixing and the vertical distribution of phytoplankton in a shallow equatorial lake (Lake George, Uganda) Oecologia 62: 611-629.

Ganf, G. G. & Blazka, P., 1974. Oxygen uptake, ammonia and phosphate excretion by zooplankton of a shallow equatorial lake (Lake George, Uganda). Limnol. Oceanogr. 19: 313-325.

Ganf, G. G. & Horne, A. J., 1975. Diurnal stratification, photosynthesis and nitrogen fixation in a shallow, equatorial lake (Lake George, Uganda). Freshwat. Biol. 5: 13-39.

Gunatilaka, A. & Senaratne, C., 1981. Parakrama Samudra (Sri Lanka) Project, a study of a tropical lake ecosystem. II. Chemical environment, with special reference to nutrients. Verh. Internat. Verein. Limnol. 21: 1000-1006.

Hale, L. J., 1965. Biological Laboratory Data. Science Paperbacks & Methuen. 147 pp.

Hofer, R. & Schiemer, F., 1983. Feeding ecology, assimilation efficiencies and energetics of two herbivorous fish: *Sarotherodon* (Tilapia) *mossambicus* (Peters) and *Puntius filamentosus* (Cuv. et Val.). In: Schiemer, F. (ed.) Limnology of Parakrama Samudra - Sri Lanka; a case study of an ancient man-made lake in the tropics. Developments in Hydrobiology (this volume). Dr W. Junk, The Hague.

Lewis, W. M., 1979. Zooplankton Community Analysis. Springer-Verlag, Berlin. 159 pp.

Moriarty, D. J. W., Darlington, J. P. E. C., Dunn, I. G., Moriarty, C. M. & Tevlin, M. P., 1973. Feeding and grazing in Lake George, Uganda. Proc. R. Soc. Lond. 184: 299-319.

Schiemer, F., 1983. The Parakrama Samudra Project - scope and objectives. In: Schiemer, F. (ed.) Limnology of Parakrama Samudra - Sri Lanka; a case study of an ancient man-made lake in the tropics. Developments in Hydrobiology (this volume). Dr W. Junk, The Hague.

Schiemer, F. & Hofer, R., 1983. A contribution to the ecology of the fish fauna of the Parakrama Samudra reservoir. In: Schiemer, F. (ed.) Limnology of Parakrama Samudra – Sri Lanka; a case study of an ancient man-made lake in the tropics. Developments in Hydrobiology. Dr W. Junk, The Hague.

Talling, J. F., 1957. Diurnal changes of stratification and photosynthesis in some tropical African waters. Proc. R. Soc. B 147: 57–83.

Authors' addresses:
A. Duncan
Department of Zoology
Royal Holloway College
Englefield Green
Surrey TW20 9TY
United Kingdom

R. D. Gulati
Vijverhof Laboratory
Limnological Institute
3631 AC Nieuwersluis
The Netherlands

9. The influence of temperature upon the duration of embryonic development of tropical *Brachionus* species (Rotifera)

A. Duncan

Keywords: development, temperature, body size, egg size, tropical, rotifers

Abstract

The duration of embryonic development of *Brachionus caudatus* from the tropical reservoir, Parakrama Samudra (Sri Lanka), was determined at several constant temperatures and compared with data on *Brachionus calyciflorus* from a neighbouring water body. The duration-temperature regressions differed in elevation, but not in slope for 28 °C. They predict a duration of embryonic development of 5.67 h for *B. caudatus* and 10.51 h for *B. calyciflorus*. The difference lay not in their responses to temperature, since the Q_{10} values are similar, but in the leftward shift of the development rate ($1/D_e$) – temperature curve for *B. caudatus* with respect to that for *B. calyciflorus*. The experimental *B. caudatus* were smaller (97–127 μm in length) than *B. calyciflorus* (295 μm) and were carrying small eggs (0.079–0.148 $\times 10^6$ μm^3 compared with 0.641 $\times 10^6$ μm^3 for *B. calyciflorus*). It is suggested that the shorter durations of *B. caudatus* are due to the smallness of the eggs produced by the small adult females which were characteristic of the planktonic populations inhabiting the lake during the period of investigation. The discussion considers these findings in relation to previously published work on rotifer embryonic development times as well as the possible causes for small eggs.

1. Introduction

Development times when combined in various ways with densities and individual weights provide the time element necessary for the calculation of instantaneous birth, death and growth rates and the production of continuously reproducing species (Edmondson 1960; Hillbricht-Ilkowska & Patalas 1967; Paloheimo 1974; Bottrell et al. 1976). In field populations of rotifers, it is difficult to obtain information on the duration of the various periods of the life cycle, which King (1969) distinguishes as (1) the embryonic period from egg formation to egg hatching, (2) the pre-reproductive period from egg hatching to first reproduction, (3) the period of reproduction and (4) the post-reproductive period from cessation of reproduction to death. If the population is subject to predation on adults, it is likely that period (4) does not exist and that period (3) is curtailed, so that the ecologically important durations would be the embryonic period, the period of immaturity and the time interval between egg depositions during reproduction.

According to Ruttner-Kolisko (1974), true growth with cleavage of cells occurs in the rotifers only during the development of the embryo, and the young rotifer emerging from the egg already possesses its adult number of nuclei. Any enlargement of the body is produced by stretching or insertion of assimilated material between the nuclei. This occurs during King's period (2), which Ruttner-Kolisko divides into two phases: I, after emergence when the size of the body does not change but active feeding takes place; and II, when ingested food has been assimilated and stretching occurs, which continues until the beginning of reproduction. Any

Schiemer, F. (ed.), Limnology of Parakrama Samudra – Sri Lanka
 ISBN 90 6193 763 9

external influence on body size such as food level would have to operate on phase I and the consequence would be seen in phase II, since during phase III, when reproduction is taking place, the adult body does not change in size.

Several workers have determined the duration of phases in the life cycle of rotifers that were reared in the laboratory either at one temperature and under excess food conditions or at different food densities (and more than one food species) or at several temperatures under optimal food levels (Bottrell et al. 1976; Halbach 1970; King 1967; Piauvaux 1977; Pourriot 1973; Pourriot & Deluzarches 1971; Pourriot & Hillbricht-Ilkowska 1969; Pourriot & Rieunier 1977; Ruttner-Kolisko 1964, 1975; Snell & King 1977). The effects of food and temperature on post-embryonic duration do not appear to have been studied simultaneously, although this is an essential requirement for ecological application, as is discussed by Bottrell et al. (1976). During their extensive studies on rotifer life cycles, Pourriot & Deluzarches (1971) determined the duration of the embryonic (D_e), post-embryonic (D_p) and generation periods ($T_g = D_e + D_p$) in 11 species at several temperatures. They concluded that, under their experimental conditions of excess food of a good quality (*Phacus pyrum*), the ratio D_p/D_e appeared to be constant and independent of temperature. This ratio ranged from 1.57 to 1.70 (14°–20 °C) for their thermophile clones of *Brachionus calyciflorus* and from 1.34 to 1.81 (10°—20 °C) for the eurythermal one. These clones exhibited different thermal characteristics as the thermophile clone did not reproduce below 12 °C and was cultured at 20°–22 °C and the eurythermal one reproduced between 5 °C and 25 °C and was reared at 10°–12 °C. The larger D_p/D_e ratios which the authors calculated from the data of Halbach (1970) for this species ranged from 3.55 to 3.99 (15°–20 °C) and were ascribed to the poor utilisation of *Chlorella* compared with *Phacus* at this temperature range as at 25 °C the ratio for Halbach's cultures dropped to 1.99. The D_p/D_e ratio, which can be calculated from Ruttner-Kolisko's (1975) data on *Hexarthra fennica* fed on *Chlorella vulgaris* at 10^6 cells ml^{-1}, were also high. This author provides some information on the time interval between egg depositions in *Hexarthra* fed on *Chlorella* and these ranged from 0.41 at 25 °C to 0.71 at 15 °C of the embryonic development time.

The differences between the two thermal clones of *Brachionus calyciflorus* fed on *Phacus* were explored by Pourriot & Deluzarches (1971) and Pourriot (1973), who found that the relationship between the development rate and temperature in the warm thermophile clone exhibited an anti-clockwise rotation (i.e. an accelerated rate) with respect to the cold eurythermal clone or, in other words, revealed a higher Q_{10} (for D_e, about 5 compared with 2.8). Pourriot (1973) was also able to show that body length and egg diameter were linearly and inversely related to temperature in both clones, but the warm clone had smaller adults and eggs at the same temperatures. Consequently, there appeared to be a linear relationship between egg diameter (L_o) and body length (L_a) for which he calculated a significant regression ($L_o = 0.384L_a + 48$ [$r = 0.897$, $n = 32$; units in μm]) based upon data from not only the two clones but also from natural populations from ponds at Gif. When the relationship between the size and duration of eggs was examined, there was a very interesting tendency for larger eggs to have a prolonged development for the same temperature, which was first commented upon in Pourriot & Deluzarches (1971). This needs testing by multi-regressional analysis since temperature is also involved and may require more experimentation to enlarge the number of data points. In general, Pourriot felt that the nutritive state of the juvenile animal was probably influential in determining the size attained by the adult and the size of eggs that she can produce (i.e. the amount of food reserves stored). Pourriot's work has been reviewed in detail because of its relevance to the present results on tropical populations of *Brachionus* spp.

2. Materials and methods

Animals were collected by plankton net from offshore stations when possible during the early morning of the day of the experiment and those individuals carrying eggs were immediately sorted into experimental dishes together with lake water filtered through 33 μm netting. This provided food of edible size and removed the larger algae. A stock of this filtered water was kept at experimental temperatures so that the water in the experimental dishes could be renewed at hourly intervals when observations were made.

During September 1979, the experimental dishes

(watch-glasses with covers greased on to prevent evaporation) were placed in a white plastic dish supported in a temperature-controlled tank by means of polystyrene foam. The watch-glasses contained 1 ml of filtered water and up to ten animals with eggs (but usually fewer). These dishes provided good conditions for the animals, but involved too much searching time in order to check every individual, by transfer to another dish, in such a large volume of water. In April 1980, the cavities of haemagglutination trays were employed; these were cut to a suitable size and floated directly on the water surface, supported by polystyrene foam for safety. The cavities contained 0.2 ml of filtered water and 1–5 animals. A stiff plastic sheet was greased on to prevent evaporation. The temperature contact was better in this system and searching required less time.

The water tank was cooled continuously by pumping its water through a refrigerator unit and its temperature was controlled by intermittent heating by means of a thermostat. The efficiency of cooling varied with the electrical supply of the town of Polunnaruwa and, when this was uninterrupted, control was adequate and within ±0.5 °C, except when a low temperature was attempted. Table 1 lists the temperature conditions of the nine experiments successfully accomplished.

The experimental procedure adopted was that described by Edmondson (1960, 1965) and Pourriot & Deluzarches (1971). If a population of rotifers is laying eggs continuously, a sample at any one moment will contain eggs at different stages of development. When the number of females with unhatched eggs is plotted against time, a linear regression, if statistically significant, will predict the time when the last youngest eggs will hatch. This represents the mean egg development duration from laying to hatching for a particular temperature. If the laying of eggs in nature is synchronised within a 24-h period and samples are collected at the same time of day, the duration of egg development estimated in this way could be too short. There is this danger in this study as samples were usually collected at about 0630 from the Milk Factory Tank and at about 0830 from Parakrama Samudra North.

Observations were made at hourly intervals from the time when a sufficient number of individuals with eggs were placed in their experimental dishes. Animals were handled individually, using a coarse-bored braking pipette which in no way damaged them. The sorting procedure usually lasted 1–2 h, but could take longer when animals were small and scarce. The larger *Brachionus calyciflorus* from the Milk Factory Tank were easier to handle than the smaller *B. caudatus* from Parakrama Samudra. The latter species was reasonably abundant during August and September 1979, though small, but was scarcer and smaller in March and April 1980. An unsuccessful attempt was made to work with an even smaller species, *Brachionus forficula*, but these tended to become trapped in the surface tension layer where the eggs did not develop.

Table 1. Experimental conditons and results[a]: regressions relating number of females with unhatched eggs (Y) with time in hours (X): Equation: Y = a + bX.

No.	Date	Species	Temperature (mean, range)	Source of animals	a	b	*df*	*F*	*P*	X when Y = 0 in hours
1.	17.9.79	*Brachionus* spp.	31.3 ± 0.3	PSN 14	22.952	-2.633	1,8	412	0.001	8.72
2.	18.9.79	*B. caudatus*	26.8	PSN 3	26.477	-3.856	1,6	65	0.001	6.87
3.	21.9.79	*B. caudatus*	26.8	PSN	14.000	-2.143	1,5	204	0.001	6.53
4.	27.9.79	*B. caudatus*	20.8 ± 0.5	PSN 3	17.232	-1.402	1,10	67	0.001	12.29
5.	27.9.79	*B. caudatus*	9.0 ± 1.0	PSN 3	14.941	-0.716	1,16	254	0.001	20.87
6.	1.4.80	*B. calyciflorus*	21.0[b]	MFT	31.550	-1.707	1,13	999	0.001	18.48
7.	2.4.80	*B. caudatus*	27.4 ± 0.4	PSN 8	15.400	-2.700	1,3	70	0.005	5.70
8.	3.4.80	*B. calyciflorus*	30.1 ± 0.1	MFT	92.218	-10.182	1,8	418	0.001	9.06
9.	7.4.80	*B. calyciflorus*	35.2[b]	MFT	20.400	-3.700	1,3	133	0.001	5.51

[a] *df* = degrees of freedom; *F* = variance ratio; *P* = probability of the regression; X when Y = 0 is the mean duration of egg development.
[b] Steady temperature throughout experiment.
[c] PSN: Parakrama Samudra North; MFT: Milk Factory Tank.

3. Results

Table 1 lists the regression of the number of females with unhatched eggs against time obtained for the nine experiments, three for *Brachionus calyciflorus* from the Milk Factory Tank (MFT) during April 1980 and six for *Brachionus caudatus* from Parakrama Samudra North (PSN) during August 1979 and April 1980. In most cases, the experimental animals were fed on natural food particles of less than 33 μm obtained from their own habitat and the food concentration, as carbon, was measured as part of the feeding studies reported in Duncan & Gulati (1983). For MFT in April 1980, this was 6 mg C l^{-1}; for PSN in August 1979 and April 1980, these were 3 mg C l^{-1} and 11 mg C l^{-1}, respectively. All the regressions were statistically significant and the last column provides the predicted mean egg development duration when there were no females left with unhatched eggs. These values have been plotted against experimental temperature in Figure 1, and Table 2 presents the exponential regressions

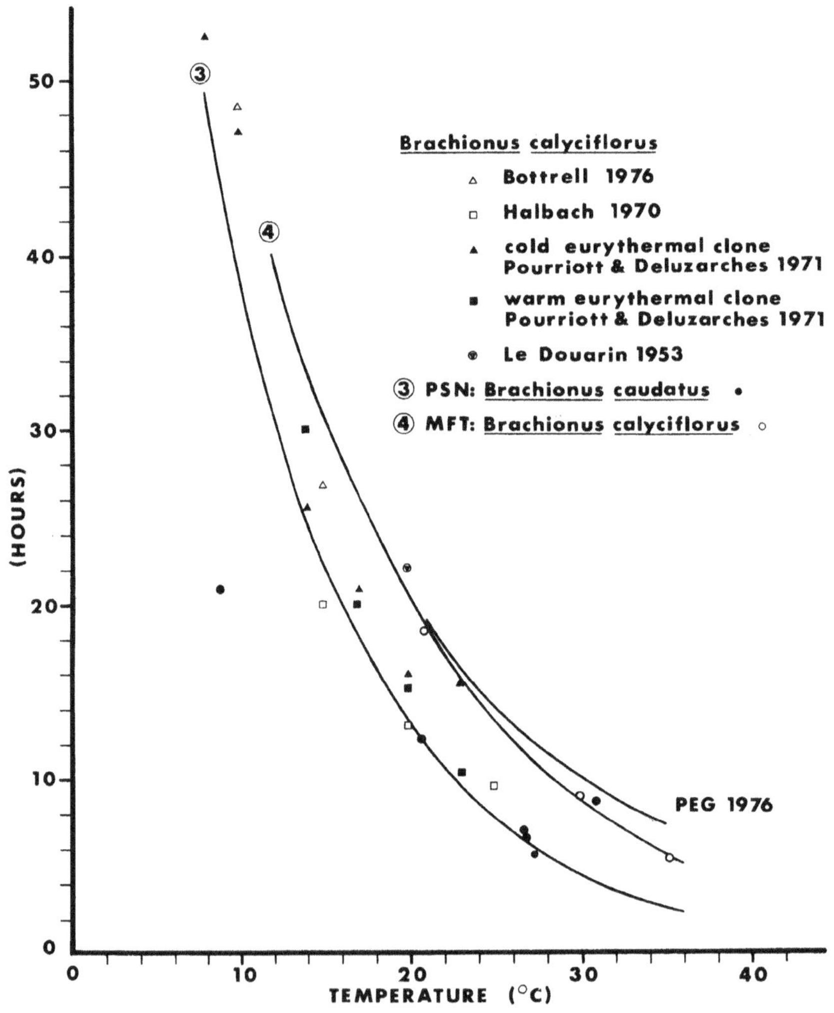

Fig. 1. The duration of egg development at various temperatures for *Brachionus caudatus* from PSN and *B. calyciflorus* from MFT together with some values from the literature for *B. calyciflorus*. The values predicted by the regression equations in Table 2 for *B. caudatus* (no. 3) and *B. calyciflorus* (no. 4) are plotted as a continuous line.

Table 2. Regressions relating mean duration of egg development in hours (Y) to experimental temperature in °C (X).[a] Equation: $Y = a \cdot e^{bX}$.

No.	Species	Source	Regressions included	a	b	*df*	*F*	*P*	Q_{10}
1.	*B. caudatus*	PSN	1, 2, 3, 4, 5, 7	32.755	-0.054	1,4	13.7	0.025	
2.	*B. caudatus*	PSN	1, 2, 3, 4, 7	24.209	-0.043	1,3	1.2	0.500	
3.	*B. caudatus*	PSN	2, 3, 4, 7	116.708	-0.108	1,2	89.6	0.025	2.94
4.	*B. calyciflorus*	MFT	6, 8, 9	110.385	-0.084	1,1	289.0	0.001	2.32
5.	*B. calyciflorus*[b]	't'		149.125	-0.116	1,2	296.0	0.000	3.19
6.	*B. calyciflorus*[b]	'e'		96.500	-0.086	1,3	33.7	0.025	2.36

[a] *df* = degrees of freedom; *F* = variance ratio; *P* = probability of the regression; Q_{10} for 1/Y versus $X = e^{+b \cdot X}$.
[b] Calculated from results published in Pourriot & Deluzarches (1971); 't' refers to the thermophile clone and 'e' to the eurythermal clone.

of egg duration on temperature, which fit the empirical data significantly. These have been derived separately for the two species of *Brachionus* and there are two significant regressions for *B. caudatus:* (1) which uses the results from all six experiments and (2) which includes the results from only four experiments (runs 2, 3, 4 and 7). Experiment 5 is omitted because 9 °C is a low temperature for a tropical species and experiment 1 because the duration obtained is long for a temperature of 31 °C and this was only the second experiment attempted in Sri Lanka.

When these two regressions for the Sri Lankan *Brachionus* species are compared, as in Table 3, in their transformed linear forms by covariance analysis (Snedecor & Cochran 1967), the residual mean squares and slopes do not differ statistically, but their elevations do, as is shown by the probabilities in Table 3. Pooling the sums of squares to derive a common slope results in a value of -0.0897, which will become positive when applied to the development rate ($1/D_e$) and represents a Q_{10} of 2.45. It is possible to calculate similarly exponential regressions for the thermophile and eurythermal clones of *B. calyciflorus* from the data published by Pourriot & Deluzarches (1971); these regressions are presented in Table 3 and are all statistically significant. An attempt was made to test for differences between these regressions, by six pairs of comparisons the results of which are given in Table 3. Table 3 reveals that there is a homogeneity of residual mean squares within all four regressions so that they are comparable. In one pair of regressions for the thermophile clone and MFT population of *B. calyciflorus* (comparison 2), the slopes are different (*P* = 0.05). This is due to the thermophile clone whose slope of -0.116 is high and represents a Q_{10} of 3.19. By pooling the sums of squares and products for the other three regressions (two Sri Lankan species and the eurythermal *B. calyciflorus*), a common slope of -0.0878 is obtained, which represents a Q_{10} of 2.41. This result supports Pourriot & Deluzarches' (1971) claim that the developmental response of the thermophile clone of *B. calyciflorus* is accelerated

Table 3. Comparison of regressions for *Brachionus* species relating mean egg duration to temperature ($Y = a \cdot e^{bX}$).[a]

Species whose regressions were compared[b]		Probabilities of			
		R.M.S.	Slopes	Elevations	
B. caudatus (PSN)	*B. calyciflorus* (MFT)	0.50	0.25	0.001	Differ in elevation
B. calyciflorus (MFT)	*B. calyciflorus* ('t')*	0.50	0.05	–[c]	Differ in slope
B. caudatus (PSN)	*B. calyciflorus* ('e')*	0.25	0.50	0.05	Differ in elevation
B. calyciflorus (MFT)	*B. calyciflorus* ('e')*	0.50	0.75	0.50	
B. calyciflorus ('t')*	*B. calyciflorus* ('e')*	0.10	0.25	0.50	
B. caudatus (PSN)	*B. calyciflorus* ('t')*	0.50	0.75	–	

[a] R.M.S. = residual mean squares; slopes = b in the above equation; elevation refers to the height of the regression.
[b] Asterisks indicate regressions calculated from results published by Pourriot & Deluzarches (1971).
[c] Not tested further as slopes were significantly different.

with respect to the others. Amongst the three other populations, it has already been demonstrated that the two Sri Lankan species have different elevations ($P = 0.001$ in comparison 1). This is also so for comparison 3 between *B. caudatus* and the eurythermal *B. calyciflorus* ($P = 0.05$). In two other comparisons (nos. 4 and 5) between the MFT and eurythermal *B. calcyciflorus* and between the thermophile and eurythermal clones of *B. calyciflorus*, the differences in elevation were not statistically significant. As the comparison of elevations between *B. caudatus* and the thermophile clone of *B. calyciflorus* resulted in a negative variance ratio (comparison 6), no conclusion can be drawn. In Figure 2, which presents plots of all four exponential regressions, but as development rate ($1/D_e$) against temperature, the curves for *B. caudatus* from Sri Lanka and the thermophile clone of *B. calyciflorus* lie close together and to the left of the MFT population and eurythermal clone of *B. calyciflorus*. In general, for the same temperature, rates ($1/D_e$) were faster and egg development (D_e) was shorter in the former pair.

There was considerable difference in the size of animals and their egg volume between the MFT *B. calyciflorus* and the PSN *B. caudatus*, which is shown in Table 4. Care was taken in April 1980 to measure the experimental animals and their eggs, but for September 1979 there is only evidence of size for the field animals. As can be seen, the egg of *B. calyciflorus* was about eight times the volume of that of *B. caudatus* used in the April experiments and four times the volume of field *B. caudatus* collected in September. The adult *B. calyciflorus*

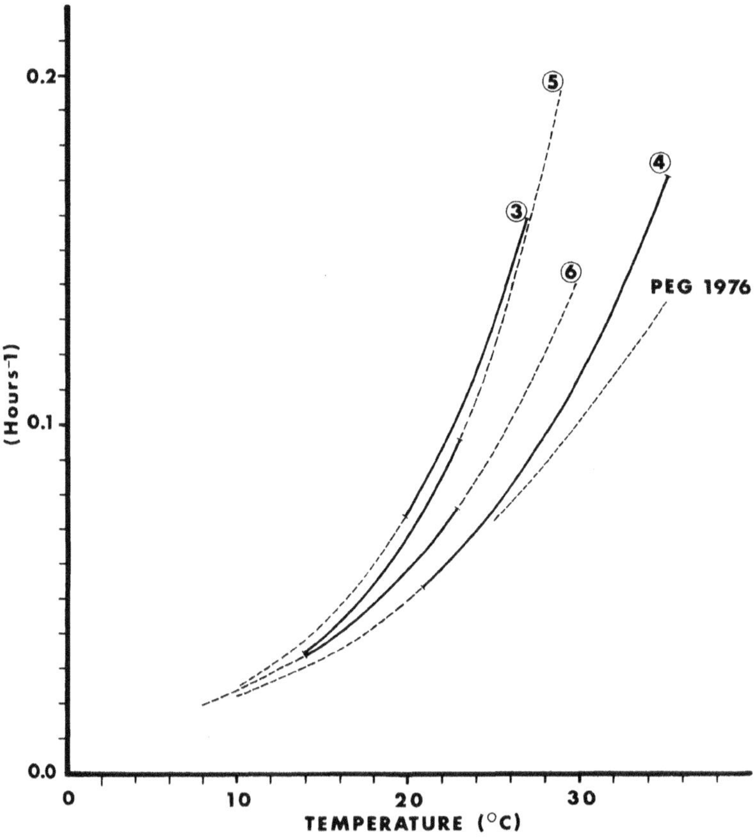

Fig. 2. The development rate ($1/D_e$) as predicted by the regression equations in Table 2 for *Brachionus caudatus* from PSN, Sri Lanka, (no. 3) and for *B. calyciflorus* from MFT, Sri Lanka (no. 4), and from France, the thermophile clone (no. 5) and the eurythermal clone (no. 6).

Table 4. Mean body length (μm) and egg volume (10^6 μm³) of experimental *Brachionus caudatus* and *Brachionus calyciflorus* in April 1980 and of field *B. caudatus* in September 1979.

Species	Stage	Date	Body length (μm)	Body volume (10^6 μm³)	Egg volume (10^6 μm³)
Brachionus calyciflorus	Adult	April	295	4.98	0.641
Brachionus caudatus	Adult	April	97	0.17	0.079
	Juvenile	April	66	0.048	-
Brachionus caudatus	Adult	Sept.	127	-	0.148

are also considerably larger than the *B. caudatus* and it may be that larger females are laying larger eggs. In order to detect whether this is so within the brachionid populations of PSN, egg volume was plotted against bodylength for 28 individuals belonging to different species of *Brachionus*. Only in *B. caudatus* were there big differences as the April animals were smaller and carried smaller eggs than those measured in September. It was possible to derive a linear relationship between egg volume (V_o, 10^3 μm³) and adult length (L_a, μm): $V_o = 1.044L_a - 12.79$ (*df* 1,7; $F = 79$; $P = 0.001$). The other species showed little variation in body length and egg size and *B. calyciflorus* is so large in comparison that it lies off the scale of this figure 3.

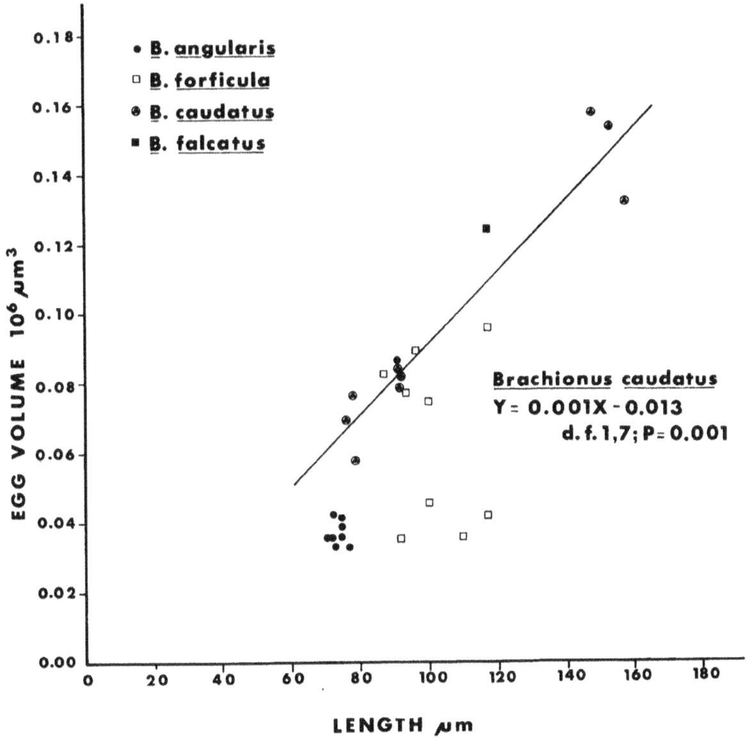

Fig. 3. The relationship between egg volume (10^6 μm³) and body length (μm) in *Brachionus* species from PSN and MFT in Sri Lanka.
Key ● *Brachionus angularis*
□ *Brachionus forficula*
Ⓐ *Brachionus caudatus*
▪ *Brachionus falcatus*
Brachionus calyciflorus from MFT had a length of 300 μm and an egg volume of 0.642×10^6 μm³.

4. Discussion

Table 3 shows that the two Sri Lankan species of *Brachionus* studied differed in the responses of their egg development to temperature. The difference lay not in their Q_{10}, which were not significantly different (2.94 versus 2.32), but in their durations. As a result of the leftward shift of the curve for *B. caudatus* from PSN with respect to that for *B. calyciflorus* from MFT, the embryonic development time of the former was much shorter (5.67 h) compared with the latter (10.51 h) for the same temperature, 28 °C. The substantially shorter egg duration in *B. caudatus* appeared to be associated with the small size of eggs being produced by the small females which were so characteristic of the PSN population. A similar difference in the sizes of eggs and females was detected by Pourriot (1973) in the warm clone of *B. calyciflorus* compared with the cold clone, as well as a tendency for larger eggs to have a prolonged development when compared for the same temperature. Pourriot suggested that small eggs and adults might be caused by poor nutritive conditions during the juvenile phase of the adults, but this is unlikely to be the case in the Sri Lankan population. The carbon concentration of their food ranged from 3 to 11 mg C l^{-1} and these are unlikely to represent limiting food levels, judging from the high fecundities observed in their population (Duncan 1983)*. Duncan & Gulati (1981) have commented upon the smallness of body size amongst the PSN rotifer species when compared with published sizes, and Duncan (1983) shows that the mean length of the rotifer planktonic community was even smaller in 1980 due to the scarcity or absence of larger species as well as larger individuals. These authors have proposed that the reduction in the body size of the planktonic rotifers during the periods when Parakrama Samudra was studied was due to the intensive predation by young fish and that the resulting small females gave birth to small eggs, which we see here have short duration times. If this is so, the consequence is that the PSN populations were developing at a rate which was approximately twice as fast as would be expected in populations with individuals with a more normal size frequency. There is evidence in Table 4 that it was mainly *B. caudatus* amongst the brachionid species of PSN which exhibited a considerable reduction in size of body and egg between September 1979 and April 1980. It is clear from the preliminary gut analysis of the young fish feeding on rotifers that it was brachionid rotifers which were taken mainly and, when identifiable, these were mostly *B. caudatus*.

Table 3 reveals that the regression for the MFT *B. calyciflorus* does not differ from that of the eurythermal clone of the species studied by Pourriot & Deluzarches (1971) and Figure 2 shows that its development rate lies close to that published by PEG (Bottrell et al. 1976) for mainly temperate species of rotifers (bearing in mind that the latter was derived from a curvilinear regression of a different formula). This suggests that the tropical *B. calyciflorus* shows similar temperature responses in egg development to that of temperate populations. In contrast, the development rate-temperature curve of *B. caudatus* (Fig. 2) lies close to that of the warm clone of *B. calyciflorus* studied by Pourriot & Deluzarches (1971) and had similar Q_{10} values (2.94 versus 3.19). The common feature in these two groups of experimental animals was the small size of the adult bodies and eggs.

* However, please read the Discussion in Duncan & Gulati (1983).

Acknowledgement

I acknowledge with great thanks the financial assistance I received from the British Council in order to carry out this work.

References

Bottrell, H. H., Duncan, A., Gliwicz, Z. M., Grygierek, E., Herzig, A., Hillbricht-Ilkowska, A., Kurasawa, H., Larsson, P. & Weglenska, T., 1976. A review of some problems in zooplankton production studies. Norw. J. Zool. 24: 419–456.

Duncan, A., 1983. The composition, density and distribution of the zooplankton in Parakrama Samudra. In: Schiemer, F. (ed.) Limnology of Parakrama Samudra – Sri Lanka: a case study of an ancient man-made lake in the tropics. Developments in Hydrobiology (this volume). Dr W. Junk, The Hague.

Duncan, A. & Gulati, R. D., 1981. Parakrama Samudra (Sri Lanka) Project – a study of a tropical lake ecosystem. III. Composition, density and distribution of the zooplankton in 1979. Verh. Internat. Verein. Limnol. 21: 1001–1006.

Duncan, A. & Gulati, R. D., 1983. Feeding studies with natural food particles on tropical, species of planktonic rotifers. In: Schiemer, F. (ed.) Limnology of Parakrama Samudra – Sri Lanka: a case study of an ancient man-made lake in the tropics. Developments in Hydrobiology (this volume). Dr W. Junk, The Hague.

Edmondson, W. T., 1960. Reproductive rates of rotifers in natural populations. Memorie Ist. Ital. Idrobiol. 12: 21–77.
Edmondson, W. T., 1965. Reproductive rate of planktonic rotifers as related to food and temperature in nature. Ecol. Monogr. 35: 61–111.
Halbach, U., 1970. Einfluss der Temperatur auf die Populationsdynamik des planktischen Rädertieres *Brachionus calyciflorus* Pallas. Oecologia (Berl.) 4: 176–207.
Hillbricht-Ilkowska, A. & Patalas, K., 1967. Methods of estimating production and biomass and some problems of qualitative calculation methods of zooplankton. Ekol. Pol. B 13: 139–172.
King, C. E., 1967. Food, age and the dynamics of a laboratory population of rotifers. Ecology 48: 111–128.
King, C. E., 1969. Experimental studies of ageing in rotifers. Exp. Geront. 4: 63–79.
Paloheimo, J. E., 1974. Calculation of instantaneous birth rate. Limnol. Oceanogr. 19: 692–694.
Piauvaux, A., 1977. Sur la durée de developpement des oeufs immédiats de quelques rotifères. Ann. Hydrobiol. Suppl. 33: 237–247.
Pourriot, R., 1973. Rapports entre la température, la taille des adultes, la longueur des oeufs et le taux de développement embryonnaire chez *Brachionus calyciflorus* Pallas (Rotifère). Ann. Hydrobiol. 4: 103–115.
Pourriot, R. & Deluzarches, M., 1971. Recherches sur la biologie des rotifères. II. Influence de la température dur la durée du développement embryonnaire et post-embryonnaire. Ann. Limnol. 7: 25–52.
Pourriot, R. & Hillbricht-Ilkowska, A., 1969. Recherches sur la biologie des quelques rotifères planctoniques. I. Resultats preliminnaires. Bull. Soc. Zool. Fr. 94: 111–118.
Pourriot, R. & Rieunier, M., 1973. Recherches sur las biologie des rotifères. III. Fécondité et durée de vie comparées chez les femelles amictiques et mictiques de quelques espèces. Ann. Limnol. 9: 241–258.
Ruttner-Kolisko, A., 1964. Über die labile Periode im Fortpflanzungsklus des Radertiere. Int. Rev. ges. Hydrobiol. 49: 473–482.
Ruttner-Kolisko, A., 1974. Plankton rotifers. Biology and taxonomy. Binnengewasser Suppl. 26: 1–146.
Ruttner-Kolisko, A., 1975. The influences of fluctuating temperature on plankton rotifers. A graphical model based on life data of *Hexarthra fennica* from Neusiedlersee, Austria. Symp. Biol. Hung. 15: 197–204.
Snedecor, G. W. & Cochran, W. G., 1967. Statistical Methods. Iowa State University Press, Ames. 593 pp.
Snell, T. W. & King, C. E., 1977. Lifespan and fecundity patterns in rotifers: the cost of reproduction. Evolution 31: 882–890.

Author's address:
Annie Duncan
Department ofZoology
Royal Holloway College
Englefield Green
Surrey TW20 9TY
United Kingdom

10. Feeding studies with natural food particles on tropical species of planktonic rotifers

A. Duncan & R. D. Gulati

Keywords: tropical, *Brachionus,* feeding, filtering, radio-active

Abstract

In order to assess the role of the planktonic rotifers in relation to the dynamics of the ecosystem of Parakrama Samudra, the filtering and feeding rates of *Brachionus caudatus* from Parakrama Samudra and *B. calyciflorus* from the Milk Factory Tank were measured on natural food particles of less than 33 μm, using a radio-tracer technique. The concentration of food was determined as carbon per millilitre. At 28 °C, the hourly filtering rates for *B. caudatus* ranged from 1.63 to 6.10 μl ind^{-1} h^{-1} in food concentrations from 6.27 to 2.21 μgC ml^{-1}. Assuming continuous feeding and uniformly labelled food, the daily feeding rates were 0.158–0.334 μgC ind^{-1} d^{-1}. *Brachionus calyciflorus* had higher filtering rates, 16.9–39.7 μl ind^{-1} h^{-1}, and higher feeding rates, 3.38–7.94 μgC ind^{-1} d^{-1} at a food concentration of 8.34 μgC ml^{-1}, but this species was about 30 times larger by volume than the small *B. caudatus* from Parakrama Samudra. These results are discussed in the light of experimental information of rotifer feeding on monospecific food available in the literature.

1. Introduction

The authors were asked to study the role of the zooplankton in the dynamics of the ecosystem of Parakrama Samudra and one aspect of this was an estimation of the daily feeding rates on natural food particles of a zooplankton consisting mainly of rotifers. Experiments on the feeding of planktonic rotifers were conducted on several occasions during the two visits in 1979 and 1980 and this chapter reports on the results obtained.

Procedures were developed on site which took into account the nature of the plankton. During August and September 1979, the most abundant taxonomic group of animals were the rotifers with some protozoans. whereas the protozoans were more numerous during March and April 1980 and planktonic crustaceans were absent at both visits (Duncan & Gulati 1981; Duncan 1983). On both occasions, the phytoplankton was characterized by blue-green algae and diatoms. During the 1979 visit, the algal crops were dominated visually by *Anabaenopsis raciborskii* and *Melosire granulata* together with an unidentified *Mougeotia* sp., but this changed after the north-east monsoonal rains so that during the 1980 visit, *A. raciborskii* and *Mougeotia* sp. provided the most important components of the algal crops by number (Rott 1983; Moniruzzaman, personal communication). However, there is other evidence for both occasions that most of the chlorophyll and the photosynthesis can be attributed to algal species smaller than these visual dominants (Dokulil et al. 1983).

2. Methods and materials

Rotifers were offered a food medium of particles small enough to be edible by them by passing lake water through a net of 33-μm or 10-μm mesh and

Schiemer, F. (ed.), Limnology of Parakrama Samudra – Sri Lanka
 ISBN 90 6193 763 9

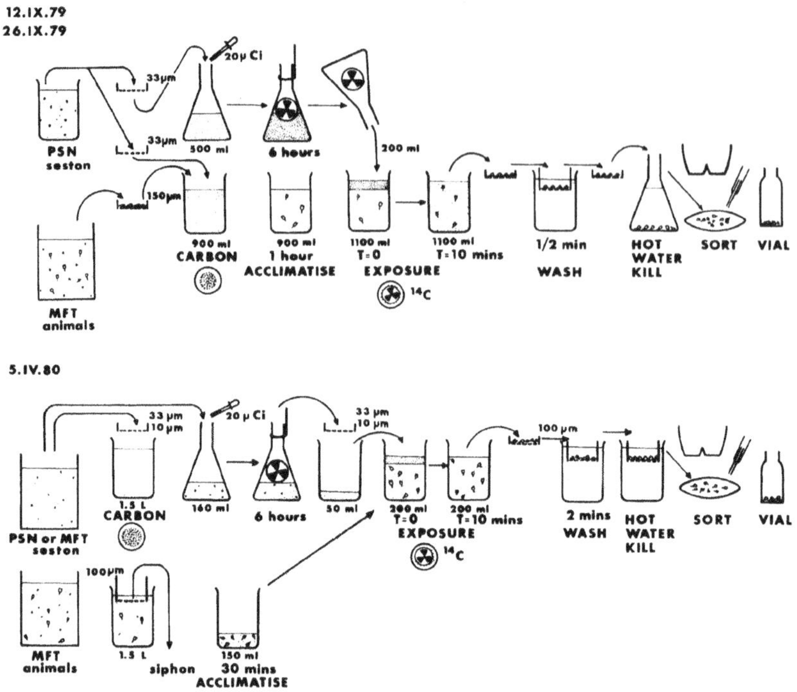

Fig. 1. The procedures adopted during the feeding experiments on different dates

this also served to reduce the possibility of radio-active contamination of the animal counts through the removal of the algal colonies which were of the same size or larger than the animals themselves. Several procedures were adopted for this two of which are illustrated in Figure 1. The small size of the main planktonic species of rotifers in Parakrama Samudra was the main cause of the difficulty experienced during both extracting live experimental animals from lake water and later sorting the preserved radio-active specimens from the final dish into scintillation vials. The only successful experiment with rotifers from this reservoir was conducted on 26 September 1979 and it proved possible to sort only the *Brachionus* spp. in numbers large enough for radio-active counting. The earlier procedures were tested using larger brachionids from a nearby reservoir, the Milk Factory Tank, where larger specimens of the main PSN brachionid species, *B. caudatus,* were present together with the much bigger *B. calyciflorus,* which was absent from PSN although once present (Fernando 1980; Fernando & Rajapaksa 1983). Advantage was taken of the presence of these larger forms in MFT in order to measure and compare their feeding rates on PSN and MFT water as food.

2.1. Procedure

Lake water to which 10 μCi of ^{14}C sodium bicarbonate per 160 ml had been added was suspended in the lake for up to 6 h. This provided a satisfactory level of specific radio-activity (0.0011–0.0394 μgC DPM^{-1}). The labelled water was filtered through a net of 33-μm or 10-μm mesh and the filtrate retained. Previously sorted or concentrated animals were kept in a known volume of lake water for a 30-min period of acclimatisation to the conditions of the experiment, in the dark and at 26°–28 °C. The labelled filtrate was added quantitatively to the water containing the animals, mixed well and left for 10 min, in the dark and at a known temperature. After exposure, the animals were removed by filtration onto a coarse net, allowed to swim in unlabelled filtered lake water as a washing process, killed by immersion in hot water and preserved in 4% formalin. The animals were then sorted into liquid scintillation vials, one by one, from a Petri

Table 1. The carbon concentration (μgC ml^{-1} ± CV [%]) of the sestonic particles in the lake water used as food medium.

Date	Site	*n*	Total concentration	*n*	Less than 33 μm	*n*	Less than 10 μm
21.9.79	PSN	1	11.11		–		–
26.9.79	PSN		–	4	11.55 ± 5.7%		–
					8.81 ± 5.9%		
4.4.80	MFT	3	9.84 ± 15.4% (100%)	3	6.24 ± 56.6% (63%)	3	6.27 ± 17.3% (64%)
5.4.80	PSN	3	3.34 ± 18.4% (100%)	3	3.15 ± 4.6% (94%)	3	2.90 ± 47.3% (87%)

dish under a binocular microscope and using a coarse braking pipette. This avoided picking up highly labelled algal material. Any variations in the detail of the procedures adopted are illustrated in Figure 1. The main variation was in the techniques used to obtain experimental animals. The technique of sorting live animals individually by braking pipette proved too time-consuming. Concentrating all the rotifers from a sample by filtering them onto a coarse net was abandoned in favour of concentrating them by siphoning off water after it had passed through a net kept suspended in the dish (Fig. 1).

2.2. *Food medium*

The specific radio-activity of the food medium was determined as μgC DPM^{-1}, calculated from estimates of its carbon concentration (μgC ml^{-1}) and radio-active load (DPM ml^{-1}). We are indebted to A. Gunatilaka for the 21 September 1979 estimate of the total particulate carbon from PSN 3, which was obtained by subtraction: particulate carbon = total organic carbon (19.11 mgC l^{-1}) minus dissolved organic carbon (8.0 mgC l^{-1}). The methods used are cited in Gunatilaka & Senaratne (1981). The other values for particulate carbon (Table 1) are based upon known volumes of food medium being filtered onto GFC (1979) or GFF (1980) pads which were then subjected to the wet combustion technique described in Strickland & Parsons (1972) and Mackereth et al. (1978), using an amperometric detection of the end-point of the back-titration. A known volume of the food medium at time = 0 h was filtered onto Millipore pads for measurement of the level of radio-activity.

2.3. *Radio-active counting*

Vials containing samples of animals or samples of the food particles were counted in a toluene-based scintillation fluid, with an efficiency of 76%–88%. Before adding toluene the animals were digested for 4 h at 55 °–60 °C in a solubiliser (Soluene 350). There were other vials containing 0.5 ml of a liquid sample of the food medium to which some alkaline solution had been added to retain the dissolved ^{14}C. These samples were counted in Instagel plus water (2:1 by volume) with an efficiency of 55%–60% and provided some information about the conditions of the labelling. Counting was continued until either 10^4 counts had been achieved or for 10 min, whichever was shorter. The 1979 samples were counted in Sri Lanka at the Radioactive Centre on the Colombo Campus of the University of Sri Lanka. The 1980 samples were counted at the Limnological Institute of the Netherlands Academy of Sciences at Nieuwersluis.

2.4. *Calculations*

$$\text{Filtering rates } (\mu\text{l ind}^{-1}\,\text{h}^{-1}) = \frac{R_e}{N \cdot t} \cdot \frac{1}{C}$$

$$\text{Feeding rates } (\mu\text{gC ind}^{-1}\,\text{h}^{-1}) = \frac{R_e}{N \cdot t} \cdot C_r$$

where: R_e = radio-active content of animals in a vial (DPM per 10 min)
N = number of animals in the vial
t = exposure time in hours
C = radio-activity of the food medium (DPM ml^{-1})
C_r = specific radio-activity of the food medium (μgC DPM^{-1}), assuming that all the particulate carbon is edible by rotifers.

The background radio-activity was subtracted from all counts.

2.5. Body size of Brachionus *spp.*

The length (a) and breadth (b) of animals from preserved net samples were measured at 400X magnification. From the mean values, the body volume was calculated as an average of the following three formulae recommended by Ruttner-Kolisko (1976):

$V = 0.12\ a^3; \quad V = 0.52\ abc;$

$$V = \frac{4}{3} \cdot \pi \cdot \left(\frac{a}{2} \cdot \frac{b}{2} \cdot \frac{c}{2}\right)$$

where c = 0.4 a and 10% is added for the foot. Body volume was converted to dry weight and to carbon content by assuming a dry weight–wet weight ratio of 10% and a carbon content of 50% of dry weight.

3. Results

The amount of food offered in the feeding experiments was estimated as the particulate carbon concentration of the filtrate of the lake water after it had been passed through a net of 33-μm or 10-μm mesh; it was hoped that most of the carbon particles in the filtrate would be edible by rotifers. The concentrations in Table 1 show that there was 3–4 times more small particulate carbon in PSN water during September 1979 than in April 1980 and that, in April 1980, MFT water had twice the amount of small particulate carbon of PSN water of the same month. Double-size fractionation of the two lake waters carried out in April 1980 reveals other differences between PSN and MFT water, since 87% of the total particulate carbon of PSN passed through the 10-μm mesh compared with only 64% for MFT water.

The species of rotifers present in sufficient numbers for a feeding experiment and which were relatively easy to handle were *Brachionus caudatus* and *B. angularis* in PSN and *B. calyciflorus* in MFT. These species differed considerably in body size and animals belonging to *B. caudatus* were much smaller in PSN than in MFT. This is shown in both range of

Table 2. The body size of *Brachionus* species from Parakrama Samudra and the Milk Factory Tank.

Species	Site	*n*	Range of length (μm)	Mean ± CV length (μm)	Mean ± CV breadth (μm)	Body volume (10^6 μm^3)	Dry weight (μg)	Carbon weight (μg)
Brachionus angularis	PSN	7	58– 77	73.9 ± 8%	56.0 ± 17%	0.063	0.006	0.003
Brachionus caudatus	PSN	20	68–119	88.4 ± 17%	80.0 ± 17%	0.126	0.013	0.006
Brachionus caudatus	MFT	20	120–250	183.9 ± 21%	147.3 ± 17%	1.036	0.104	0.052
Brachionus calyciflorus	MFT	16	210–350	297.9 ± 14%	263.3 ± 13%	4.740	0.474	0.237

Table 3. Clearance rates and feeding rates on natural food in Sri Lankan species of *Brachionus*.

Date	Species	Habitat[a]	Food level (μgC ml^{-1})	Food source	Nos. of indiv.	*n*	Clearance rates (μl ind^{-1} h^{-1})	Feeding rates (μgC ind^{-1}) Hourly	Daily
12.9.79	*Brachionus calyciflorus*	MFT	8.34	PSN	50	3	39.7 ± 14.5%	0.331	7.94
12.9.79	*Brachionus calyciflorus*	MFT	8.34	PSN	100	3	16.9 ± 7.0%	0.140	3.36
12.9.79	*Brachionus calyciflorus*	MFT	8.34	PSN	100	3	17.0 ± 10.1%	0.142	3.40
26.9.79	Small *Brachionus* spp.[b]	PSN	2.17	PSN	64	1	4.66	0.010	0.243
26.9.79	*Brachionus caudatus*	PSN	2.21	PSN	4	1	5.8	0.0128	0.308
26.9.79	Small *Brachionus* spp.[b]	PSN	2.24	PSN	34	1	7.7	0.0173	0.414
5.4.80	*Brachionus caudatus*	MFT	6.27[c]	MFT	93	1	1.968	0.0123	0.296
5.4.80	*Brachionus caudatus*	MFT	6.27[c]	MFT	20	1	1.299	0.0081	0.195
5.4.80	*Brachionus caudatus*	MFT	3.15	PSN	47	1	1.578	0.0049	0.119
5.4.80	*Brachionus caudatus*	MFT	3.15	PSN	100	1	2.604	0.0082	0.197
5.4.80	*Brachionus caudatus*	MFT	2.90[c]	PSN	100	1	3.541	0.0103	0.246
5.4.80	*Brachionus caudatus*	MFT	2.90[c]	PSN	100	1	6.060	0.0176	0.422

[a] MFT: Milk Factory Tank; PSN: Parakrama Samudra North.
[b] Probably *Brachionus angularis*.
[c] Food medium contained particles of less than 10 μm; all other food media were less than 33 μm.

body lengths and in the average body volumes calculated from mean length and mean breadth according to the formulae given earlier (Table 2). Although the calculation of body volumes from linear dimensions is subject to some error, the differences in size are quite large: the *B. calyciflorus* and *B. caudatus* from MFT have an average volume which is 38 and 8 times greater than that of *B. caudatus* from PSN. The average dry weight and carbon content reflect the same differences because they have been derived from body volumes, using the same ratios for conversion which are given earlier. Small PSN *Brachionus* spp. were used in feeding experiments with PSN food in 1979, but in 1980 it was decided to use the larger MFT *B. caudatus*, fed on PSN and MFT food. This was because of the greater handling difficulties associated with small individuals and because, in April, PSN *Brachionus* spp. were even smaller and not very abundant.

The hourly clearance rates and daily feeding rates of these species of *Brachionus*, which were collected from PSN or MFT and fed on food from PSN or MFT, are listed in Table 3 and come from 12 feeding runs carried out on three occasions. The calculations employed assume that (a) all the ^{14}C-labelled particles and all the particulate carbon is available for ingestion by rotifers, (b) the food is uniformly labelled and (c) feeding is continuous and constant for 24 h. *Brachionus calyciflorus* in PSN water filtered at much higher rates than *B. caudatus* coming from either water body or when fed on food from either source. In one group of *B. calyciflorus*, the filtering rate is almost double that of the other two groups, although all three were fed on PSN food with the same level of radioactivity ($212 \pm 3.3\%$ DPM ml^{-1}); this value of 39.7 μl ind^{-1} h^{-1} is high, but not the highest recorded for this species and was measured at a high temperature (28 °C). In general, the range of filtering rates measured on *B. caudatus* from both reservoirs (1.3–7.7 μl ind^{-1} h^{-1}) falls within that reported in the relevant literature (Starkweather 1980). The narrow range of natural food levels available to offer in the feeding experiments does not permit us to extract a clearance rate–food density relationship, but there does appear to be an inverse trend between these two variables.

It was hoped that feeding rates would provide some estimate of feeding conditions for rotifers in PSN since the amount of food ingested is influenced by both food level and by the effect particle density has on filtering rates. However, the mean daily feeding rates per individual in Table 4 show bigger differences between species than between food levels. The ten times greater feeding rates ($3.38 \pm 0.8\%$ μgC ind^{-1} h^{-1}) of *B. calyciflorus* compared with other species (0.158–0.334 μgC ind^{-1} h^{-1}) is more likely to be associated with its greater body size than with other factors. A more comparable expression is the relative daily food ration obtained by dividing the daily feeding rate in carbon by the carbon content of the body (Table 2) of the consumer. This has units of μgC μgC^{-1} d^{-1} or, since carbon is the common unit, d^{-1}. Table 4 shows a wide spread of values from 3.0 to 6.4 d^{-1} for MFT *B. caudatus*, 14.3 to 33.5 d^{-1} for MFT *B. calyciflorus* to 53.8 d^{-1} for PSN *B. caudatus*. Some of these values are quite high.

Table 4. Mean clearance rates and feeding rates on natural food of Sri Lankan species of *Brachionus*.

Species	Habitat[a]	Food level (μgC ml^{-1})	Food source	n	Clearance rates (μl ind^{-1} h^{-1})	Daily feeding rates[b] (1) (μgC ind^{-1} d^{-1})	Daily feeding rates[b] (2) (d^{-1})
Brachionus calyciflorus	MFT	8.34	PSN	3	39.7 ± 14.5%	7.94	33.5
Brachionus calyciflorus	MFT	8.34	PSN	6	16.9 ± 7.8%	3.38	14.3
Brachionus caudatus & *Brachionus angularis*[c]	PSN	2.21	PSN	3	6.1 ± 25%	0.323	53.8
Brachionus caudatus	MFT	6.27[d]	MFT	2	1.63 ± 29%	0.246	4.7
Brachionus caudatus	MFT	3.15	PSN	2	2.09 ± 35%	0.158	3.0
Brachionus caudatus	MFT	2.90[d]	PSN	2	4.80 ± 37%	0.334	6.4

[a] MFT: Milk Factory Tank; PSN: Parakrama Samudra North.

[b] (1) absolute feeding rates; (2) relative feeding rates (μgC μgC^{-1} d^{-1})

[c] Probably *B. angularis*.

[d] Food medium contained particles of less than 10 μm; all other food media had particles of less than 33 μm.

Values for relative daily food rations for *Brachionus* spp. can be calculated from data available in the literature (Doohan 1973; Pilarska 1977; Leimeroth 1982), but for a lower temperature (20 °C). These attain values of up to 10 d^{-1}, but only in relatively high food levels (more than 15 μgC ml^{-1}). The actual value of the relative daily food ration varies with the units employed (cal cal^{-1} d^{-1} or μgC μgC^{-1} d^{-1}), but are similar in order of magnitude although Pilarska (1977) lists values in units of cal μgDW^{-1} which give different values. None of the species involved *(B. plicatilis, B. rubens* and *B. calyciflorus)* was as small in size as PSN *B. caudatus*. Bearing in mind small size and high temperature, a high relative daily ration may be expected for PSN *B. caudatus*, although probably not so high as 53.8 d^{-1}. The values for *B. calyciflorus* from Sri Lanka are probably too large.

4. Discussion

For temperate waters with a cladoceran plankton, the food concentrations given in Table 1 would be considered quite high since threshold values for reproduction in species of *Daphnia* are as low as 0.05 μgC ml^{-1} or lower and maximal fecundity can be attained at food densities of 0.2–0.6 μgC ml^{-1} (Lampert & Schober 1980; Rocha, unpublished; Duncan, unpublished). Chalk (1981) gives critical food concentrations for the feeding of *Daphnia magna* which ranged between 0.2 and 14.0 μgC ml^{-1} and which depended largely upon the size of the food cell. The critical food concentrations were determined for a 3-mm *D. magna* fed on 13 kinds of monospecific food which ranged from a bacterium, a yeast, algae to a ciliate. There was a tendency for there to be a decrease in the critical concentration with increase in food size and she was able to derive a series of significant log-linear relationships between these two variables, irrespective of how the cell size was measured: diameter, projected area, cell volume or carbon content. Her relationship for critical numerical density (cells ml^{-1}) on cell volume [μm^3] was

$$N_k = 1.1963 \times 10^6 \cdot V^{-0.676} \quad (P = 0.001) \qquad (1)$$

where N_k is the cells ml^{-1} at which the feeding rate is 71% of its maximal rate and V is volume.

Recent work (Gilbert & Starkweather 1977, 1978; Starkweather & Gilbert 1977a and b, 1978) has explored in detail the feeding behaviour of *B. calyciflorus* in response to the size and shape of algal cells offered at various cell densities either singly or with more than one species. When fed on monospecific cultures of food cells of different size (*Aerobacter aerogenes*, 1.8–3.1 μm; *Rhodotorula glutinis*, 2.7–8.7 μm; *Euglena gracilis*, 5.0–8.7 μm), this species of rotifer showed different patterns of feeding rates against food density. With the two smaller food species, feeding rates increased continuously with food concentration up to food levels of 100 μg dry weight per ml whereas ingestion rates on the larger *Euglena gracilis* remained constant at food densities greater than 5 μg ml^{-1}. In the case of *Aerobacter aerogenes* with a cell weight of 0.094 pg dry weight, 100 μg dry weight per ml represents a cell density of 1.06×10^9 cells ml^{-1}, a numerical concentration which is well below the cited critical concentration of 10^{12} cells ml^{-1}. It seems that temperate *Brachionus* species may require much higher concentrations of cells smaller than about 1 000 μm^3 than daphnids in order to attain their maximal feeding rates.

Starkweather (1980) summarises for *Brachionus* species what is known in the literature about the level of the critical food concentration above which feeding rates remain constant. There are seven experiments where the critical concentrations are given as dry weight or can be converted to dry weight from the cells sizes given in the paper. For *B. calyciflorus*, these are *Lagerheimia ciliata* (= *Chodatella ciliata*), 100 μg ml^{-1}; *Euglena gracilis*, 5–10 μg ml^{-1}; *Rhodotorula glutinis*, about 100 μg ml^{-1}; *Aerobacter aerogenes*, more than 100 μg ml^{-1}; *Anabaena flos-aquae*, 3.0 μg ml^{-1} (Erman 1962; Starkweather & Gilbert 1977a; Starkweather et al. 1979; Starkweather 1981). For *B. plicatilis*, there is a record of *Dunaliella tertiolecta*, 47.5 μg ml^{-1} (Theilacker & McMaster 1971) and for *B. rubens*, of *Chlorella vulgaris*, 45 μg ml^{-1} (Pilarska 1977). Apart from the values for *E. gracilis* and *Anabaena*, these appear to be rather high food concentrations for a critical level and higher than comparable values for *Daphnia* spp. When the cell concentration is plotted against cell size as volume, there appears to be a similar trend for the critical numerical concentration to decrease with increase in cell size as is reported by Chalk (1981) and the following regression can be derived:

$N_c = 1.93 \times 10^{11} \cdot V^{-2.430}$ (df 1,4; $F = 308$; $P = 0.001$) (2)

where N_c is the minimal cell density at which the feeding rate is maximal and constant. The most important difference between the daphnid and rotifer regressions is the steepness of the slope of the regression for *Brachionus* spp. The two regressions overlap at a cell volume of 1 000 μm^3 and a numercial density of 10^4 cells. ml^{-1}, but for cell sizes less than 1 000 μm^3, the rotifer critical numerical concentration shows a much greater increase per unit volume than that for *D. magna*. It seems therefore that critical concentrations of 100 μg dry weight per ml are associated with small size of food cell. The need for such high densities of small particles may be associated with both the feeding mechanism and the energetic situation since the metabolic demands of a rotifer of 0.1 μg dry weight will be 3–6 times greater than those of a daphnid of 10 μg or 100 μg body weight. In another aschelminthe, the benthic nematode *Plectus palustris*, the critical concentration of bacterial food for maximum fecundity was also very high at 2 000 μg DW ml^{-1} (Duncan et al. 1974; Schiemer et al. 1980). Although the log-linear relationship between the critical cell density and food particle size for *Brachionus* species was derived from only a few published experiments (see eq. 2), it seems useful to use it to predict the critical food density at maximal feeding rates for cells between 1 μm^3 and 1 000 μm^3. The results for this are presented in Table 5 as cells ml^{-1}. An unpublished regression between cell carbon and cell volume of algae (Chalk et al., unpublished; see Table 5) has been used to first convert cell volume to carbon and then cell ml^{-1} to μgC ml^{-1}. From these predicted critical food weights associated with food cells of different size, it appears that brachionids rarely meet in the nature the food concentrations required for them to attain maximal feeding rates on cells of 100 μm^3 or smaller, although bacterial densities of 10^6–10^7 per ml can be found in activated sludge liquor of which rotifers form part of the fauna (Pike & Curds 1971). Of the seven abundant algal species in PSN contributing to the algal biomass which fall between 1 μm^3 and 1 000 μm^3, there are (in order of abundance) *Anabaena raciborski* (520 μm^3), *Synedra acus* (405 μm^3), *Scenedesmus* spp. (860 μm^3) *(S. acuminatus, S. arcuatus, S. opoliensis), Mougeotia* sp. (1 005 μm^3) and *Monoraphidium* spp. (18 μm^3) *(M. irregularis, M. setiforme)* (Rott 1983; Moniruzzaman, unpublished; Dokulil et al. 1983). Judging from Starkweather's (1981) results on the feeding behaviour of *B. calyciflorus* on *Anabaena flos-aquae, B. caudatus* is probably capable of capturing, processing and ingesting algal filaments such as *A. raciborski*. Rott (1983) lists many other small algae which have not been counted and which may also contribute to the stock of algal cells of a size available to rotifers. This is supported by the finding of Dokulil et al. (1983) that 57%–94% of the carbon assimilation and of the chlorophyll biomass is contributed by the less than 33-μm fraction. Nonliving organic particles probably colonised by bacteria are also available as food to certain rotifers although this has not been demonstrated for *Brachionus* (Starkweather & Bogdan 1980). Some estimate of how much is available can be obtained as the difference between sestonic and algal carbons. The carbon concentration of the whole algal sample from PSN, obtained by converting algal volumes, was approximately 3.2 μgC ml^{-1} on 1 September 1979 and 2.1 μgC ml^{-1} on 8 March 1980. When this is compared with the values in Table 1 for the total sestonic carbon (11.11 μgC ml^{-1} on 21 September

Table 5. For *Brachionus* species maximal feeding rates, the predicted critical food concentrations, by number and by weight, for food cells of different volume.

Cell volume (μm^3)	Cell carbon (pgC)[a]	Predicted critical food concentrations (cells ml^{-1})[b]	(μgC $^{-1}$)[c]
1	0.16	1.94×10^{11}	31 000[d]
10	1.7	7.20×10^{8}	1 224
100	18	2.68×10^{6}	48
200	38	2.40×10^{5}	9.2
400	76	9.22×10^{4}	7.0
600	118	3.44×10^{4}	4.1
800	160	1.71×10^{4}	2.7
1 000	200	9.95×10^{3}	2.0

[a] Predicted from Chalk, Duncan & Rocha (unpublished): $C = 0.152\ V^{1.039}$ ($df = 1{,}15$; $F = 447$; $P = 0.001$), where C = pgC $cell^{-1}$; V = μm^3 $cell^{-1}$).

[b] Predicted from eq. 2 in this paper: $N_c = 1.93 \times 10^{11}\ V^{-2.43}$ (df 1,4; $F = 308$; $P = 0.001$).

[c] Calculated from cell ml^{-1} × pgC $cell^{-1}$.

[d] In activated sludge liquor, rod-shape bacteria (1.5 × 0.7 μm) of density 1.1 g cm^{-3} attain counts of 6.6×10^9 cells.ml^{-1} (Pike & Curds 1971). From this, the calculated cell volume, cell carbon and carbon concentration equal 0.58 μm^3, 0.32 pg C and 2110 μg C ml^{-1}, respectively.

1979 and 3.34 μgC ml^{-1} on 5 April 1980). It appears that this was a larger source of food particles in 1979 (about 8 μgC ml^{-1}) than in 1980 (about 1 μgC ml^{-1}), both quantitatively and relatively.

Acknowledgements

We acknowledge with thanks the financial assistance received from the British Council and the Dutch Organisation of Tropical Research to carry out this work.

References

Chalk, E. A., 1981. Cladoceran filter feeding in a Thames Valley Reservoir. PhD thesis, CNAA (Central London Polytechnic/Thames Water Authority). 266 pp.

Doohan, M., 1973. An energy budget for adult *Brachionus plicatalis* Muller (Rotatoria). Oecologia (Berl.) 13: 351–362.

Dokulil, M., Bauer, K. & Silva, I., 1983. An assessment of the phytoplankton biomass and primary productivity of Parakrama Samudra, a shallow man-made lake in Sri Lanka. In: Schiemer, F. (ed.) Limnology of Parakrama Samudra – Sri Lanka: a case study of an ancient man-made lake in the tropics. Developments in Hydrobiology (this volume). Dr W. Junk, The Hague.

Duncan, A., 1983. The composition, density and distribution of the zooplankton in Parakrama Samudra in March and April 1980 compared with August and September 1979. In: Schiemer, F. (ed.) Limnology of Parakrama Samudra – Sri Lanka: a case study of an ancient man-made lake in the tropics. Developments in Hydrobiology (this volume). Dr W. Junk, The Hague.

Duncan, A. & Gulati, R. D., 1981. Parakrama Samudra (Sri Lanka) Project – a study of a tropical lake ecosystem. III. Composition, density and distribution of the zooplankton in 1979. Verh. Internat. Verein. Limnol. 21: 1001–1008.

Duncan, A., Schiemer, F. & Klekowski, R. Z., 1974. A preliminary study of feeding rates on bacterial food by adult females of a benthic nematode, *Plectus palustris* de Man 1880. Pol. Arch. Hydrobiol. 21: 249–258.

Erman, L. A., 1962. The quantitative aspects of nutrition and food selectivity in the planktonic rotifer *Brachionus calyciflorus* Pall [in Russian: JPRS 19: 894]. Zool. Zh. 41: 31–48.

Fernando, C. H., 1980. The freshwater zooplankton of Sri Lanka, with a discussion of tropical freshwater zooplankton composition. Int. Revue ges. Hydrobiol. 65: 85–125.

Fernando, C. H. & Rajapaksa, R., 1983. Some remarks on long term and seasonal changes in the zooplankton of Parakrama Samudra. In: Schiemer, F. (ed.) Limnology of Parakrama Samudra – Sri Lanka: a case study of an ancient man-made lake in the tropics. Developments in Hydrobiology (this volume). Dr W. Junk, The Hague.

Gilbert, J. J. & Starkweather, P. L., 1977. Feeding in the rotifer *Brachionus calyciflorus*. I. Regulatory mechanisms. Oecologia (Berl.) 28: 125–131.

Gilbert, J. J. & Starkweather, P. L., 1978. Feeding in the rotifer *Brachionus calyciflorus*. III. Direct observations on the effects of food type, food density, changes in food type and starvation on the incidence of pseudotrochal screening. Verh. Internat. Verein. Limnol. 20: 2382–2388.

Gunatilaka, A. & Senaratne, C., 1981. Parakrama Samudra (Sri Lanka) Project, a study of a tropical lake ecosystem. II. Chemical environment, with special reference to nutrients. Verh. Internat. Verein. Limnol. 21: 994–1000.

Lampert, W. & Schober, U., 1980. The importance of 'threshold' food concentrations. Kerfoot, W. C. (ed.) Evolution and ecology of zooplankton communities, pp. 265–267. Univ. Press New England, Hanover.

Leimeroth, N., 1980. Respiration of different stages and energy budgets of juvenile *Brachionus calyciflorus*. Hydrobiologia 73: 195–197.

Mackereth, F. J. H., Heron, J. & Talling, J. F., 1978. Water Analysis. Freshwater Biological Association. Scientific Publ. no. 36. 120 pp.

Pike, E. B. & Curds, C. R., 1971. The microbial ecology of the activated sludge process. In: Microbial aspects of pollution. Symp. Ser. 1 Soc. Appl. Bact. AP: 123–147.

Pilarska, J., 1977. Eco-physiological studies on *Brachionus rubens* Ehrbg (Rotatoria). I. Food selectivity and feeding rates. Pol. Arch. Hydrobiol. 24: 319–354.

Rott, E., 1983. A contribution to the phytoplankton species composition of Parakrama Samudra, an ancient man-made lake in Sri Lanka. In: Schiemer, F. (ed.) Limnology of Parakrama Samudra – Sri Lanka: a case study of an ancient man-made lake in the tropics. Developments in Hydrobiology (this volume). Dr W. Junk, The Hague.

Ruttner-Kolisko, A., 1976. In: Bottrell, H., Duncan, A., Gliwicz, Z. M., Grygierek, E., Herzig, A., Hilbricht-Ilkowska, A., Kurasawa, H., Larsson, P. & Weglenska, T., 1976. A review of some problems in zooplankton production studies. Norw. J. Zool. 24: 419–456.

Schiemer, F., Duncan, A. & Klekowski, R. Z., 1980. A bioenergetic study of a benthic nematode, *Plectus palustris* de Man 1880, throughout its life cycle. Oecologia (Berl.) 44: 205–212.

Starkweather, P. L., 1980. Aspects of the feeding behaviour and trophic ecology of suspension-feeding rotifers. Hydrobiologia 73: 63–72.

Starkweather, P. L., 1981. Trophic relationships between the rotifer *Brachionus calyciflorus* and the blue-green alga *Anabaena flos-aquae*. Verh. Internat. Verein. Limnol. 21: 1507–1514.

Starkweather, P. L. & Bogdan, K. G., 1980. Detrital feeding in natural zooplankton communities: discrimination between live and dead algal foods. Hydrobiologia 73: 83–85.

Starkweather, P. L. & Gilbert, J. J., 1977a. Feeding in the rotifer *Brachionus calyciflorus*. II. Effect of food density on feeding rates using *Euglena gracilis* and *Rhodotorula glutinis*. Oecologia (Berl.) 28: 133–139.

Starkweather, P. L. & Gilbert, J. J., 1977b. Radiotracer determinates of feeding in *Brachionus calyciflorus:* the importance of gut passage times. Arch. Hydrobiol. Ergeb. Limnol. 8: 261–263.

Starkweather, P. L. & Gilbert, J. J., 1978. Feeding in the rotifer *Brachionus calyciflorus*. IV. Selective feeding on tracer particles as a factor in trophic ecology and in situ technique. Verh. Internat. Verein. Limnol. 20: 2389–2394.

Starkweather, P. L., Gilbert, J. J. & Frost, T. M., 1979. Bacterial feeding by the rotifer *Brachionus calyciflorus:* clearance and ingestion rates, behaviour and population dynamics. Oecologia (Berl.) 44: 26–30.

Strickland, J. D. H. & Parsons, T. R., 1972. A practical handbook of seawater analysis, 2nd edn. Bull. Fish. Res. Bd. Can. 167. 310 pp.

Theilacker, G. H. & McMaster, M. F., 1971. Mass cultivation of the rotifer *Brachionus plicatilis* and its evaluation as a food for larval anchovies. Mar. Biol. 10: 183–188.

Author's addresses:
A. Duncan
Department of Zoology
Royal Holloway College
Englefield Green
Surrey TW20 9TY
United Kingdom

R. D. Gulati
Vijverhof Laboratory
Limnological Institute
3631 AC Nieuwersluis
The Netherlands

11. Sediment characteristics and benthic community oxygen uptake rates of Parakrama Samudra, an ancient man-made lake in Sri Lanka

P. Newrkla

Keywords: tropical limnology, man-made lakes, sediments, community oxygen uptake

Abstract

Investigations on Parakrama Samudra sediments were carried out during August and September 1979. Main characteristics of the sediment were determined, oxygen uptake rates of undisturbed sediment cores and oxygen consumption rates of the lake water were measured. The results provide evidence that most of the primarily produced organic matter is mineralized in the pelagic zone. Decomposition processes are distinctly higher in the pelagic zone compared to the benthic zone (ratio 9:1). This is explained by: (a) wind- and thermally induced currents resuspending the uppermost layers of sediment and (b) the high temperature of the lake water enhancing mineralization processes of easily decomposable organic matter.

1. Introduction

Decomposition of sediment organic matter can be important for nutrient recycling in lakes (Mortimer 1941/42). The rate of nutrient release by the sediment will depend on the metabolic activity of the benthic community, which has been estimated in terms of oxygen-uptake rates of undisturbed sediments (Odum 1957; Edwards & Rolley 1965; Hargrave 1973, 1976; Pamatmat 1977; Granéli 1978; etc.). Conversion of oxygen consumption rates into the equivalent of carbon released allows comparison between primary production and benthic mineralization (Hutchinson 1957; Hargrave 1973; Rich 1975; Granéli 1979).

However, total benthic oxygen consumption is the sum of biological aerobic metabolism and inorganic chemical oxydation of reduced metabolic end-products. If oxygen supply to the sediment becomes insufficient, anaerobic metabolism will increase and the neglect of other terminal electron acceptors will lead to an underestimation of total metabolism (Rich 1979).

The present work was part of a multidisciplinary investigation of an ancient, man-made lake in Sri Lanka (Schiemer 1983). The lake, Parakrama Samudra, is situated in the 'dry zone' of Sri Lanka (7°55′N; 81°E; elevation: 59.1 m a.s.l.) and consists of three basins (which are further referred to as PSN, PSM and PSS) with a total area of 22.6 km². Investigations were carried out during August and September 1979, the season of low water level. On account of its shallowness (max. depth of the northern basin in August 1979: 2.8 m), turbulent mixing of the water by strong winds together with thermic convectional currents caused sediment resuspension and high turbidity. For further details, see Schiemer (1981).

This chapter describes the characteristic parameters of Parakrama Samudra sediments and presents a first approximation of sediment metabolism in order to evaluate its share in mineralization processes in the whole lake, as compared to pelagic primary production.

Schiemer, F. (ed.), Limnology of Parakrama Samudra – Sri Lanka
 ISBN 90 6193 763 9

2. Methods

Sediment samples were taken with Perspex tubes (inner diameter = 5 cm, outer diameter = 6 cm, length = 40 cm) by hand from 11 sampling sites (Fig. 1). A silicone-sealed slot along the side of the tubes allowed measurements of redox potential (platinum against Calomel electrodes) and pH values (Ingold insert electrode, diameter = 3 mm) with minimal disturbance of stratification. Subsequently, the uppermost layer of mud (the first 3 cm) was analysed for water content (difference between wet and constant dry weight, after drying at 105 °C for 24 h) and for weight loss on ignition (12 h at 450 °C).

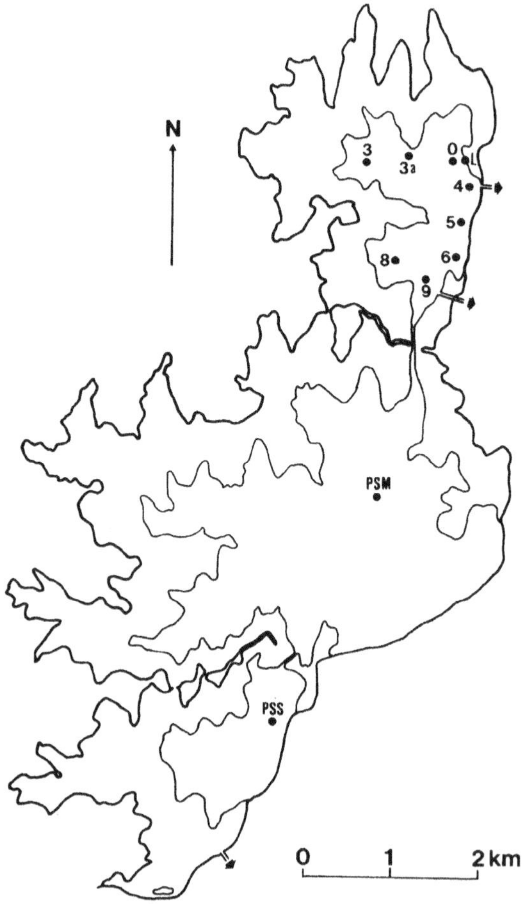

Fig. 1. Parakrama Samudra: outer contour line corresponds the max. filling level (59.1 m a.s.l.); inner contour the water level at 53.6 m a.s.l. Sampling stations are indicated by black dots; arrows indicate outlets.

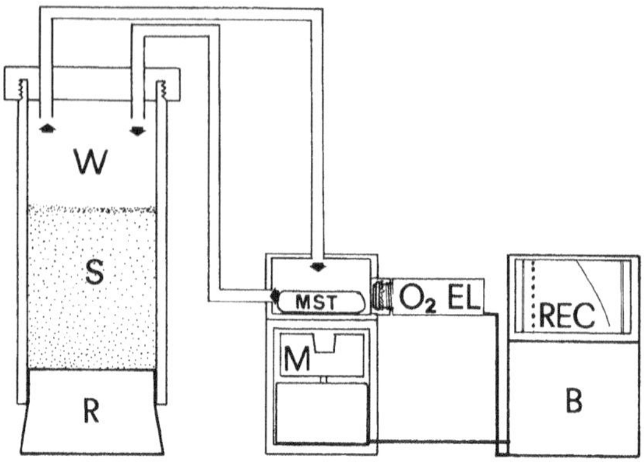

Fig. 2. Experimental set-up for measuring benthic oxygen-uptake rates: B, battery supply; M, magnet + motor; MST, magnetic stirring bar; O_2EL, oxygen sensor; R, rubber stopper; REC, recorder; S, sediment core; W, water; arrows indicate direction of water flow.

Oxygen uptake rates of the benthic community were measured using Perspex tubes of the same diameter, but only 12 cm length. Sediment cores were brought to the laboratory and adapted to simulated in situ temperature (26.8 °C; ±0.5). The water overlying the sediment was connected via glass tubes to a stirring chamber containing a polarographic oxygen electrode (Fig. 2). The system was carefully tightened to avoid leakage. Decreasing oxygen concentrations in the dark were recorded for 2–3 h. The oxygen consumption of water drawn from above the sediment was determined using Winkler bottles and subtracted from the total core uptake.

3. Results

The littoral sediment is dominated by fine sand, while silty clay material is characteristic for the open lake area. For the distribution pattern and thickness of soft sediment layers, see Figure 3.

The uppermost layer of the soft sediment consists of a fine silty material, rich in organic matter. Its water content varied from 89% to 92%, the weight loss on ignition was between 18.6% and 20.5% of the dry weight (Table 1). With increasing depth, sediment compaction resulted in decreasing water content.

Redox profiles indicate the position of the discontinuity layer and depth of oxygen penetration

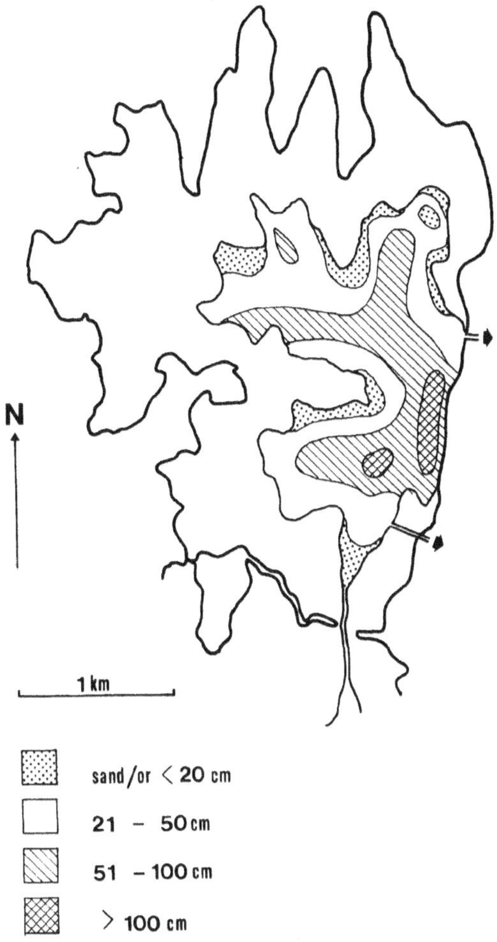

Fig. 3. Distribution of soft sediments in PSN below the 53.6 m a.s.l. contour line. Outer contour corresponds to the max. filling level; arrows indicate outlets. Depth of soft sediment in centimeters.

Table 1. Water content and weight loss on ignition as % of dry weight of Parakrama Samudra sediments. For positions of sampling sites, see Figure 1.

Station	Sediment depth (cm)	Water content (% wet weight)	Weight loss on ignition
PSN 3	0- 4	91.3	20.5
PSN 4	0- 4	85.4	19.2
PSN 5	0- 4	89.6	18.6
PSN 5	4- 8	85.1	20.6
PSN 5	12-13	81.4	22.7
PSN 8	0- 4	91.3	20.1
PSN 9	0- 4	90.6	19.6
PSM	0- 4	91.5	19.2
PSS	0- 4	-	17.6

Table 2. pH values in the water-sediment interphase, measured in undisturbed sediment columns.

Depth (cm)	Sampling stations PSN 3	PSN 8	PSN 9	PSN 4	PSM
+4	8.7	8.7	8.8	9.0	8.3
±0	7.9	8.0	7.9	8.2	7.9
-1	7.2	7.5	7.4	7.4	7.4
-2	6.9	7.2	7.0	7.2	7.0
-4	6.8	6.8	6.6	6.8	6.6
-10	6.5	6.5	6.5	6.6	6.5
-14	6.3	6.4	6.4	-	6.4

into the sediment (Fig. 4). The sandy littoral sediment is aerobic down to a depth of 4 cm. In soft mud, oxygen was limited to the uppermost millimetres, except at PSN 6, where anaerobic conditions extended above the sediment–water interface. The improvement of oxygen supply to the sediment by wind action is demonstrated by the deeper position of the discontinuity layer during windy conditions (Fig. 5). In the middle and southern basin of Parakrama Samudra, oxygen supply to the sediment was sufficient to maintain oxic conditions in the uppermost millimetres.

The pH of the water above the sediment was alkaline with slight variations between the different sampling sites (Table 2). Within the sediment, pH values decreased with increasing depth due to the accumulation of metabolic end-products.

The average oxygen-uptake rate of the benthic community of the northern basin was 49.3 mgO_2 m^{-2} h^{-1} ± 10.4% (95% conf. limits) (Table 3). Sediment oxygen-uptake rates were 69.3 mgO_2 m^{-2} h^{-1}

Table 3. Benthic oxygen-uptake rates (mgO_2 m^{-2} h^{-1}) of replicate undisturbed sediment samples, measured at 26.8 °C, from different sites of Parakrama Samudra (see Fig. 1).

Station	Oxygen-uptake rates
PSN 0	61.7, 64.0, 68.6
PSN 3	49.4, 45.8
PSN 3a	50.7, 34.6
PSN 4	35.9, 35.9
PSN 5	68.0, 54.0
PSN 6	28.8, 46.7
PSN 8	44.8, 52.0
PSN 9	55.8, 54.7
PSN L	45.1, 45.5, 43.2
PSM	66.0, 72.0, 69.3
PSS	72.0, 66.0, 66.0

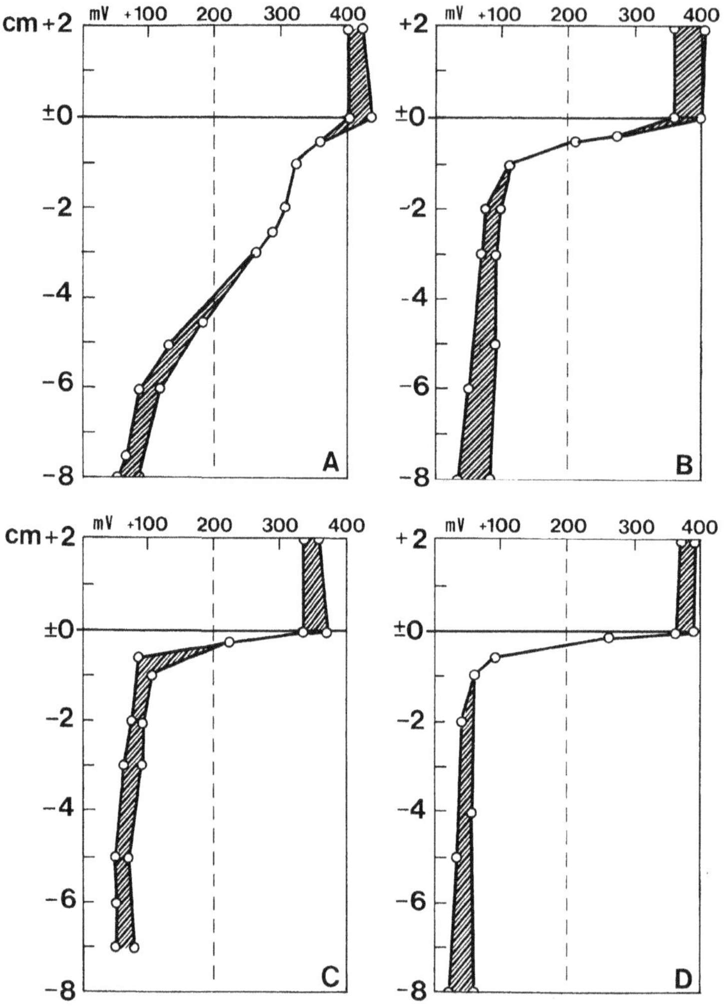

Fig. 4. Redox-potential profiles measured in the sediment–water interphase: striped area = range of redox values; dashed line = threshold between aerobic and anaerobic conditions; A = PSN littoral; B = PSN 3; C = PSM; D = PSS (see Fig. 1).

in the middle basin and 68.0 mgO_2 m^{-2} h^{-1} in the southern basin. The amount of oxygen consumed by the lake water was found to vary with time and depth, obviously depending on the suspended sediment load which varied with wind action.

4. Discussion

The importance of wind-induced turbulences for resuspension of sediments and nutrient recycling was pointed out by Viner (1975). In shallow tropical lakes, convective currents produced by thermic events appear to play a similar important role (Bauer 1983). Strong water currents will erode the uppermost layers of mud so that part of the benthic mineralization processes will be shifted into the pelagic zone. At the same time, oxygen supply to the sediment stimulates benthic community metabolism, as noted by several workers (Edwards & Rolley 1965; Hargrave 1969; Edberg 1976; Granéli 1977; Newrkla 1982). Increased metabolic rates im-

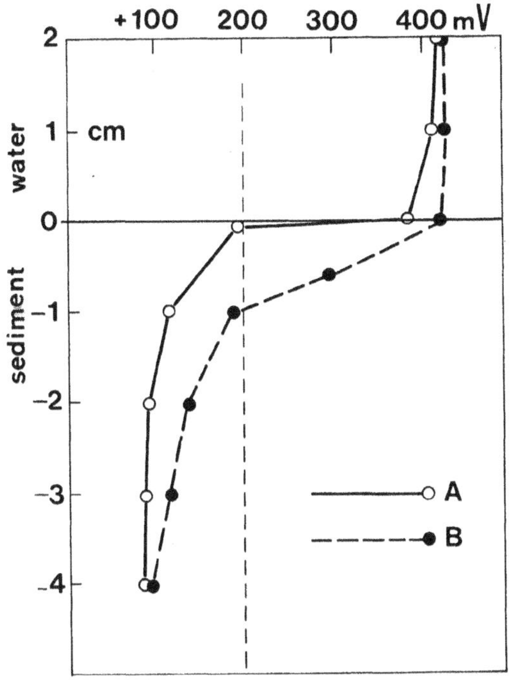

Fig. 5. Redox-potential profiles of Parakrama Samudra sediments from site PSN 5, measured during calm (A) and windy (B) periods: dashed line indicates threshold of oxygen.

ply faster decomposition of organic matter and recycling of nutrients for primary production.

Organic matter as the primary source of energy for the benthic community may govern metabolic processes. Rybak (1969) found a correlation between oxygen-uptake rates and organic content of the sediments of various Polish lakes differing in their trophic conditions. However, due to unknown effects of other variables, a simple correlation between benthic oxygen-uptake rates and organic content of the sediment often fails to appear (Johnson & Brinkhurst 1971; Pamatmat & Bhagwat 1973; Edberg & Hofsten 1973). In the present study, no correlation between benthic oxygen consumption and organic matter content of the sediment was found.

Based on the assumption that organic matter reaching the sediment surface is mainly autochthonous, Hargrave (1973) related oxygen-uptake rates of undisturbed sediments to primary production and to mixed-layer depth by the formula:

$C_o = a \cdot (\frac{C_i}{z_m})_b$, where C_o = carbon equivalent of annual benthic oxygen consumption, C_i = annual primary production, z_m = mixed-layer depth, and a and b are constants.

Proceeding from a mean primary production rate of 3.9 gC m^{-2} day^{-1} measured in PSN (Dokulil et al. 1983), and assuming a mixed-layer depth of about 1 m, benthic oxygen-uptake rate should amount to 106.5 mgO_2 m^{-2} h^{-1}. The difference of this theoretical value to the actual measured rate (49.3 mgO_2 m^{-2} h^{-1}) indicates that only a minor fraction of photosynthetically produced matter is

Table 4. Comparison of pelagic primary production and benthic oxygen-uptake rates of some shallow lakes from the temperate region. Literature values were converted if necessary into daily C-production rates and hourly oxygen-uptake rates, respectively.

Author	Lake	Temp. (°C)	Depth tot.	z_m	Primary prod. (gC m^{-2} d^{-1})	Benthic resp. (mgO_2 m^{-2} h^{-1})	% C resp. in sediment
Hargrave							
(1973)	Tuel	8	17	6	1.63	31.3	20
	Bagsvaerd	10	2	2	1.43	34.7	25
	Frederiksborg	10	3	3	1.59	30.1	20
Granéli							
(1978)	Trummen	9-10	2.5	1.9	0.7	15.4	26
	Växjösjön	9-10	6.3	3.5	0.97	12.1	13
	Vombsjön	9-10	15	6	1.43	19.0	13.8
	Södra Bergundasjön	9-10	5.4	2.3	1.57	17.2	11.4
Newrkla							
(1979)	Jeserzer	13°	6.6	4.3	0.73	23.8	34
Hunding (1979)	Mývatn	12 (Summer)	0.5	-	No direct comparison possible, but gives a ratio between pelagic and benthic respiration of		70:30

mineralized on the sediment surface. Resuspension of the uppermost sediment layers seems to be responsible for the shift of mineralization processes into the pelagic zone. Correspondingly, BOD values of the lake water were high, reaching values of 7.2 mgO_2 l^{-1} d^{-1}. Thus, 90% of the primary production is compensated for by pelagic decomposition processes, while the remaining 10% is respired at the sediment surface. However, we are not able to distinguish which fraction of the total pelagic oxygen consumption must be attributed to an actual pelagic decomposition and which part is caused by resuspension of sediments.

We lack measurements of benthic primary production, which could play a considerable part within the total budget. In Castle Lake, benthic oxygen-uptake rates were found to equipoise almost exactly the maximal benthic primary production (Naeme 1975), implying that organic matter produced in the pelagic zone is almost completely mineralized within the free water body while a minor part is fossilized within the sediment. The fraction of organic matter produced by pelagic algae which reaches the sediment surface is variable, being mainly dependent upon depth, morphometry and temperature of the lake. For example, only 0.8% of net primary production was left in 25-m depth of a temperate lake (Ohle 1965). In two shallow Danish lakes, 20%–25% of primary production is mineralized by the benthic community (Hargrave 1973). Table 4 gives some information about the relation between pelagic primary production and benthic oxygen consumption measured in some shallow lakes. In Parakrama Samudra, the ratio of pelagic to benthic mineralization (9:1) is determined by two factors: (a) Resuspension of sediment, shifting mineralization processes from the benthos to the pelagic zone, where greater oxygen availability favours higher metabolic rates; and (b) fast mineralization of seston because of the high temperature, as compared to lakes of similar morphometry but situated in the temperate climate.

Acknowledgments

I am grateful for financial support given by the city of Vienna (MA 7, no.: 3221/79). For the invitation to join the project, I am indebted to Dr. F. Schiemer, as well as to my colleagues in the team of the Parakrama Samudra Project for critical discussions. I am obliged to Prof. M. Pamatmat for revising the manuscript.

References

Bauer, K., 1983. Thermal stratification, mixis, and advective currents in the Parakrama Samudra Reservoir, Sri Lanka. In: Schiemer, F. (ed.) Limnology of Parakrama Samudra – Sri Lanka: a case study of an ancient man-made lake in the tropics. Developments in Hydrobiology (this volume). Dr W. Junk, The Hague.

Dokulil, M., Bauer, K. & Silva, I., 1983. An assessment of the phytoplankton biomass and primary productivity of Parakrama Samudra, a shallow man-made lake in Sri Lanka. In: Schiemer, F. (ed.) Limnology of Parakrama Samudra – Sri Lanka: a case study of an ancient man made lake in the tropics. Developments in Hydrobiology (this volume). Dr W. Junk, The Hague.

Edberg, N., 1976. Oxygen consumption of sediment and water in certain selected lakes. Vatten 1: 1–12.

Edberg, N. & Hofsten, B.v., 1973. Oxygen uptake of bottom sediments studied *in situ* and in the laboratory. Wat. Res. 7: 1285–1294.

Edwards, R. W. & Rolley, H. L. J., 1965. Oxygen consumption of river muds. J. Ecol. 53: 1–19.

Granéli, W., 1977. Measurement of sediment oxygen uptake in the laboratory using undisturbed sediment cores. Vatten 3: 1–15.

Granéli, W., 1978. Sediment oxygen uptake in south Swedish lakes. Oikos 30: 7–16.

Granéli, W., 1979. The influence of *Chironomus plumosus* larvae on the exchange of dissolved substances between sediment and water. Hydrobiology 66: 149–160.

Hargrave, B. T., 1969. Similarity of oxygen uptake by benthic communities. Limnol. Oceanogr. 14: 801–805.

Hargrave, B. T., 1973. Coupling carbon flow through some pelagic and benthic communities. J. Fish Res. Bd. Can. 30: 1317–1326.

Hargrave, B. T., 1976. Metabolism at the benthic boundary. In: McCave, I. (ed.) The Benthic Boundary Layer. Plenum, New York. 323 pp.

Hunding, C., 1979. The oxygen balance of Lake Mýratn, Iceland. Oikos 32: 139–150.

Hutchinson, G. E., 1957. A Treatise on Limnology. Wiley, New York. 1015 pp.

Johnson, M. G. & Brinkhurst, R. O., 1971. III. Benthic community metabolism in Bay of Quinte and Lake Ontario. J. Fish. Res. Bd. Can. 28: 1715–1725.

Mortimer, C. H., 1941/42. The exchange of dissolved substances between mud and water in lakes. J. Ecol. 29: 280–329 and J. Ecol. 30: 147–201.

Naeme, P. A., 1975. Oxygen uptake of sediments in Castle Lake, California. Verh. Internat. Verein. Limnol. 19: 792–799.

Newrkla, P., 1979. Sedimentverhältnisse und Sauerstoffaufnahme der benthischen Gemeinschaft Jeserzer oder Saisser See, Kärnten. Carinthia II, 169/89: 331–335.

Newrkla, P., 1982. Methods for measuring benthic community respiration rates. In: Gnaiger, E., Forstner, H. (eds.) Handbook on Polarographic Oxygen Sensors: Aquatic and Physiological Applications. Springer, Berlin.
Odum, H. T., 1957. Trophic structure and productivity of Silver Springs, Florida. Ecol. Monogr. 27: 55–112.
Ohle, W., 1965. Primärproduktion des Phytoplanktons und Bioaktivität holsteinischer Seen, Methoden und Ergebnisse. Helsinki Helsingfors, Limnoligisymposion (1964), Suomen Limnologinen Yhdistys Limnologiska Föreningen, Finland, 24.43.
Pamatmat, M. M., 1977. Benthic community metabolism: a review and assessment of present status and outlook. In: Coull, B. C. (ed.) Ecology of Marine Benthos. Belle W. Baruch Library in Marine Science 6. University of South Carolina Press, Columbia.
Pamatmat, M. M. & Bhagwat, A. M., 1973. Anaerobic metabolism in Lake Washington sediments. Limnol. Oceanogr. 18: 611–627.
Rich, P. H., 1975. Benthic metabolism of a soft-water lake. Verh. Internat. Verein. Limnol. 19: 1023–1028.
Rich, P. H., 1979. Differential CO_2 and O_2 benthic community metabolism in a soft water lake. J. Fish. Res. Bd. Can. 36: 1377–1389.
Rybak, J. I., 1969. Bottom sediments of the lakes of various trophic type. Ekol. Pol. 17: 611–662.
Schiemer, F., 1981. Parakrama Samudra (Sri Lanka) Project, a study of a tropical lake ecosystem. I. An interim review. Verh. Internat. Verin. Limnol. 21: 993–999.
Schiemer, F., 1983. The Parakrama Samudra Project – scope and objectives. In: Schiemer, F. (ed.) Limnology of Parakrama Samudra – Sri Lanka: a case study of an ancient man-made lake in the tropics. Developments in Hydrobiology (this volume). Dr W. Junk, The Hague.
Viner, A. B., 1975. The sediments of Lake George (Uganda). I: Redox potential, oxygen consumption and carbon dioxide output. Arch. Hydrobiol. 76: 181–197.

Author's address:
P. Newrkla
Institute of Zoology
University of Vienna
Althanstr. 14
A-1090 Vienna
Austria

12. A contribution to the ecology of the fish fauna of the Parakrama Samudra Reservoir

F. Schiemer & R. Hofer

Keywords: tropics, reservoir, fish ecology, trophic relationships, seasonal changes

Abstract

The chapter provides a survey on the fish fauna of the Parakrama Samudra reservoir; 23 species were encountered during two research visits, using different sampling techniques. According to their habitat preferences and feeding biology, the numerically abundant species can be grouped in

a) Littoral 'Aufwuchs' feeders (e.g. *Puntius sarana, P. filamentosus, Labeo dussumieri, Etroplus suratensis, Rasbora daniconius*),
b) zoobenthivorous species (e.g. *Puntius dorsalis* and *P. chola*), and
c) limnetic species feeding on phytoplankton (*Sarotherodon mossambicus, Amblypharyngodon melettinus*) and zooplankton (*Ehirava fluviatilis*).

High food-overlap values have been encountered in the species pairs *P. sarana* – *R. daniconius* and *P. filamentosus* – *E. suratensis* in the littoral zone, and in *P. dorsalis* – *P. chola* in the offshore zone. An analysis of the diel feeding pattern of the latter two species indicates the mechanisms by which food competition between them is relaxed. *Puntius chola* feeds throughout the 24-h cycle, predominantly in the offshore areas. *Puntius dorsalis* feeds mainly during the night and its nocturnal feeding activity is connected with inshore migrations.

Observed seasonal changes in the species composition and population structure of the fish fauna are discussed with regard to (a) discontinuous breeding activity and spawning migrations of species, (b) seasonal changes in habitat and resource availability and (c) predation effects by fish and fish-eating birds.

1. Introduction

A knowledge of habitat and food requirements of common species of fish and their trophic interactions are a prerequisite for an understanding of tropical reservoir ecosystems and their resource management (Schiemer 1983). Within the framework of the Parakrama Samudra Project, ecological studies on several abundant species have been carried out, including small-sized economically unexploited species. The latter are due to their high population densities, of great importance for the trophic structure of the lake ecosystem. An analysis of their habitat and food requirements can indicate food niches which could be economically exploitable and also outline their possible effects as competitors and predators when introducing 'exotic' fish species, e.g. *Sarotherodon mossambicus.*

Details on the feeding ecology of *S. mossambicus* and *Puntius filamentosus* are given in separate chapters (Hofer & Schiemer 1983; Dokulil 1983). The digestive physiology of several species encountered in Parakrama Samudra has been discussed by Hofer & Schiemer (1981). The present chapter pro-

Schiemer, F. (ed.), Limnology of Parakrama Samudra – Sri Lanka
 ISBN 90 6193 763 9

vides a survey on the overall abundance of fishes in different parts of the reservoir, their distribution pattern, diel migrations and food relationships.

2. Methods

Experimental fishing was carried out mainly in March/April 1980, using monofilament gill nets mesh sizes, knot to knot 6, 12, 15, 20, 28, 40, 50, 60, 70, 80, 90 mm; 25 m long, 2 m high). Catches as individuals per unit effort (CPUE), i.e. per unit exposure time and per net. In addition to gill nets, other catch techniques have been used: for small-sized fishes within the littoral zone, a 1 × 1 m lift net (1.5-mm mesh size); long-line fishing and angling. Most fishing was carried out in the northern basin of the lake (PSN); a small series of catches was also obtained from the middle basin (PSM) and the southern basin (PSS). In August/September 1979, some material for gut analysis was obtained from professional cast-net fishermen from the littoral zone of PSN.

Gut analyses were carried out on several species of fish both in September/October 1979 and March/April 1980. The portion of different food items in terms of volume of the total gut content of individual fish was estimated using a scale from 0 to 5 (following the plant sociological classification). For calculation of mean values of food composition in a sample of fish, the classes 1–5 were weighted as 0.5, 1, 2, 4 and 8, respectively.

The diel feeding cycle in *Puntius chola* and *P. dorsalis* was studied by determining the relative gut filling throughout the 24-h cycle. The gut content was squeezed out of the intestine, its fresh weight determined and expressed as a percentage of the body weight (with empty intestine) of the fish.

3. Results

3.1. Composition of the fauna

In the course of both visits (August/September 1979 and March/April 1980), 23 species of fish were encountered in the lake (Table 1) out of a total of 54 species known from freshwater habitats of the is-

Table 1. List of species found in March/April 1980 in different habitats of the Parakrama Samudra Reservoir (det. R. Hacker).[a]

Species	1	2	3	4	5
Ehirava fluviatilis Deraniyagala	++	++	+++		++
Amblypharyngodon melettinus (Val.)	++	+++	+		+++
Danio aequipinnatus (McClelland)		++			+++
Esomus danrica thermoicos (Val.)		++			
Labeo dussumieri (Val.)		+	+		+
Puntius chola (Ham.-Buch.)	+	+	+++	+	++
Puntius dorsalis (Jerdon)	+++	+++	+++	++	++
Puntius filamentosus (Val.)	+++	++	++	+++	+++
Puntius sarana (Ham.-Buch.)		++			++
Rasbora daniconius (Ham.-Buch.)	++	+++	++	+	+++
Ompok bimaculatus (Bloch)		+	+	+	+
Heteropneustes fossilis (Bloch)		+	+		
Mystus keletius (Val.)		+		++	
Mystus vittatus (Bloch)	+	+	++	+	+
Anquilla bicolor bicolor McClelland			+		
Anquilla nebulosa nebulosa McClelland			+		
Hemiramphidae sp.	++	+	++	++	
Ophiocephalus striatus Bloch		+			
Sarotherodon mossambicus (Peters)	+++	+++	+++	+++	++
Etroplus maculatus (Bloch)	+				
Etroplus suratensis (Bloch)	++	+		+	
Glossogobius giuris (Ham.-Buch.)	++	++	+	+	++
Mastacembelus armatus (Lacepede)		+	+	+	+

[a] + = rare, ++ = common, +++ = abundant; 1 = PSN, bay with macrophyte cover, 2 = sandy littoral, 3 = PSN, open lake, soft mud bottom, 4 = PSM, open lake with dead, submerged trees, 5 = outflow channel of PSN.

land. According to their origin, this fauna consists of three elements:

a) Brackish water species, colonizing the lake from estuarine habitats (e.g. *Ehirava fluviatilis*, Hemiramphidae sp., *Etroplus suratensis*).
b) Introduced, 'exotic' species; e.g. *Sarotherodon mossambicus* was introduced in Sri Lanka in 1952 and is at present economically the most important freshwater fish of the island. Its success in Parakrama Samudra reservoir is well documented (e.g. De Silva & Fernando 1980). Several other species have been introduced into the lake,but apparently with less success (e.g. *Tilapia rendalli*).
c) The remaining species belong to a freshwater ichthyofauna with a wide distribution in different habitat types (rivers and different types of stagnant water bodies) in the South-East Asian region. However, a truly lacustrine fauna is lacking (Fernando & Indrasena 1969).

3.2. Habitat preferences and distribution pattern

Parakrama Samudra is a large, shallow impoundment (area = 25.5 km^2, z_{max} = 12.7 m) with an extensive littoral fringe and large areas of a water depth of less than 1 m (Schiemer 1983). Hence littoral habitats are of importance for many fish species in the lake. The water level shows strong seasonal fluctuations with an average annual amplitude of 52.5–56.2 m above sea level. This implies a constant change in the ecological character of the marginal habitats. Above the 53-m contour line, sediments consists mainly of sand or a sand–silt mixture. During the dry season, these areas are covered with terrestrial vegetation which is of importance after flooding as food and as habitat structure which in turn provide an additional food source in the form of an Aufwuchs community. The importance of true aquatic macrophytes in the lake is low. At low water level in August/September 1979, no macrophytes were encountered; while in February– April 1980, submerged macrophytes were found exclusively in one bay in PSN in the form of a dense cover of *Ceratophyllum* sp. Some littoral areas in PSN and PSM are covered by gravel and rocks. The whole area of PSM is an inundated jungle, the submerged trees of which provide a structure for a rich Aufwucns community.

The bottom of the open lake, approximately below the 53-m contour line, consists of soft sediments with a high content of organic matter (20% of sediment dry weight; Newrkla 1982).

In March/April 1980, experimental fishing concentrated on the following locations (Fig. 1a):

PSNA Littoral zone covered densely with *Ceratophyllum*, 1.0–1.5 m deep.
PSN B: 30 m offshore, 1.0–1.5 m deep, sandy bottom with sparse cover of terrestrial plants.
PSN C: 10–30 m offshore, 1.0–1.5 m deep, sandy and rocky bottom (in front of the Polonnaruwa Resthouse).
PSN D: 100–150 m offshore (east of Resthouse), 3 m deep, soft mud bottom.
PSN E: 700 m offshore, 4 m deep, soft mud bottom.
PSM: offshore station in the middle basin of the lake, 8 m deep, soft mud bottom, dense stands of submerged trees.

PSN A, PSN B and PSN C represent different types of littoral habitats; PSN D and PSN E form a transect to offshore locations.

In August/September 1979, gill-netting was carried out at various offshore stations in PSN (soft mud bottoms, 2 m deep) and in PSM (soft mud bottom, 6 m deep).

From catch statistics obtained with various sampling techniques and gill nets of different mesh sizes, a broad pattern of habitat preferences of the common species can be recognized (Table 1 and 2). Species with inshore predominance: The marginal zone of the lake is the preferred habitat of fry and juveniles of several species of fish. *Sarotherodon mossambicus* fry was abundant both in September/October 1979 and in March/April 1980. During the latter period,numerous schools (~50–200 individuals of 15–30 mm length) patrolled along the shoreline and were heavily predated by several species of fish and birds (see below). Besides *S. mossambicus*, the juveniles of *Ehirava fluviatilis* and *Rasbora daniconius* were abundant in sheltered littoral areas (Duncan 1983).

The littoral fish fauna includes a number of smaller-sized species, some of which (*Danio aequipinnatus*, *Esomus danrica thermoicos*, *Glossogobius giu-*

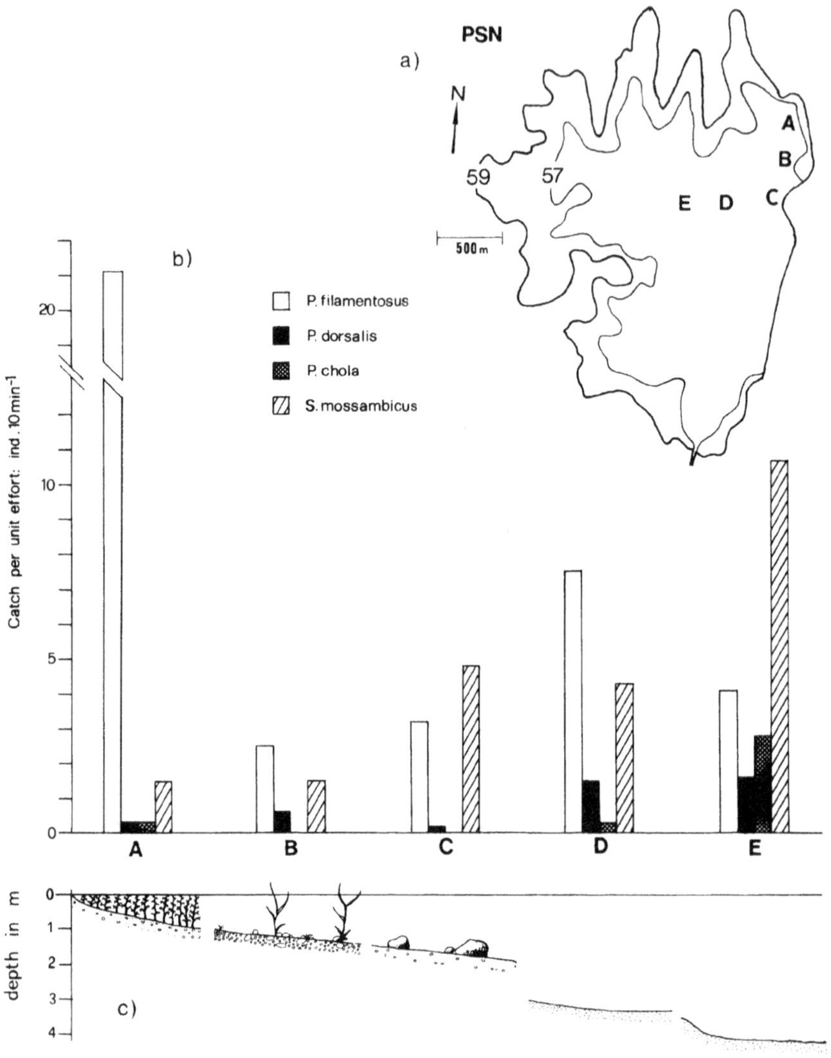

Fig. 1. (a) contour map (m above sea level) of Parakrama Samudra North (PSN) with position of sampling stations; (b) CPUE (20 mm gill nets) at different localities during daytime; (c) diagrammatic representation of habitat types.

ris) occur preferably in the inshore zone, while others (viz. *Rasbora daniconius, Ehirava fluviatilis, Amblypharyngodon melettinus* and Hemiramphidae sp. are common both in the littoral and the offshore areas.

Among the middle- and larger-sized species of the lake, several feed predominantly on the epigrowth of macrophytes, gravel and rocks in littoral habitats. This group contains three economically important fish, viz. *Labeo dussumieri, Puntius sarana* and *Etroplus suratensis.* None of them was abundant during the two research periods but all three species are reported to be caught in larger numbers during certain periods of the year (De Silva & Fernando 1980). An economically less important species with a similar habitat preference is *Puntius filamentosus.* It was the dominant species in the 20-mm gill-net catches from inshore localities in PSN (Fig. 1b) as well as in the open lake area of PSM, which is structured by submerged trees. The

Table 2. Total catch will gill nets of various mesh sizes in different habitats in the Parakrama Samudra reservoir, March/April 1980.

Locality	Mesh size of gill nets	No. of catches	Total exposure time (min)	*Puntius filamentosus*	*Puntius dorsalis*	*Puntius chola*	*Puntius sarana*	*Labeo dussumieri*	*Danio aequipinnatus*	*Rasbora daniconius*	*Amblypharyngodon melettinus*	*Esomus danrica thermoicos*	*Sarotherodon mossambicus*	*Etroplus suratensis*	*Etroplus maculatus*	*Mystus vittatus*	*Mystus keletius*	*Ompok bimaculatus*	*Glossogobius giuris*	*Mastacembelus armatus*
PSN A	12 mm	3	15	11		2					8								2	
PSN A	20 mm	20	260	540	75	5							35	3	9				4	1
PSN B	20 mm	9	120	53	54	3		1	1				22	1		2			1	
PSN C	6 mm	4	20	1		1			7	5	2	8								
PSN C	12 mm	1	30	3	1	2				9	1								5	
PSN C	20 mm	17	370	123	24	1	2	1		1			157	9		2	1	1		1
PSN D	20 mm	25	520	177	93	13		5					208	17		9			3	1
PSN D	28 mm	9	330	3	16			3					74							
PSN E	6 mm	11	145	1		2				39	45									
PSN E	12 mm	7	95		21	224				13	5		7			19		1	9	
PSN E	15 mm	9	111	12	10	91				2			24	1		27			2	
PSN E	20 mm	30	1683	245	166	226		2					704	2		17		1	1	
PSN E	28 mm	1	35										71							
PSN E	40 mm	13											18							
PSN E	50 mm	20											29							
PSN E	60 mm	20											59							
PSN E	70 mm	15											1							
PSN E	80 mm	15											1							
PSN E	90 mm	15											0							
PSM	12 mm	3	94	4		6				1	1					6			3	
PSM	15 mm	3	75	50	2	7	1				1		7			10			3	
PSM	20 mm	5	176	157	6	2							15	1		3	3			

Table 3. Catch per unit effort (±SD) of 20 # mm gill nets (20 m long, 2 m high) calculated for 10-min exposure times during daytime (0603–1800 hours) and at night (1800–0630 hours). At water depths below the height of the net, the nets were set at the bottom.

		n	*Puntius filamentosus*	*Puntius dorsalis*	*Puntius chola*	*Sarotherodon mossambicus*
PSN A. Macrophyte Bay	Day	15	21.1 (±10.2)	0.3 (±0.5)	0.03 (±0.6)	1.5 (± 3.4)
	Night	4	34.0 (±19.6)	17.3 (±7.1)	0	2.2 (± 1.7)
B. Sandy Littoral	Day	4	2.5 (± 1.0)	0.6 (±0.7)	0	1.5 (± 0.9)
	Night	5	7.0 (± 9.7)	8.2 (±3.6)	0.6 (±0.8)	5.0 (± 5.6)
C. Resthouse Bay	Day	10	3.2 (± 2.4)	0.2 (±0.4)	0	4.8 (± 8.6)
	Night	7	6.1 (± 4.0)	2.0 (±1.3)	0.1	7.8 (± 4.9)
D. 100 m offshore	Day	16	7.5 (± 7.4)	1.5 (±2.0)	0.3 (±0.8)	4.3 (± 5.6)
	Night	8	3.9 (± 3.7)	5.8 (±9.2)	0	8.8 (±11.4)
E. 500 m offshore	Day	23	4.1 (± 5.0)	1.6 (±1.8)	2.8 (±2.3)	10.7 (±20.6)
	Night	7	2.5 (± 2.6)	2.0 (±1.8)	2.2 (±1.9)	5.4 (± 9.7)
PSM offshore	Day	5	8.9 (± 9.3)	0.3 (±0.6)	0.1 (±0.3)	0.8 (± 1.0)

whole group of 'Aufwuchs' feeders are prospective candidates for new reservoirs with submerged trees and vegetation (e.g. Uda Walawe in the southern province).

Open lake fauna: at times, high densities of planktonic larvae of Hemiramphidae sp. and *E. fluviatilis* have been encountered (see Duncan, 1982). Among the small-sized species *R. daniconius, A. melettinus, E. fluviatilis* and *Hemiramphidae* sp. are abundant in the limnetic zone. The 20-mm gill-net catches from the open lake area of PSN in March/April 1980 contained *S. mossambicus* (TL 85–130 mm), *P. filamentosus, P. chola, P. dorsalis* and *Mystus vittatus* (ranked according to their abundance). Gill nets with mesh sizes from 28 to 90 mm yielded exclusively adults of *S. mossambicus.*

The catch per unit effort of *S. mossambicus, P. chola* and *P. dorsalis* during daytime is higher in the offshore areas compared to the littoral zone (Fig.

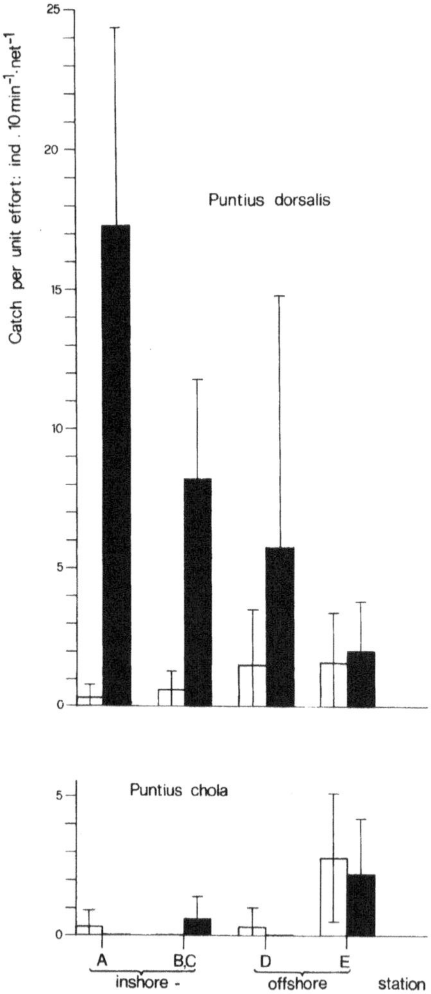

Fig. 2. Catch per unit effort of *P. dorsalis* with 20 mm gill nets. The catch statistics (arithmetic means and standard deviations) refer to different sampling stations (see Fig. 1). Catches obtained during daytime = white; during nighttime = black.

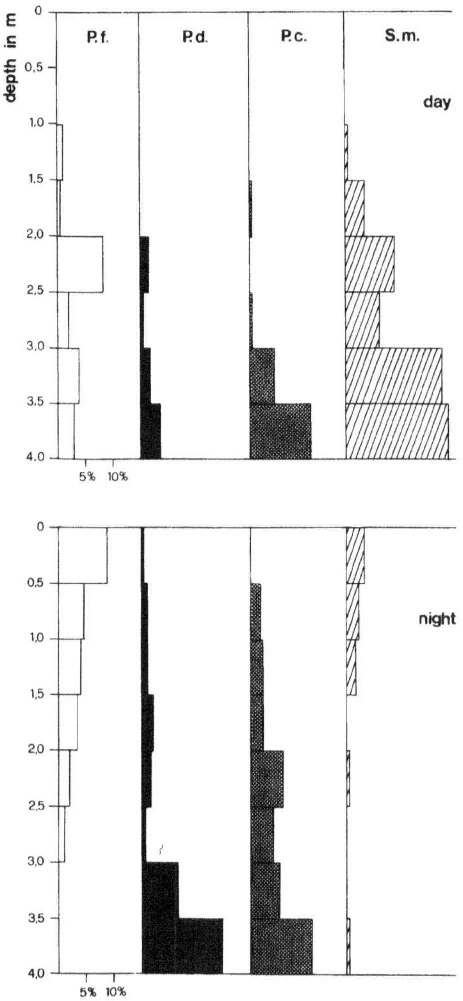

Fig. 3. Vertical distribution of four common species in the offshore zone of PSN during day- and nighttime. The occurrence in different strata is expressed in percent of the total catch with 20 mm gill nets. Pf = *Puntius filamentosus,* Pd = *P. dorsalis,* Pc = *P. chola,* Sm = *Sarotherodon mossambicus.*

1b and Table 3). While the distribution pattern and catch statistics of *P. filamentosus* and *P. chola* remain similar during day and night, considerable diurnal differences were observed in *P. dorsalis*. Catch statistics reveal strongly increased activity during the night combined with inshore migrations (Fig. 2). Increased CPUE in inshore localities obtained shortly after sunset indicate that these inshore migrations may already begin in the late afternoon.

The vertical distribution in the water column was assessed at the offshore station of PSN from the position of the fishes in the vertically exposed 20-mm gill nets. During daytime, more than 90% of the fish occurred in the lower half of the water column. In the night, *S. mossambicus* and *P. filamentosus* show an upward shift with maximal occurrence near the surface while the two benthivorous species *P. dorsalis* and *P. chola* remain near the bottom (Fig. 3). The same pattern holds good for the smaller size classes (12- and 15-mm gill nets) of the latter two species.

3.3. Feeding ecology

In the following, the dietary habits of some of the species are discussed, based on gut analysis (Tables 4 and 5). A detailed study including the diel feeding cycle was carried out for *P. dorsalis, P. chola, P. filamentosus* and *S. mossambicus*. Results on the two latter are only summarized here and treated in greater detail in a separate paper (Hofer & Schiemer 1983).

3.3.1. Puntius chola and P. dorsalis. The two species show a remarkable similarity in their feeding habits. In the offshore stations, where both occur in high densities, their food consists of zoobenthos and sedimented material, which is sucked in together with animal food mainly chironomids, cyclopids, cladocerans and ostracods. This food represents the soft mud zoobenthic community, but reveals a high selectivity towards larger-sized components (especially chironomids and cyclopids, whilst ostracods are usually under-represented). Frequently setae of oligochaetes are found in the guts of both species, but their contribution to the diet is difficult to quantify.

Table 6 reveals the striking food overlap of the two species when occurring in the same habitats. The food resources are used with a very similar selectivity. However, while *P. chola* is restricted to offshore localities, *P. dorsalis* is common in several habitat types of the lake and its food varies according to the availability in these different localities. Gut contents of specimen caught in the littoral zone contained, besides chironomids and crustaceans, other aquatic insects and plant seeds (see Table 5).

Table 4. Relative importance of different food items in the diet of fishes in August/September 1979.[a]

	Porifera	Bryozoa	Gastropoda	Entomostraca	Chironomidae	Trichoptera	Other aqu. insects	Terrestrial insects	Fish	Filamentous algae	Macrophytes	Seeds	Coarse detritus	Fine detritus
Puntius filamentosus														
PSN offsh. $n = 7$ TL 130–150				0.1	0.1						8.0			3.1
PSM offsh. $n = 10$ TL 135–150	0.8				0.6		0.7			0.9	0.8		3.3	6.0
P. dorsalis														
PSN offsh. $n = 8$ TL 150–180				4.1	2.0		0.1						0.3	6.5
PSM offsh. $n = 6$ TL 140–190	0.2			2.7	2.0		0.5						1.3	7.3
P. chola														
PSN offsh. $n = 11$ TL 115–130				5.0	0.8								0.6	7.7
Etroplus suratensis														
PSN litt. $n = 10$ TL 140–225	0.7	2.2		0.1	1.0					0.1	2.0		3.3	4.0
PSM litt. $n = 4$ TL 155–225		0.5	2.5	0.5	4.0	0.2	0.2			0.4			3.0	3.0

Table 5. Relative importance of different food items in the diet of fishes in March/April 1980.[a]

	Porifera	Bryozoa	Gastropoda	Entomostraca	Chironomidae	Trichoptera	Other aqu. insects	Terrestrial insects	Fish	Filamentous algae	Macrophytes	Seeds	Coarse detritus	Fine detritus
Puntius filamentosus														
PSN A $n = 10$ TL 125–148						0.1			1.0	8.0				
PSN C $n = 10$ TL 135–150			0.1	0.1		0.1	0.4		6.5	1.4				
PSN E $n = 10$ TL 130–150			0.9	0.6		1.5	2.1		1.7	4.6				
OF_{canal} $n = 11$ TL 117–132	7.5				0.4		0.2		1.1					
P. dorsalis														
PSN A $n = 6$ TL 130–155			5.0	4.4							2.4	4.0		
PSN B $n = 8$ TL 130–170			0.7	4.6	1.6	0.1					3.6	7.5	4.0	
PSN E $n = 13$ TL 135–170			2.1	8.0							0.2		8.0	
OF_{canal} $n = 12$ TL 125–150	2.8	2.8	2.0	3.0	1.0	1.1	0.1							
P. chola														
PSN E $n = 14$ TL 120–135			3.9	5.3							0.6		7.6	
P. sarana														
PSN C $n = 7$ TL 130–200				0.3			3.7	2.3	0.1	3.4	3.4	3.4		
Rasbora daniconius														
PSN C $n = 3$ TL 100–110							2.7		3.3	2.7				
PSN E $n = 6$ TL 100–115				2.8*			4.3	5.3						* pupae
Mystus vittatus														
PSN E $n = 9$ TL 100–145			5.4	2.5	0.1	1.0					0.4	6.0		
Etroplus suratensis														
PSN C $n = 12$ TL 70–90	1.3	0.7	0.1	0.2	0.1	0.1			6.8	1.2		2.3		

[a] The values represent the mean for a number of specimen (n) from various sampling stations in PSN and the outflow canal (see above). The scale range from 0 to 8. TL = total length of fishes in mm.

Specimens from the outflow channel of the lake showed a varied diet with a high proportion of bryozoans and gastropods in addition to chironomids and entomostracans.

3.3.2. Diel cycle of feeding. The average relative gut filling of *P. chola* specimens from the offshore station of PSN remains at a similar level throughout the 24-h cycle, but shows considerable individual variation at any time of the day (Table 7). The pattern of *P. dorsalis* obtained at the same locality is clearly different: gut contents are low in the late afternoon and at sunset, increase strongly from sunset until midnight and remain high until sunrise (Fig. 4). These feeding rhytms agree well with the activity patterns of the two species as indicated by

Table 6. Relative abundance of different taxa (in % of numbers) in the diet of *Puntius chola* and *P. dorsalis* caught in offshore stations of PSN and PSM.[a]

	Cladocera	Cyclopidae	Ostracoda	Chironomida
PSN, 1979:				
P. chola $n = 11$; TL 115–130	58.8	38.2	0.3	2.6
P. dorsalis $n = 9$; TL 150–175	44.4	47.5	0	8.1
PSN, 1980:				
P. chola $n = 14$; TL 115–136	15.1	23.0	2.8	58.9
P. dorsalis $n = 13$; TL 140–160	17.6	4.4	4.9	73.1
PSM, 1979:				
P. dorsalis $n = 6$; TL 140–190	0	25.0	40.9	34.1

[a] n = number of fishes analysed, TL = total length of fishes in mm.

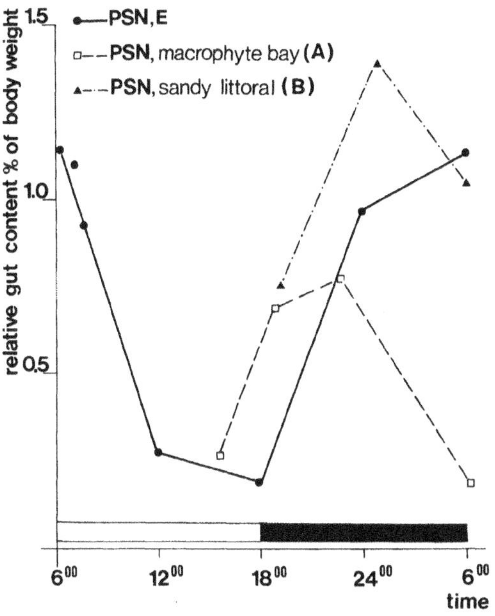

Fig. 4. Diurnal pattern of relative gut content (in % of body weight) in *P. dorsalis*, caught at various habitats in PSN. Circles, full line = PSN E; (offshore station) squares, broken line = PSN, macrophyte bay (Station A); Crosses, dottet line = PSN, sandy littoral (Station B). For details see Table 8.

the catch statistics (Fig. 3, see above), viz. nocturnal activity of *P. dorsalis* and continuous activity of *P. chola*.

However, the diel feeding rhythm of *P. dorsalis* varies considerably in different habitats in the lake (Table 8). For example, specimens caught in the littoral zone showed a high degree of gut filling already shortly after sunset, at a time when the CPUE in the littoral zone increased strongly. This increase in catch connected with high gut filling immediately after sunset indicate that migrations start already in the late afternoon and are connected with a high feeding activity. Satiation in those animals is attained earlier in the night compared to fishes remaining in the offshore localities, which is indicated by a decrease in relative gut filling already during the night.

Table 7. Relative gut content (in % of body weight) of *Puntius chola* at different times of the day in the open lake of PSN (PSN E). Arithmetic mean ($\bar{x}$), standard deviation (SD) and range of values.

Time	n	$\bar{x}$	SD	Range
6.3.80, 0600	18	1.40	0.42	0.7–2.1
26.3.80, 0700	5	1.20	0.42	0.8–1.8
5.3.80, 1200	20	1.41	0.48	0.6–2.6
5.3.80, 1800	13	1.20	1.04	0.0–3.4
5.3.80, 2400	17	1.18	0.41	0.1–1.8

Table 8. Relative gut content (in % of body weight) of *Puntius dorsalis* at different times of the day and in various habitats. Arithmetic mean ($\bar{x}$), standard deviation (SD) and range of values.

Locality	Time	n	$\bar{x}$	SD	Range
PSN A	27.3.80, 0620	20	0.18	0.40	0.0–1.7
PSN A	26.3.80, 1840	11	0.69	0.39	0.2–1.6
PSN A	25.3.80, 2300	12	0.78	0.33	0.0–1.2
PSN B	2.4.80, 0600	18	1.06	0.66	0.1–2.1
PSN B	1.4.80, 1830	18	0.76	0.25	0.3–1.3
PSN B	2.4.80, 2450	12	1.38	0.46	0.5–2.1
PSN E	6.3.80, 0600	20	1.15	0.32	0.5–1.7
PSN E	26.3.80, 0700	11	1.11	0.47	0.5–1.8
PSN E	29.3.80, 0730	10	0.92	0.24	0.6–1.3
PSN E	5.3.80, 1200	19	0.28	0.26	0.0–0.8
PSN E	31.3.80, 1530	11	0.27	0.34	0.0–0.9
PSN E	5.3.80, 1800	9	0.19	0.25	0.0–0.7
PSN E	5.3.80, 2400	10	0.97	0.18	0.8–1.3
PSN outflow	31.3.80, 0620	12	0.55	0.46	0.0–1.6
PSN outflow	30.3.80, 1230	14	0.76	0.40	0.0–1.5
PSN outflow	30.3.80, 1800	10	0.53	0.29	0.2–0.9

Again a different pattern of feeding activity was observed in the outflow channel of PSN, i.e. a riverine situation, where *P. dorsalis* occurred at high densities in March/April 1980. The gut filling at this locality remained essentially constant throughout the day, but at a relatively lower level compared to the diurnal peak values attained in the reservoir. This reveals that not only the type of food, but also the diel feeding cycle in *P. dorsalis* varies with the habitat.

3.3.3. Puntius filamentosus. This species shows a broad range of diet, with a predominance of littoral and 'Aufwuchs' food. Specimens from offshore stations in PSN contained littoral vegetation and various aquatic and terrestrial insects in their gut. The food at PSN C (littoral, with sandy and rocky bottom) consisted mainly of filamentous algae, macrophytes, littoral fauna and terrestrial insects. At station PSN A, covered densely with *Ceratophyllum* sp., food consisted almost entirely of macrophytes, while in the outflow channel bryozoans were the predominant food.

We could show that food uptake in this species

occurs predominantly during daytime and that a purely macrophytic diet is not sufficient to cover the energetic demand of the species (Hofer & Schiemer 1983).

3.3.4. Puntius sarana. The food of a small number of specimens (TL 130–200 mm) caught at PSN C in March/April 1980 consisted of plant material (mainly submerged terrestrial macrophytes and seeds), terrestrial insects, larger-sized aquatic insects (e.g. Hemiptera) and fish (e.g. fry of *S. mossambicus*).

3.3.5. Rasbora daniconius. Adult specimens (TL 100–115 mm) caught in the offshore zone in March/April 1980 had fed on terrestrial insects, chironomid pupae and fish fry. These food items indicate that food uptake occurs in the open water body and at the water surface. Similar-sized specimens caught within the littoral zone had fed on algal epigrowth, macrophytes, terrestrial insects and fish. Observations in March 1980 showed that *R. daniconius* is one of the most important predators of *S. mossambicus* fry in the littoral zone.

3.3.6. Amblypharyngodon melettinus. The diet of specimens (TL 90–100 mm), both from the limnetic and the littoral zones, was comprised of the typical phytoplankton assemblage with a quantitative (in terms of biomass) predominance of *Mougeotia, Microcystis* and *Anabaenopsis reciborski* in PSN, March 1980. The relative length of the gut (6.25 according to Hofer & Schiemer 1981) indicates the herbivorous nature of this species.

3.3.7. Mystus vittarus. This small-sized catfish (TL 100–144 mm) was frequently caught in the near bottom stratum of the 12- to 20-mm gill nets in the open areas of the lake. Food items are mainly of animal origin: Cyclopidae, Chironomidae, Hemiptera, Cladocera, Trichoptera, *Chaoborus* larvae, terrestrial insects, Hydracarina, plant seeds (listed according to their quantitative importance). Several of these components are of littoral origin (e.g. *Chaoborus, Diaphanosoma,* cyclopids in PSN), indicating that inshore migrations must take place.

3.3.8. Etroplus suratensis. The composition of gut contents of adult *E. suratensis* (TL 140–225 mm) caught in September 1979 showed striking differences in fishes from PSN and PSM. Gut filling in specimens from PSN was generally low. Their diet consisted mainly of detritus and terrestrial plant material, indicating that fishes were feeding on recently inundated areas. The portion of larger-sized 'Aufwuchs' organisms (bryozoans, sponges, chironomids, *Caridina* sp.) was low. In specimens from PSM, the gut content was distinctly higher and consisted mainly of littoral chironomid species and gastropods. These differences indicate that the low water level in August/September 1979 deprived the species of littoral feeding grounds in PSN (see below) while in PSM a high macrobenthos density in a belt of dead mollusc shells and a rich 'Aufwuchs' community on the numerous submerged tree trunks provided more suitable trophic conditions.

Despite intensive experimental fishing in March/April 1980, only a few small-sized specimens (TL 68–90 mm) were obtained, mainly at station PSN C. The diet of these fishes was dominated by filamentous algae, followed by sponges, macrophytes, bryozoans, trichopternas, chironomids and entomostracans. A few specimens caught at PSN A (covered with *Ceratophyllum*) contained mainly macrophytes, epiphytic algae, trichopterans and chironomids.

This food analysis reveals that *E. suratensis* is feeding preferably on hard substrates and on macrophytes in the littoral zone, while soft mud habitats with a predominance of meiobenthos (Schiemer, in preparation) are not providing for a usable diet. These findings allow one to predict that Sri Lankan reservoirs with a rich littoral structure (e.g. macrophytic vegetation) and newly flooded reservoirs with inundated jungle provide good trophic conditions for this species.

3.3.9. Sarotherodon mossambicus. The feeding beology was studied for medium-sized stages (TL 85–130 mm; see Hofer & Schiemer 1983). A study of the relative gut filling and the pH of the stomach content revealed that feeding continues throughout the 24-h cycle. The vertical distribution in the water column, in the lower half of the water column during daytime, near the surface during night, indicates that food is taken up both by sucking up the detrital aggregate of the upper sediment layers (see Bowen 1980) and by filtering suspended phytoplankton (e.g. Fish 1955). A certain preference for the ingestion of sediments is indicated by the high ash frac-

Table 9. Major food resources of fish in PS.

	Limnetic				Airborne	Littoral items			
	Phytoplankton	Zooplankton	Sediment & detritus	Zoobenthos		Algal epigrowth	Macrophytes	Invertebrates	Fish
Ehirava fluviatilis		–						–	
Amblypharyngodon melettinus	–								
Labeo dussumieri						–			
Puntius chola			–	–					
P. dorsalis			–	–				–	
P. filamentosus					–	–	–	–	
P. sarana					–	–	–	–	–
Rasbora daniconius		–			–	–	–	–	–
Mystus vittatus									
Sarotherodon mossambicus (ad.)	–		–						
Etroplus suratensis						–	–	–	

tion of the stomach content (Hofer & Schiemer 1983). Dokulil (1983) ascertained that a part of the ingested alive algae passes the intestine without losing its viability. This may hold especially good for the material which passes the intestine without entering the stomach. An energy budget calculated for the species showed favorable trophic conditions for the species in the lake (Hofer & Schiemer 1983).

Table 10. Food overlap values of fish species in the littoral zone (PSN C, *P. dorsalis* from PSN B) and the open lake (PSN E) in March/April 1980. The calculations are based on data given in Table 6.[a]

Littoral zone					
	Pf	Pd	Ps	Rd	Es
Pf	XX	0.03	0.28	0.59	0.82
Pd		XX	0.23	0	0.05
Ps			XX	0.86	0.15
Rd				XX	0.50
Es					XX
Open lake					
	Pf	Pd	Pc	Rd	Mv
Pf	XX	0.14	0.14	0.24	0.25
Pd		XX	0.76	0	0.50
Pc			XX	0	0.72
Rd				XX	0
Mv					XX

[a] Pf = *Puntius filamentosus*, Pd = *P. dorsalis*, Pc = *P. chola*, Ps = *P. sarana*, Rd = *Rasbora daniconius*, Mv = *Mystus vittatus*, Es = *Etroplus suratensis*.

3.4. Trophic interactions

According to their habitat preferences and their feeding biology (see Table 9), the common species of fish in Parakrama Samudra can be grouped as:

a) Littoral 'Aufwuchs' feeders, with a high portion of epiphytic algae, macrophytes and littoral fauna in their diet. This group contains *Puntius sarana*, *P. filamentosus*, *Labeo dussumieri* and *Etroplus suratensis*.
b) Soft mud zoobenthivors: *P. chola* is bound to offshore stations; *P. dorsalis* and *Mystus vittatus* exhibit inshore migrations with food uptake in the littoral zone.
c) Limnetic species feeding on phytoplankton (*Sarotherodon mossambicus*, *Amblypharyngodon melettinus*), zooplankton (*Ehirava fluviatilis*, *R. daniconius?* Hemiramphidae sp. and air-borne material (*R. daniconius*, Hemiramphidae sp.).

To outline possible competitive interactions, food-overlap coefficients for several pairs of species, both in the inshore and the offshore localities, have been assessed (Table 10). Food-overlap coefficients have been calculated from estimates of food composition (Table 5) by Schoener's formula: $O = 1 - 0.5\ (\sum_n |\ p_a - p_b\ |\ )$ (p_a, p_b = fraction of a food component (n) in the diet of the species a and b) (Schoener 1970).

In the inshore zone, high overlap values (>50%) occur between *P. filamentosus, E. suratensis* and *Rasbora daniconius.* The diet of *P. sarana* is similar only to that of *R. daniconius,* but differs from that of the other two species. Overlap between *P. dorsalis* and the other species compared is low.

In the open lake area, the similarity in diet is particularly high between the two zoobenthivorous species *P. chola* and *P. dorsalis.* The results discussed above indicate the mechanisms by which competitive interactions between these two species are relaxed: A simultaneous use of the same food resources occurs mainly in the offshore areas during the night. The nocturnal migrations of *P. dorsalis* into the littoral zone correspond with the period of its highest feeding activity. Thus, these inshore migrations lead to a considerable extension of the resource availability of this species in the lake. Hence, resource partitioning exists both with regard to the time pattern of feeding and spatial segregation at periods of highest feeding activity.

The overlap between the two *Puntius* species and *Mystus vittatus* is actually lower than expressed by the coefficient, since littoral and offshore food items have been classified under the same categories. No overlap was found between these zoobenthivorous species and *Rasbora daniconius,* which feeds in the open water column and at the water surface. *Puntius filamentosus* exhibits only a week overlap with all the species compared. The diet of *S. mossambicus* and *A. melettinus* consists of phytoplankton, which in the former may be mainly taken up in form of sedimented ooze, while the latter feeds probably in the limnetic zone.

Predation effects by fish influence both the zooplankton and the zoobenthos community. Duncan (1983) explains the predominance of small-sized rotifers and the virtual absence of crustaceans in the zooplankton by the predation effects of fish, exerted by species like *Ehirava fluviatilis,* juveniles of Hemiramphidae sp. and, possibly, juveniles of *Rasbora daniconius.* Similarly the predominance of meiobenthos and the paucity of macrobenthos, e.g. larger-sized chironomids and oligochaetes in the soft mud bottoms (Schiemer, in preparation), are a result of the intensive predation pressure exerted on zoobenthos by the dense population of *P. chola* and *P. dorsalis.*

The importance of piscivorous fish for the species and size distribution of the fish fauna in the lake and its dynamics is not well analysed. Several species feed on fish fry, especially those of *S. mossambicus* in the marginal zone of the lake. *Rasbora daniconius* and *Cottogobius giuris* have been observed to be particularly important in this respect, but a number of other species feed on littoral fry, too (e.g. *Ophiocephalus striatus.* Hemiramphidae sp., *Puntius sarana* and *Mystus vittatus*). No observations have so far been carried out on the feeding ecology of larger-sized predators in the lake, especially the eels and catfish.

Large populations of fish-eating birds, especially the Indian Shag (*Phalacrocorax fuscicollis*) and the Little Cormorant (*P. niger*), exhibit a strong predation pressure upon the fish population of the reservoir. Winkler (1983) estimated the fish consumption by cormorants to be 254 metric tons per year. Thus their feeding habits and the size and species composition of their food are of great practical relevance. Winkler (1983) found middle-sized (standard length 130 mm) cichlids (both *S. mossambicus* and *Etroplus suratensis*) as main food components of the Indian Shag. The Little Cormorant contained smaller-sized specimens (SL 30–70 mm) of the two cichlids and of *P. filamentosus* in their diet.

3.5. Seasonal changes in the fish fauna

Species composition and population structure of fish exhibited some striking differences during the two visits. These fluctuations can be a result of (a) discontinuous breeding activity and spawning migrations, (b) changes in habitat and resource availability and (c) predation effects by fish and fish-eating birds. It remains open as to whether the observed changes reflect a normal seasonality or are a result of the atypical, prolonged low water level before August/September 1979 caused by a dam break in November 1978.

Strong differences between the two visits were observed in the breeding activity of *S. mossambi-*

cus. While in September/October breeding activity was high (numerous nests in the inshore zone of PSN), no breeding sites were found in March/April 1980 although some of the characteristic breeding localities were specifically searched. Discontinuous reproduction in this species is further indicated by striking differences in the catch of the 20-mm gill-net size class of the species, e.g. complete absence in August/September, high abundance in March/April. These observations agree with data on the seasonality of the gonadosomatic index of the species in Parakrama Samudra (De Silva & Chandrasoma 1980) and the high percentage of spent females in the period of low water level in August.

Periodic spawning and spawning migrations could explain the strong difference observed in the composition of the medium-sized *Puntius* species (*P. chola, P. dorsalis, P. filamentosus*) at the two visits. In some of the indigenous cycprinids, e.g. *P. sarana* (Chandrasoma & De Silva 1981) and *P. dorsalis* (De Silva 1983), the maximal spawning activity was observed during the rainy season in connection with migrations into riverine habitats, a phenomenon well known from various tropical freshwater species (Lowe-McConnell 1975).

Seasonal changes in resource availability can induce fluctuations in the abundance and composition of the fish fauna. It appears that in the northern basin of the reservoir, especially the littoral 'Aufwuchs' feeders are subject to seasonal limitations caused by water level fluctuations. For example, the complete absence of adult *E. suratensis* in PSN in March/April 1980 (in contrast to August/September 1979, where they formed a considerable portion of the catch of cast-net fishermen) may be explained by food limitations during the prolonged period of low water level in the foregoing summer, during which 'Aufwuchs' communities were reduced.

The possible effects of predators, mainly fish and birds, have been outlined above. The fish-eating birds feeding at the Parakrama Samudra reservoir belong to a large population, exploiting the various aquatic habitats, rice fields, channels, villus, reservoirs etc. in the Polonnaruwa area. The actual number of birds feeding on the lake undergoes strong short-term fluctuations caused by the particular conditions of food availability (growth of a cohort of fish into an eatable size class, water level situation).

References

Bowen, S. H., 1980. Detrital non protein amino acids are the key to rapid growth of *Tilapia* in Lake Valencia, Venezuela. Science 207: 1216–1218.

Chandrasoma, J. & De Silva, S. S., 1981. Reproductive biology of *Puntius sarana*, an indigenous species, and *Tilapia rendalli* (*melanopleura*), an exotic, in an ancient man-made lake in Sri Lanka. Fish. Mgmt. 12: 17–28.

De Silva, S. S., 1983. Reproductive strategies of some major fish species in Parakrama Samudra reservoir and their possible impact on the ecosystem – a theoretical consideration. In: Schiemer, F. (ed.) Limnology of Parakrama Samudra – Sri Lanka: a case study of an ancient man-made lake in the tropics. Developments in Hydrobiology (this volume). Dr W. Junk, The Hague.

De Silva, S. S. & Chandrasoma, J., 1980. Reproductive biology of *Sarotherodon mossambicus*, an introduced species, in an ancient man-made lake in Sri Lanka. Env. Biol. Fish 5: 53–60.

De Silva, S. S. & Fernando, C. H., 1980. Recent trends in the fishery of Parakrama Samudra, an ancient man-made lake in Sri Lanka. In: Furtado, J. (ed.) Tropical Ecology and Development, pp. 927–937. Society for Tropical Ecology, Kuala Lumpur, Malaysia.

Dokulil, M., 1983. Aspects of gut passage of algal cells in *Sarotherodon mossambicus* (Pisces, Cichlidae). In: Schiemer, F. (ed.) Limnology of Parakrama Samudra – Sri Lanka: a case study of an ancient man-made lake in the tropics. Developments in Hydrobiology (this volume). Dr W. Junk, The Hague.

Duncan, A., 1983. The composition, density and distribution of the zooplankton in Parakrama Samudra. In: Schiemer, F. (ed.) Limnology of Parakrama Samudra – Sri Lanka: a case study of an ancient man-made lake in the tropics. Developments in Hydrobiology (this volume). Dr W. Junk, The Hague.

Fernando, C. H. & Indrasena, H. H. A., 1969. The freshwater fisheries of Ceylon. Bull. Fish. Res. Stn. Ceylon, 20: 101–134.

Fish, G. R., 1955. The food of *Tilapia* in East Africa. Uganda J. 19: 85–89.

Hofer, R. & Schiemer, F., 1981. Proteolytic activity in the digestive tract of several species of fish with different feeding habits. Oecologia (Berl.) 48: 342–345.

Hofer, R. & Schiemer, F., 1983. Feeding ecology, assimilation efficiencies and energetics of two herbivorous fish: *Sarotherodon* (*Tilapia*) *mossambicus* (Peters) and *Puntius filamentosus* (Cuv. et Val.). In: Schiemer, F. (ed.) Limnology of Parakrama Samudra – Sri Lanka: a case study of an ancient man-made lake in the tropics. Developments in Hydrobiology (this volume). Dr W. Junk, The Hague.

Lowe-Mc Connell, R. H., 1975. Fish communities in tropical freshwaters. Longman, New York.

Newrkla, P., 1983. Sediment characteristics and bethic community oxygen uptake rates of Parakrama Samudra, an ancient man-made lake in Sri Lanka. In: Schiemer, F. (ed.) Limnology of Parakrama Samudra – Sri Lanka: a case study of an ancient man-made lake in the tropics. Developments in Hydrobiology (this volume). Dr W. Junk, The Hague.

Schiemer, F., 1983. The Parakrama Samudra Project – scope and objectives. In: Schiemer, F. (ed.) Limnology of the Parakrama Samudra – Sri Lanka: a case study of an ancient man-made lake in the tropics. Developments in Hydrobiology (this volume). Dr W. Junk, The Hague.
Schoener, T. W., 1970. Nonsynchronous spatial overlap of lizards in patchy habitats. Ecology 51: 408–418.

Authors' addresses:
F. Schiemer
Institute of Zoology
University of Vienna
Althanstr. 14
A-1090 Vienna
Austria

R. Hofer
Institute of Zoophysiology
University of Inssbruck
Peter Mayerstr. 1a
A-6020 Innsbruck
Austria

Winkler, H., 1983. The ecology of cormorants (genus *Phalacrocorax*). In: Schiemer, F. (ed.) Limnology of the Parakrama Samudra – Sri Lanka: a case study of an ancient man-made lake in the tropics. Developments in Hydrobiology (this volume). Dr W. Junk, The Hague.

Appendix

(Drawings by P. Newrkla)

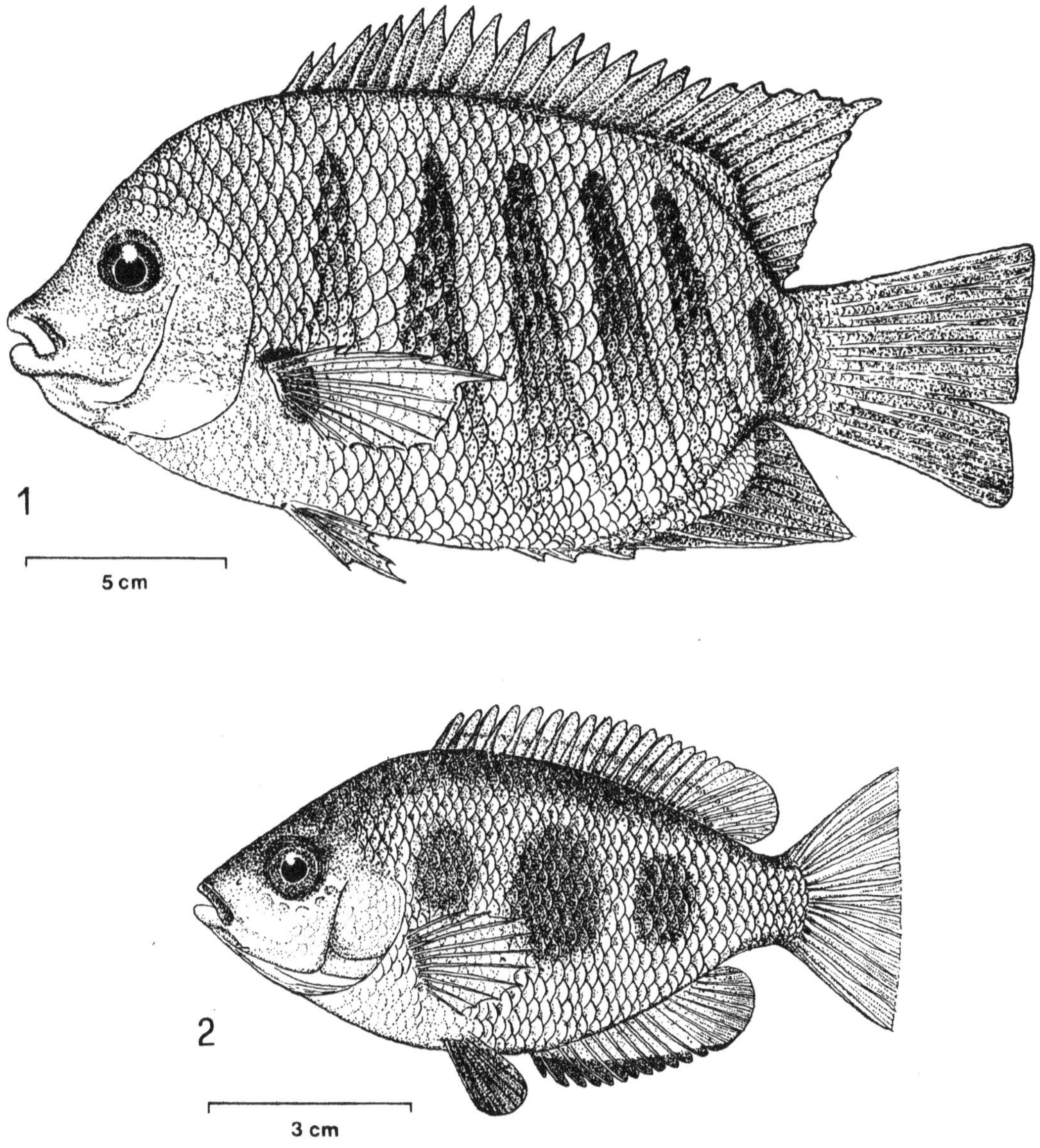

Plate I. 1. *Etroplus suratensis* (Bloch); 2. *Etroplus maculatus* (Bloch).

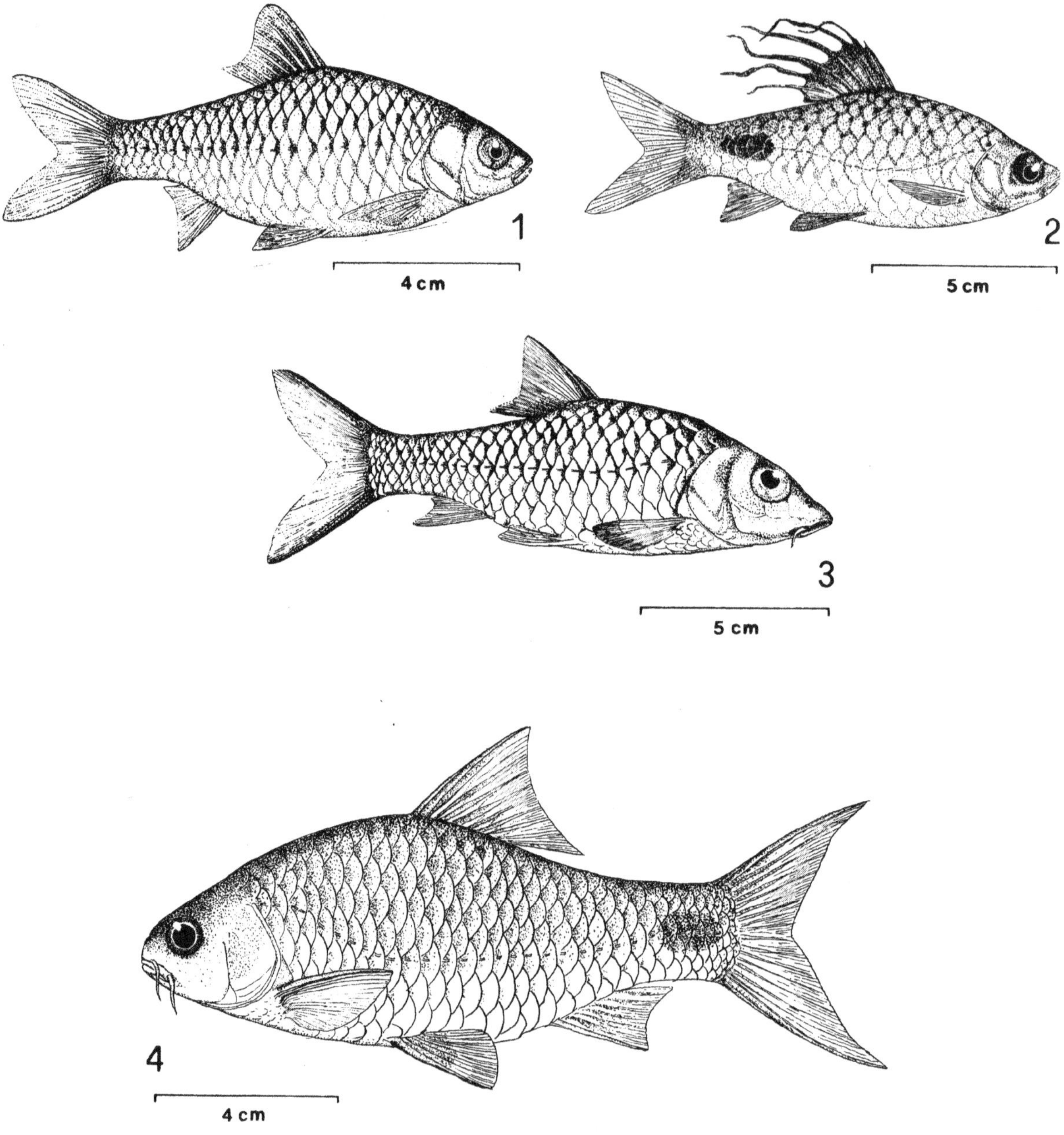

Plate II. 1. *Puntius chola* (Ham.-Buch.); 2. *Puntius filamentosus* (Val.); 3. *Puntius dorsalis* (Jerdon); 4. *Puntius sarana* (Ham.-Buch.).

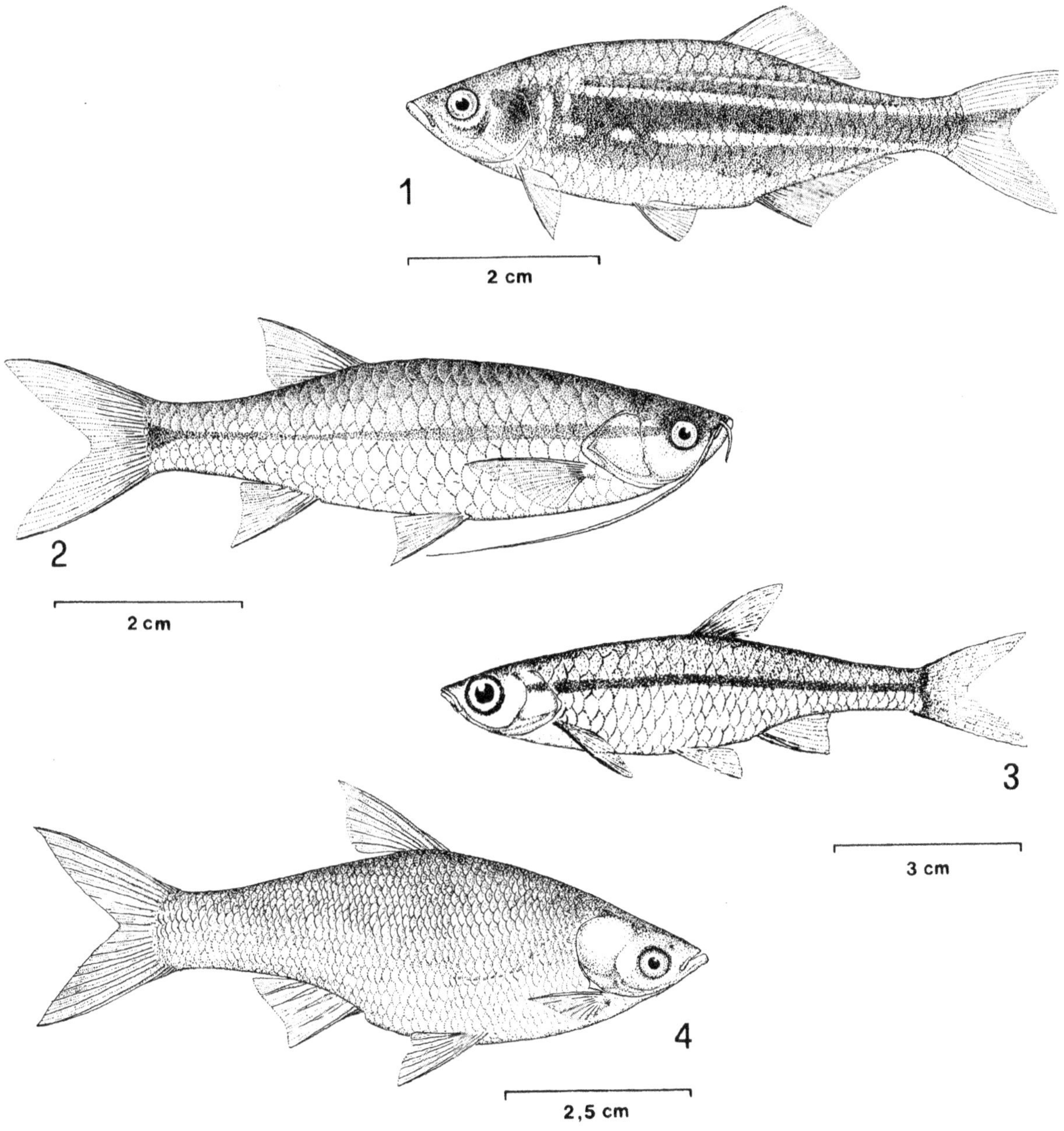

Plate III. 1. *Danio aequipinnatus* (McClelland); 2. *Esomus danrica thermoicos* (Val.); 3. *Rasbora daniconius* (Ham.-Buch.), juv.; 4. *Amblypharyngodon meletinus* (Val.).

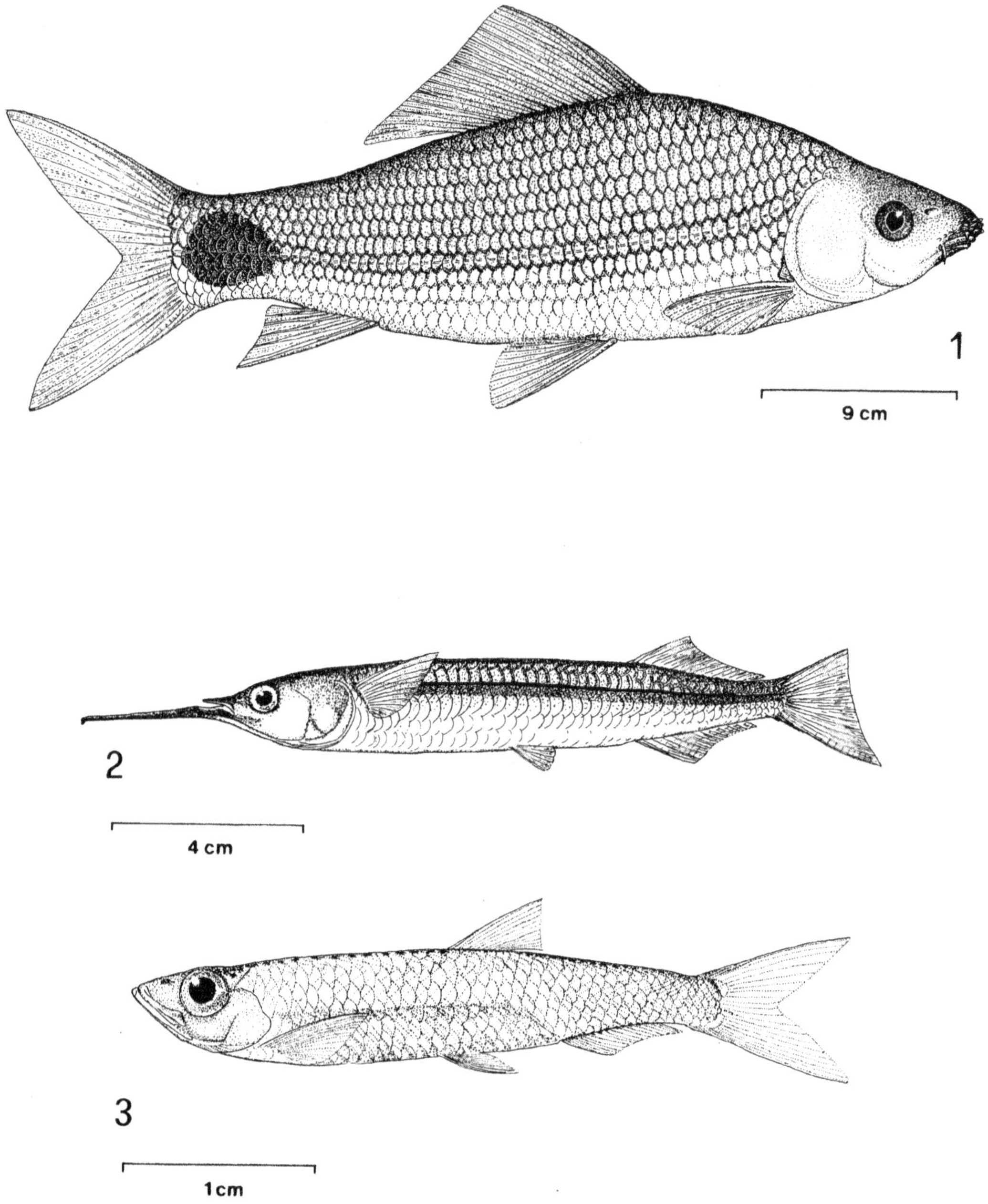

Plate IV. 1. *Labeo dussumieri* (Val.); 2. Hemiramphidae sp.; 3. *Ehirava fluviatilis* Deraniyagala.

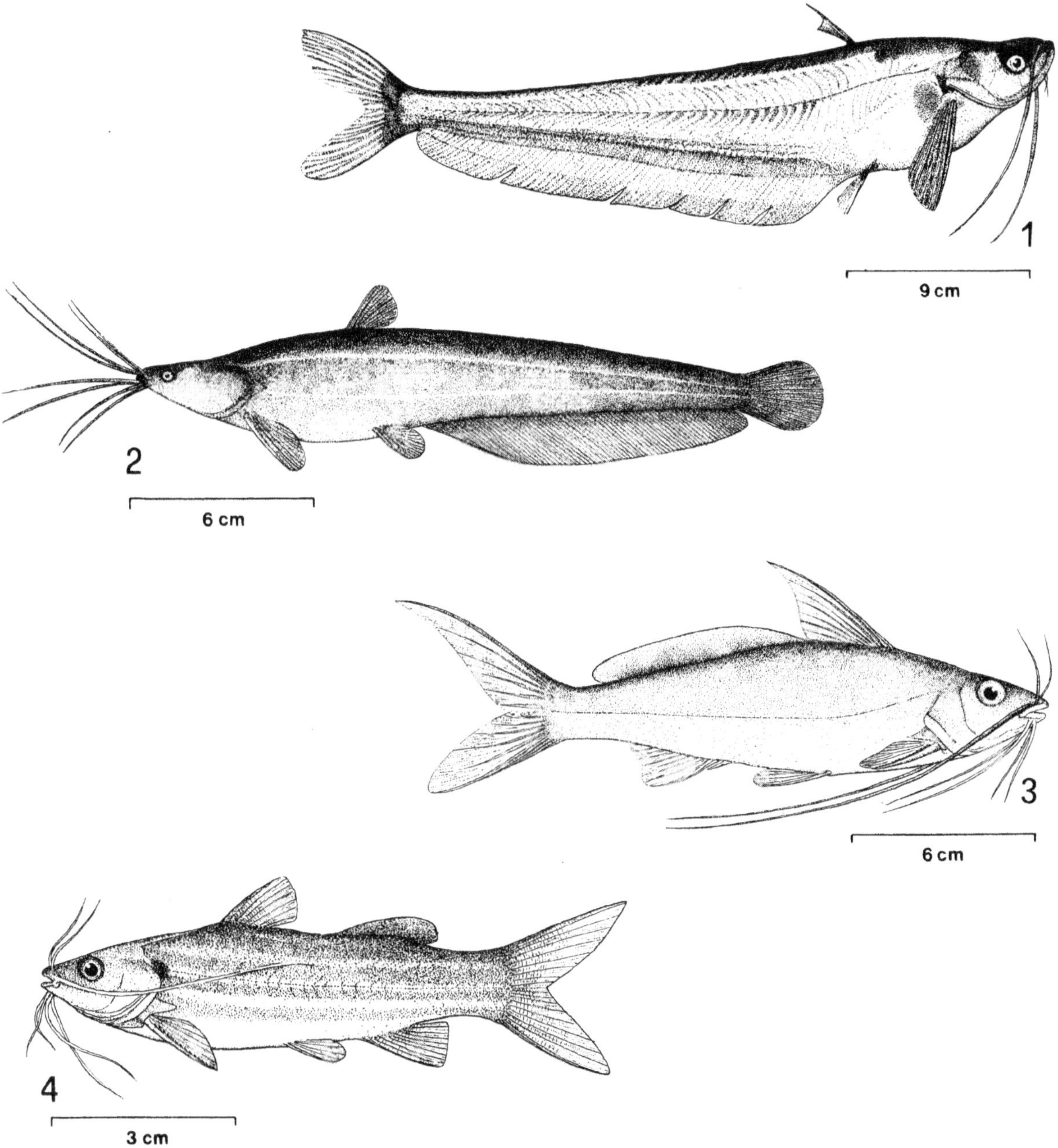

Plate V. 1. *Ompok bimaculatus* (Bloch); 2. *Heteropneustes fossilis* (Bloch); 3. *Mystus keletius* (Val.); 4. *Mystus vittatus* (Bloch).

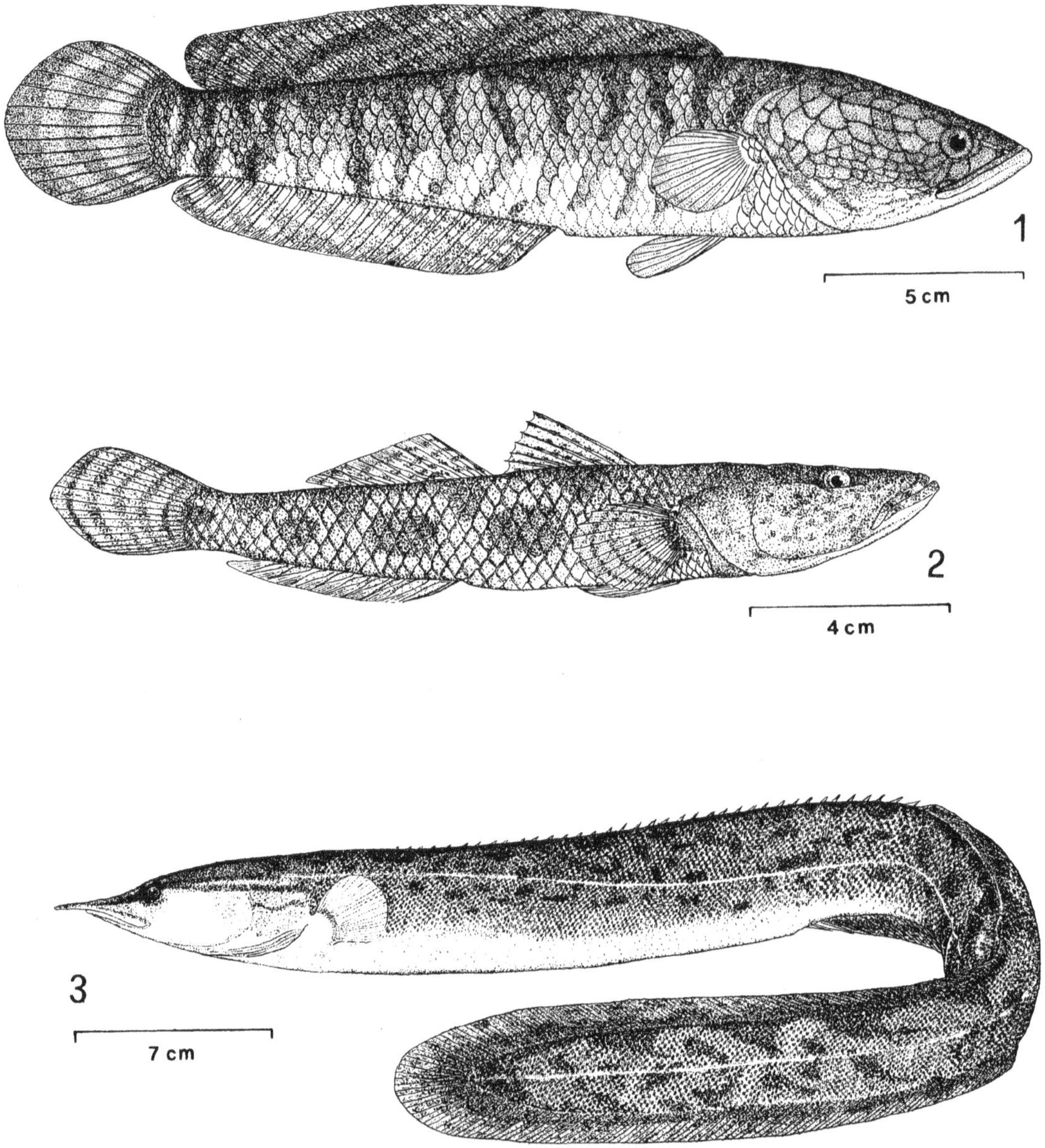

Plate VI. 1. *Opchiocephalus striatus* (Bloch); 2. *Glossogobius giuris* (Ham.-Buch.); 3. *Mastacembelus armatus* (Lacepede).

13. Feeding ecology, assimilation efficiencies and energetics of two herbivorous fish: *Sarotherodon* (Tilapia) *mossambicus* (Peters) and *Puntius filamentosus* (Cuv. et Val.)

R. Hofer & F. Schiemer

Keywords: Feeding ecology, energetics, herbivorous fish, tropical lake, *Sarotherodon mossambicus*, *Puntius filamentosus*

Abstract

The feeding habits, food consumption, assimilation efficiencies and energy budget of two dominant species of fish in Parakrama Samudra, a shallow artificial basin in the dry zone of Sri Lanka, were studied. *Puntius filamentosus*, a cyprinid, is omnivorous, but in dense standing crops of macrophytes it feeds exclusively on plants. Food is consumed only during the day time. The utilization of macrophytic matter is very low, resulting in a negative energy budget. Energetic problems of this species are discussed.

The diet of juvenile *Sarotherodon mossambicus* (mean body weight: 19.1 g), a cichlid, is very uniform, consisting of phytoplankton and mineral sediment. A constant gut filling indicates a more or less continuous diurnal feeding pattern. In spite of the high ash content of the diet the energy budget is highly positive and results in a daily production rate of 1.4% of body weight. This value coincides with growth rates found in feeding experiments reported in the literature.

1. Introduction

Sarotherodon mossambicus (Peters) and *Puntius filamentosus* (Val.) are two of the most abundant species in Parakrama Samudra (PS), an artificial lake in the dry zone of Sri Lanka. In both species plant material forms an important component of the diet (phytoplankton for *S. mossambicus* and macrophytes for *P. filamentosus*). Considerable differences exist in the structure of the digestive tract. In *S. mossambicus*, as in other herbivorous cichlids (Moriarty 1973), the hydrochloric acid of the stomach causes lysis of algal cells. Subsequently these cells are subjected to digestive enzymes in a long coiled intestine, which measures 6.8 times the body length (Hofer & Schiemer 1981).

In cyprinids without a distinct stomach, like *P. filamentosus*, digestive processes are restricted solely to enzymatic hydrolysis in an alkaline environment. Plant cells are to some extent broken down mechanically by the action of pharyngeal teeth (Prejs 1976). The digestive tract of *P. filamentosus* is comparatively short, about 2.5 times the body length (Hofer & Schiemer 1981). In both species, chemically *(S. mossambicus)* or mechanically *(P. filamentosus)* undamaged cells are not, or only very slightly subjected to enzymatic hydrolysis.

Specific feeding habits and utilization of food in *S. mossambicus* and *P. filamentosus* were studied within the framework of an ecosystem study of Parakrama Samudra (Schiemer 1981). The possible importance of the two species within the ecosystem as well as their food limitations are shown by preliminary calculations of energy budgets.

2. Materials and methods

The present study was carried out in March–April 1980. *P. filamentosus* and *S. mossambicus* were caught with gill nets of 20 mm (knot to knot) mesh size. In *P. filamentosus* the gut content was

Schiemer, F. (ed.), Limnology of Parakrama Samudra - Sri Lanka
 ISBN 90 6193 763 9

squeezed out of the intestinal tract and weighed. The long, coiled intestine of *S. mossambicus* rendered preparation of its gut content very difficult and tedious and even precise preparations involved a certain loss of matter. In this species gut content was determined by weighing the whole digestive tract and subtracting from this value the mean specific weight of the empty digestive tract of starved (24 h) *S. mossambicus.* In both species, gut content was expressed as a percentage of the body weight (with empty intestine). Microscopic analyses of the gut content were carried out in order to determine the relative frequency of different food components (see Schiemer & Hofer 1983).

The duration of the gut passage of *P. filamentosus* was determined in field experiments (Hofer et al. 1982) by the following technique: Approximately 1 kg of artificial food, labelled with red-dyed cellulose powder was distributed over an area of 50 × 100 m in a shallow bay covered by submerged macrophytes. The artificial food consisted of fish meal, blood meal, glucose and the red-dyed cellulose powder. The mixed powder was melted in an oven and cast to form thin layers, which were then broken up into small pieces. The dyed cellulose is inert and can thus be recognized in the gut. Between 1.7 and 4.5 h after distribution of the food fish were netted at approximately 30-min intervals. From a total of 138 specimen caught, 24 contained labelled food in the gut. In these specimen, the distance of the marker from the anterior end of the gut was measured and expressed in percentage of total gut length. Assuming a constant gut passage, duration of passage through the whole intestine could be calculated. Furthermore, the weight of the intestinal content ingested between feeding of artificial diet and catching the fish was considered as the consumption rate within this time.

For chemical analyses faeces were collected as follows: in *P. filamentosus* the content of the last centimeter and in *S. mossambicus* the content of the last three centimeters of hindgut were carefully removed and immediately dried in an oven (80 °C). *Ceratophyllum sp.* for chemical analysis was collected from the locality of the field experiment with *P. filamentosus.* Only tips of the branches were used.

2.1. Chemical analysis

Protein was extracted with sodium hydroxide and determined colorimetrically with Folin reagent (Lowry et al. 1951). Lipids were extracted with mixtures of chloroform with methanol and petroleum ether and determined gravimetrically. Both procedures were described by Dowgiallo (1975). The soluble carbohydrate fraction easily hydrolysable in 0.5 N sulphuric acid as well as the crude fibre (cellulose and hemicellulose) fraction solubilized in 28 N sulphuric acid were determined using the phenolsulphuric acid reagent with photometer reading at 486 nm (Dubois et al. 1956) as modified by Dowgiallo (unpublished data). Dry weight was measured at 80 °C, ash weight at 480 °C. A Phillipson microcomb calorimeter was employed for determinations of energy values (Prus 1975).

2.2. Assimilation efficiency

Assuming that ash (Conover 1966) or crude fiber (Buddington 1979) is not assimilated, the assimilation efficiency (U^{-1}) can be calculated from the percentage of ash and/or crude fiber in food and faeces by using the following formula:

$$U^{-1} = 100 - 100 \frac{\%\ \text{food fiber or ash} \times \%\ \text{faecal nutrients}}{\%\ \text{faecal fiber or ash} \times \%\ \text{food nutrients}}$$

Using ash as a marker, U^{-1} is expressed in % of weight, whereas U^{-1} based on crude fiber is expressed in % of energy value. The pH of the stomach contents was measured with a glass electrode.

3. Results and discussion

3.1. Puntius filamentosus

3.1.1. Distribution pattern. The species was recorded both from the open lake and the littoral zone. However, by far the highest population density occurred in a littoral area with dense cover of *Ceratophyllum sp. P. filamentosus* was the dominant species in the 20 mm gill net catches. Day-night catches did not indicate a diurnal inshore-offshore migration pattern such as was found in other species (for details see Schiemer & Hofer 1983).

3.1.2. Feeding. The diet of *P. filamentosus* is very heterogeneous, but vegetable components often dominate (Table 1). Feeding habits are influenced by the food availability. In a macrophyte-covered

Table 1. Analyses of gut content of *Puntius filamentosus* at various localities in the northern basin of Parakrama Samudra (PSN). Relative amount of food components was estimated: 1: 0–20%; 2: 20–40%; 3: 40–60%; 4: 60–80%; 5: 80–100% of total gut content.

Food	PSN littoral without macro-phytes	Outflow of PSN	PSN pelagial (station 3)	PSN macrophyte-covered littoral
Porifera	0.4	–	–	–
Bryozoa	–	4.4	–	–
Oligochaeta	1.6	–	–	–
Entomostraca	0.2	–	0.6	–
Hemiptera	0.1	–	1.4	0.1
Chironomidae	0.2	–	1.0	–
Trichoptera	–	0.7	–	–
Terrestr. insects	0.7	0.4	1.8	–
Filam. algae	4.2	1.2	1.3	–
Macrophytes	1.3	–	3.4	5.0
Sand	2.7	2.5	–	–

bay it fed mainly on *Ceratophyllum sp.*, whereas in the outflow of the northern basin of Parakrama Samudra bryozoa were ingested predominantly. The diurnal pattern of feeding activity was studied at several localities but in most detail in the macrophyte covered bay (Fig. 1A). Before sunrise the gut is completely empty or only small residues from the previous days meal are present at the posterior end of the gut (see Fig. 1A).

Feeding starts at sunrise, and from two hours after sunrise until sunset food uptake rates appear to be constant. During the night the relative gut content decreases, which indicates a stop in feeding activity after sunset. A similar but less pronounced diurnal pattern was found at two other localities (Fig. 1B). The degree of gut filling and, in consequence, the consumption rates are positively correlated to the proportion of plants in the diet: In the evening the gut filling of specimens feeding on *Ceratophyllum* is 3.6 times higher than that of mainly carnivorous *P. filamentosus* in the outflow of PSN.

3.1.3. Food consumption. The duration of gut passage in *P. filamentosus* was measured in the 'macrophyte bay' between 11 a.m. and 4 p.m. The transport of the marker (red cellulose powder) is shown

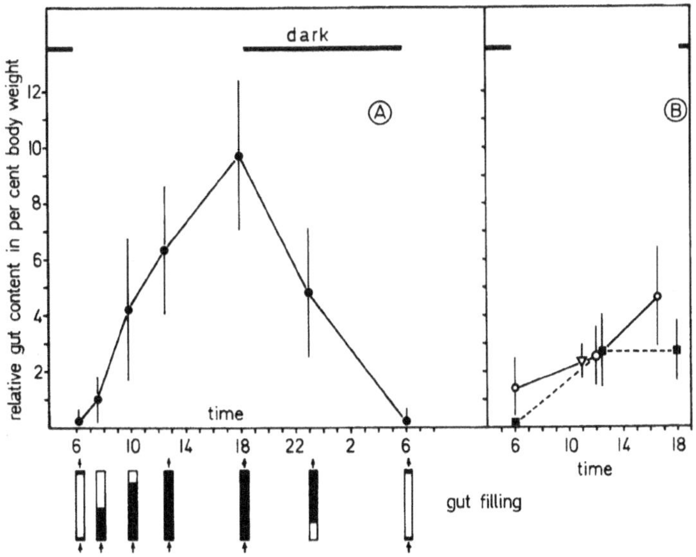

Fig. 1. Diurnal gut filling and relative gut content in per cent of body weight (with empty gut) in *P. filamentosus* at various localitities in Parakrama Samudra (PS). A: Macrophyte-covered bay of northern basin of PS (PSN), $N = 116$; B: circles: offshore of PSN (station 3), $N = 44$, squares: outflow of PSN, $N = 49$, triangles: offshore of PSM (middle basin), $N = 10$. Means and standard deviations are given.

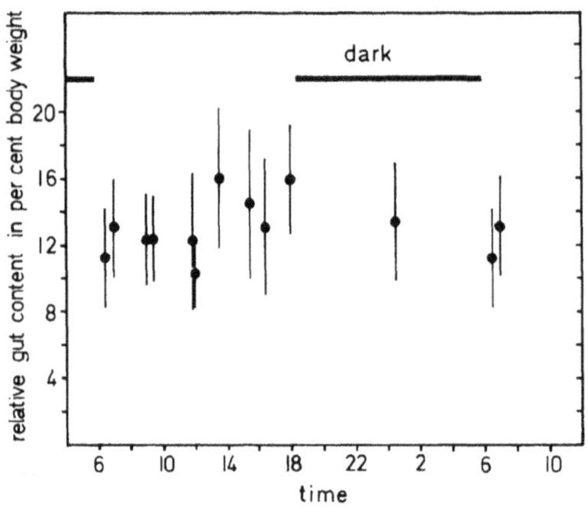

Fig. 2. Diurnal pattern of relative gut content in per cent of body weight (with empty gut) in juvenile *S. mossambicus* (body weight: 19.1 ± 5.8 g). Means and standard deviations are given. Average gut content of all measurements: 13.0 ± 3.8% of body weight, $N = 164$.

in Figure 3. The total passage time was calculated as 4.73 ± 0.87 h and the relative consumption rate is 1.86 ± 0.94% of body weight per hour. Daily relative consumption amounts to 21.5% fresh weight and 1.7% dry weight (dry weight of *Ceratophyllum* = 7.9% of fresh weight).

3.1.4. Chemical analysis. The results of the chemical analysis of *Ceratophyllum* and the gut contents in different sections of the intestinal tract are presented in Table 2. The value for *Ceratophyllum* represents the mean of four samples. For the analysis of gut contents the material of several specimens had to be pooled. In all samples remarkably high ash weights were found. In *Ceratophyllum* a high percentage (15%) remained unaccounted for by the analysis.

3.1.5. Assimilation efficiency. Despite the rather uniform diet of *P. filamentosus* caught in the macrophyte bay, our knowledge of the mode of food uptake is insufficient for the investigator to collect a fully representative sample of material ingested by the fish. This limitation has to be borne in mind when considering the calculated values. Differences could be expected especially in the amount of sediment (see large variations of ash weight in *Ceratophyllum;* Table 2) and animal components. However, in the large volume of plant material found in the intestine, only very occasionally were small animal remains detected.

The assimilation efficiency, based on ash and on crude fiber as the inert matter of a natural diet, is listed in Table 3. The results from the two methods are identical.

Cellulose does not seem to be digested by herbivorous cyprinids. In contrast to zooplanktivorous and detritivorous specimens of roach (*Rutilus rutilus* L.), only very low cellulolytic activities (using soluble carboxymethylcellulose) were found in gut fluid of herbivorous specimens (Niederholzer & Hofer 1980; Prejs & Blaszczyk 1977). Insoluble cellulose (cotton) was not hydrolysed, not even by the detritivorous *R. rutilus* (unpublished results).

These results conform with the absence of utiliza-

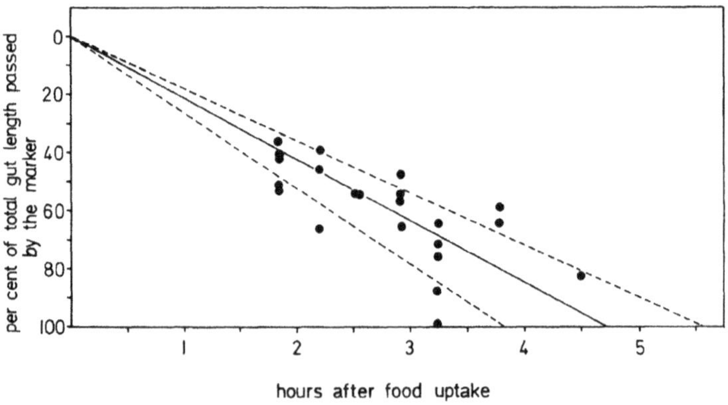

Fig. 3. The passage of red stained artificial food through the intestine of *P. filamentosus* feeding on *Ceratophyllum*. The food was offered at 11 a.m. Gut passage: 4.73 ± 0.87 h. Number of fish caught: 138, 17.4% had ingested artificial food.

Table 2. Chemical and energetic analyses of *Ceratophyllum sp.* and food samples in several parts of the digestive tract in *P. filamentosus* and *S. mossambicus*. Diet of *P. filamentosus: Ceratophyllum sp.*, diet of *S. mossambicus:* phytoplankton and mineral sediments. In *Ceratophyllum* 4 different samples were analysed, in all other cases pooled samples of several specimens were used. Mean values are given, and, if more than three parallel analyses were made, standard deviations as well.

Species	Sample	Ash (mg/g)	Crude fiber (mg/g)	Soluble carbohydrates (mg/g)	Protein (mg/g)	Lipids (mg/g)	Calorific values (J/g)	Ashfree calorific value (J/g)
Puntius filamentosus	Ceratophyllum (food)	279.0 ±47.0	147.9 ±13.8	183.8 ±28.2	152.8 ±11.9	85	12 827 ±844	17 797
	Content of the foregut	–	72.6 ±10.4	180.4 ±3.2	154.7 ±8.4	–	–	–
	Faeces	321.0	171.2	192.8	100.1 ±7.8	–	11 524 ±265	16 972
Sarotherodon mossambicus	Content of:							
	stomach	455.0 ±1.3	35.7 ±1.8	106.1 ±2.4	145.4 ±7.2	68	9 704 ±119	17 806
	Intestine -1	–	22.8 ±1.3	94.9 ±14.7	105.3 ±5.5	–	–	–
	Intestine -2	–	31.4 ±1.6	49.5 ±1.4	89.6 ±3.2	–	–	–
	Faeces	570.6 ±13.2	35.9 ±1.5	61.1 ±1.0	83.1 ±5.8	–	7 739 ±262	18 024

tion of crude fiber in *P. filamentosus* feeding on *Ceratophyllum*. Perhaps the great difference in assimilation efficiency of protein and soluble carbohydrates indicates a much better utilization of animal 'Aufwuchs' than of plant material itself. Considering the condition of the faeces (Fig. 4) in which most of the leaves of *Ceratophyllum* are cut into large pieces between 2 and 5 mm length, an assimilation efficiency of 43% for plant proteins seems quite impossible.

3.1.6. Energy budget. Table 5 presents a preliminary energy budget, based on ingestion rates measured in the field and the assimilation efficiency of a *Ceratophyllum* diet.

A specimen weighing 31.8 g consumes about 0.54 g dry *Ceratophyllum* or 6 671 J per day (Table 5). In this species crude fiber is not utilized and should be subtracted from ingested organic matter. Of the remaining 5 265 J only 24.6% are assimilated, resulting in a daily assimilation rate of only 1 295 J.

Oxygen consumption rates have not been measured for the species. Assuming a similar oxygen consumption as in *S. mossambicus*, corrected to the specific weight of the fish, daily respiratory costs amount to 2 506 J (see p. 163). This means a highly negative energetic budget of -1 211 J per day. To maintain its metabolism under such conditions the fish would have to provide 30.8 mg lipids or 67.3 mg protein. This discrepancy requires a consideration of possible experimental errors and of ecological alternatives for the species.

a) Possible errors in measuring consumption rates and assimilation efficiency. Significant mistakes in the determination of consumption rate can be excluded. Hofer et al. (1982) have shown with *Rutilus rutilus* that results obtained with the field method applied in the present study correspond exactly with the results of parallel experiments in the laboratory. In order to balance the energetic budget, the relative daily consumption rate of *P. filamentosus* would have to be doubled. A balanced budget could also be obtained by doubling the assimilation efficiency. Considering the condition of *Ceratophyllum* in the faeces (Fig. 4) such a high assimilation of plant matter appears to be impossible. On the other hand, a high utilization of proteins and low utilization of carbohydrate (Table 3) indicates a more efficient utilization of animal 'Auf-

Table 3. Assimilation efficiencies (U^{-1}) in *S. mossambicus* and *P. filamentosus*. Calculations of assimilation efficiencies based on the assumption that ash is not utilized by the fish (Conover 1966) are given in per cent of weight. Calculations based on crude fiber (Buddington 1979) are presented in percentage of energy value.

Assimilation efficiency (U^{-1})	*Sarotherodon mossambicus*	*Puntius filamentosus*	
	based on ash	based on ash	based on crude fiber
Organic matter	37.2	18.1	–
Non crude fiber organic matter	38.4	23.0	24.6
Soluble carbohydrates	54.2	8.8	9.4
Crude fiber	19.8	-0.6	–
Protein	54.4	43.1	43.4

wuchs' than of plant material. One explanation for the unbalanced energy budget might be that *P. filamentosus* picks up selectively a greater amount of small animals than is indicated by microscopic analysis of the gut contents and by hand collection for determination of chemical components of the food. An additional consumption of about 0.58 g animal food would be necessary to balance the energy budget. (Calorific value of chironomids and tubificids: 21 800 J/g dry weight – Johnson & Brinkhurst 1971; dry weight of tubifex: 16.1% of fresh weight – Pandian & Raghuraman 1972; average assimilation efficiency of animal food: ~60% – Fisher 1979). Taking a 2.5-fold daily renewal of total gut content, at least 0.23 g of animals should be found in each gut content. As mentioned above, residues of larger animals, like chironomid larvae, were only occasionally observed. On the other hand, very small, thin-skinned animals, like rotifers, nematodes or other small worms, which are digested within a short time without leaving readily detectable residues, are not found. But it is open to question whether *P. filamentosus* selectively collects these organisms from the surface of *Ceratophyllum* leaves.

b) Ecological alternatives – migrations and alternative food: The density of *P. filamentosus* was significantly higher in the macrophyte covered lit-

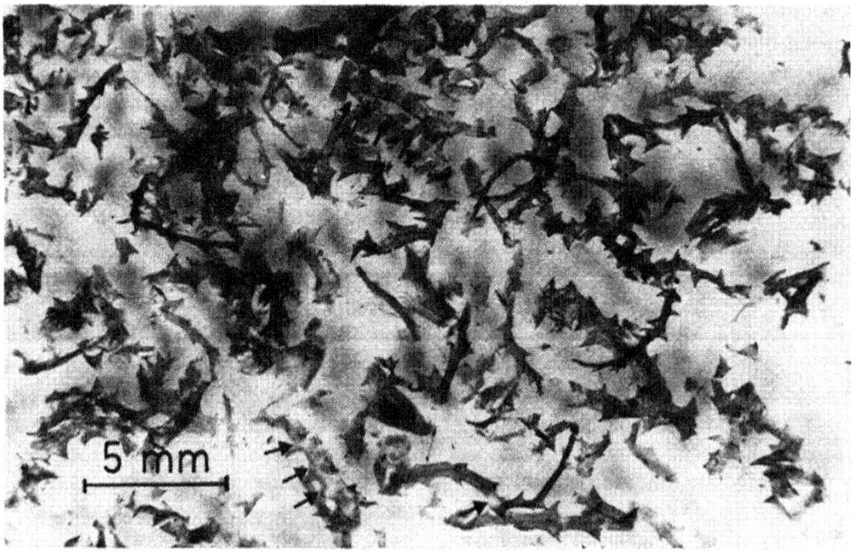

Fig. 4. Condition of *Ceratophyllum* in the faeces of *P. filamentosus*. The arrows indicate the action of pharyngial teeth.

toral than elsewhere in the lake (Schiemer & Hofer 1983), and also the relative gut content of these specimens was twice that found in *P. filamentosus* at other places in the lake. It is therefore surprising that with this macrophyte diet the fish can cover only little more than half of the respiratory costs. On the other hand, this species lives in schools and migrates across the whole lake. Therefore the time spent in the macrophyte zone may be limited (perhaps to a few days) and a mixed diet (plant and animals), which guarantees a higher assimilation rate, may be consumed (Table 1) outside the closed *Ceratophyllum* crops.

In conclusion it can be said that in spite of all inaccuracies involved in estimating bioenergetic parameters in the field it is obvious that *P. filamentosus* is not able to cover its energetic demands solely by feeding on macrophytes. This has already been suggested for some other cyprinids, e.g. for grass carp (Kraupauer 1967; Fischer & Lyakhnovich 1973) or roach (Hofer & Niederholzer 1980). In contrast to other cyprinid species (Hofer, unpublished results and Prejs 1976), *P. filamentosus* can only grind macrophytes with its pharyngeal teeth to a very small extent, which results in a low assimilation efficiency. Furthermore, the digestive system of *P. filamentosus* is unable to attack crude fiber (Table 3). With a 20% utilization of this matter, as was found in *S. mossambicus*, *P. filamentosus* would gain about 280 J per day.

In the light of the results concerning the energetics of *P. filamentosus*, the high population densities of this species in the area of Parakrama Samudra covered with macrophytes and the feeding on macrophytic material is surprising. *P. filamentosus*, like many other omnivorous cyprinids, accepts all kinds of food and appears to prefer dominating food sources, like *Ceratophyllum* in dense standing crops. *P. filamentosus* is essentially a substrate feeder, and often animal, bacterial and algal 'Aufwuchs' has a greater nutritive value than the substrate itself. The high utilization of protein and low utilization of carbohydrates confirm this assumption.

3.2. *Sarotherodon mossambicus*

3.2.1. Distribution pattern. The size class of fish (19.1 g body weight) obtained with the 20 mm gill nets occur over the whole lake area. During daytime the highest catches were obtained in the offshore areas. Inshore migrations appear to occur during the night. A distinct diurnal shift was observed in the vertical distribution pattern in the water column: during daytime more animals are found near the bottom, during nighttime more in the surface stratum. For details of the distribution pattern see Schiemer & Hofer (1983).

3.2.2. Feeding habits and food consumption. S. mossambicus feeds by sucking up the upper layer of sediment which contains a high concentration of sedimented phytoplankton. A great part of the sand ingested together with the organic matter is ejected through the opercular openings and the mouth (Bowen 1979). Additionally, like other cichlids, *S. mossambicus* is also able to filter phytoplankton suspended in the water (Moriarty 1973; Fish 1951). Both feeding habits occur in PS as judged by the vertical distribution pattern of the species in the water column. The results of a microscopic analysis of the food of *S. mossambicus* in PS are discussed by Dokulil (1983).

The rather constant mean values of gut filling indicate that the feeding activity continues throughout the 24-h cycle (Fig. 2). The average gut content was 13.0 ± 3.8% of the body weight. Because of the predominant occurrence of *S. mossambicus* in the open lake, its lower population densities and aggregated distribution pattern (see Schiemer & Hofer 1983), the field method for determination of gut passage time and food consumption which was applied for *P. filamentosus* could not be used.

Therefore, the values of the relative gut content of *S. mossambicus* in PS had to be combined with published data on the gut passage time in *Tilapia nilotica* (10 h; Moriarty & Moriarty 1973) for calculations of food consumption rates. *T. nilotica* and *S. mossambicus* have similar feeding habits and their intestines are also similar in length (Al-Hussaini & Kholy 1954). Since the gut of juvenile *S. mossambicus* is constantly full, the daily relative consumption rate was calculated by multiplying the mean relative gut content (13.0 ± 3.8%; Fig. 2) by the number of daily gut fillings (2.4). During a 24-h period *S. mossambicus* consumes 31.2% fresh matter and 4.29% dry matter of its body weight (dry weight of gut content: 13.75% of fresh weight).

3.2.3. pH in the stomach. The pH of the mixed stomach content of *S. mossambicus,* measured in the early morning and evening, was found to be very constant (1.99 ± 0.37) which is in accordance with the continuous feeding observed. This value corresponds with available data on *S. mossambicus* and other cichlid species (*Tilapia nilotica:* 1.4–2.0, Moriarty (1973); *Tilapia rendalli;* 2–4 and *Tilapia guineensis:* 1–4, Payne (1978) and *S. mossambicus:* 1.25–1.5, Bowen (1976)). Moriarty (1973) found little digestion in *T. nilotica* when algal cells were subjected to pH values above 1.8. Low values of about 1.4 were found only at the bottom of the stomach.

3.2.4. Chemical analysis of the gut content. The results of the chemical analysis of the gut contents in different sections of the intestinal tract are presented in Table 2. As in *P. filamentosus* the ash fraction is high. A considerable fraction (19%) of the phytoplankton-seaiment samples remained unaccounted for, i.e. was not detected in the analysis. These losses of organic matter could partly be explained by the 'unaccounted-for fraction' of detrital samples, which are neither protein, carbohydrate nor lipid (Bowen 1979).

3.2.5. Assimilation efficiency. S. mossambicus feeds both on phytoplankton and the uppermost sediment layers and thus it is impossible to collect the specific food for chemical analysis. Therefore ingested phytoplankton and sediment in the stomach were taken for calculations. Principally, secretions of organic matter (enzymes, mucus) into the stomach by the fish itself should be considered. Little or no digestion occurs in the stomach of

Table 4. Assimilation efficiencies in herbivorous and detritivorous cichlids, taken from the literature.

Species	Food		U^{-1} in %	Author
Tilapia	Microcystis		70	
nilotica	Anabaena		75	Moriarty &
	Nitzschia	carbon	79	Moriarty
	Chlorella		49	(1973)
	lake phytoplankton		43	
Tilapia	Najas guadalupensis	total	29	
zillii		non-cellulose		Buddington
		organic matter	56	(1979)
		protein	75	
Sarotherodon	plant	total	35–39	Mironowa
mossambicus	algae	total	28–52	(1974)
Sarotherodon	Chlorella	protein	47	Kirilenko
mossambicus	blue green algae	protein	61	et al. (1975)
Sarotherodon	detritus	organic matter	32–43	
mossambicus		energy	39–45	
		protein	44–48	Bowen (1979)
		soluble		
		carbohydrates	33–38	
		crude fiber	30–42	
Sarotherodon	detritus	protein-amino	72	
mossambicus		acids		Bowen (1980)
		non protein		
		amino acids	64	

S. mossambicus (Fish 1951; Bowen 1976; Hofer, unpublished), but during algae filtration mucus is transported into the stomach (Fish 1951). Therefore, a somewhat increased organic proportion in the content is also lost with the faeces.

The calculation of assimilation efficiency of *S. mossambicus* is based on the ash content. In contrast to *P. filamentosus* this species can utilize crude fiber. A similar result was obtained by Bowen (1979), who found in detritivorous *S. mossambicus* an assimilation efficiency of between 30 to 40% for crude fiber (see Table 4). Nevertheless, the low amount of crude fiber in the diet of *S. mossambicus* (3.6% of dry weight) renders its utilization less important. Generally, assimilation efficiencies found in *S. mossambicus* of PS correspond with the results of herbivorous and detritivorous *S. mossambicus* and other cichlids in the literature (Table 4).

3.2.6. Energy budget. In juvenile *S. mossambicus* of about 19.1 g body weight, daily food consumption amounts to about 0.82 g dry matter of phytoplankton and mineral sediment, giving an energetic value of 7 951 J (Table 5). 37.2% or 2 958 J are assimilated. For estimating respiratory costs, data of Mironowa (1976) were used. Daily oxygen consumption of *S. mossambicus* (body weight about 19 g) reaches 6.785 mg O_2 per g body weight and 129.59 mg O_2 for the total fish. Using an oxygen equivalent of 13.585 J per mg O_2 (Brett & Groves 1979), this results in total daily respiratory costs of 1 757 J. For *S. mossambicus* of the same size Caulton (1978) found a somewhat lower daily respiratory rate of 1 530 J. Comparing respiratory experiments of other tropical species (Brett & Groves 1979) and considering that oxygen consumption of starving fish in the respiratory chamber is significantly lower than in fish feeding under natural conditions, the higher respiratory values of Mironova (1976) were preferred. This results in a daily production of 1 201 J or 0.268 g fresh weight = 1.4% of body weight (1 g fresh weight of *S. mossambicus* = 4 479 J; Meakins 1976) which coincides exactly with production rates of *Tilapia aurea* (Stanley & Jones 1976): in specimens weighing 23.8 g with a daily relative feeding rate of 2.84% dry matter of *Spirulina platensis*, a daily relative production of 1.44% was found. *Tubifex*-feeding *S. mossambicus* of about the same size showed a relative daily growth rate of 1.35% (Mironova 1976). In both cases the production rate was measured and not calculated. Our analyses indicate good growth conditions for *S. mossambicus* in PS, at least for the size class under consideration.

Table 5. Roughly calculated energy budget of *S. mossambicus* and *P. filamentosus* in Parakrama Samudra.

	Sarotherodon mossambicus	*Puntius filamentosus*
Diet	Phytoplankton and sediment	*Ceratophyllum*
Wet weight of fish in g	19.1	31.8
Daily consumption in g wet weight	5.96	6.84
Daily consumption in g dry weight	0.82	0.54
Daily consumption in J	7 951	6 671
Daily consumption of non-crude fiber organic matter in J	–	5 265
Assimilation efficiency of organic matter in per cent of weight	37.2	–
Assimilation efficiency of non-crude fiber organic matter in per cent of calorific value	–	24.6
Daily assimilation in J	2 958	1 295
Daily respiratory costs in J data from Mironowa (1976)	1 757	2 506
Daily energy balance in J	+1 201	–1 211
Daily production rate in per cent of body weight	1.4	–

Acknowledgements

This investigation was carried out within the framework of the Parakrama Samudra Limnology Project, sponsored by the 'Österreichischen Bundeskanzleramt', the UNESCO, the 'Österreichischen Bundesministerium für Wissenschaft und Forschung' and by the 'Fonds zur Förderung der wissenschaftlichen Forschung in Österreich', project no. 3307/1. We acknowledge the critical review of the manuscript by S. De Silva & Z. Fischer.

References

Al-Hussaini, A. H. & Kholy, A., 1954. On the functional morphology of the alimentary tract of some omnivorous teleost fish. Proc. Egypt. Acad. Scienc. 12: 17–39.

Bowen, S. H., 1976. Mechanism for digestion of detrital bacteria by the cichlid fish *Sarotherodon mossambicus* (Peters). Nature 260, 5547: 137–138.

Bowen, S. H., 1979. A nutritional constraint in detritivory by fishes: The stunted population of *Sarotherodon mossambicus* in Lake Sibaya, South Africa. Ecological Monographs 49: 17–31.

Bowen, S. H., 1980. Detrital non protein amino acids are the key to rapid growth of Tilapia in Lake Valencia, Venezuela. Science 207: 1216–1218.

Brett, J. R. & Groves, T. D. D., 1979. Physiological energetics. In: Hoar, W. S., Randall, D. J. & Brett, J. R. (eds.) Fish Physiology VIII, pp. 279–352. Academic Press, New York, San Francisco, London.

Buddington, R. K., 1979. Digestion of an aquatic macrophyte by *Tilapia zillii* (Gervais). J. Fish Biol. 15: 449–455.

Caulton, M. S., 1978. The effect of temperature and mass on routine metabolism in *Sarotherodon (Tilapia) mossambicus* (Peters). J. Fish Biol. 13: 195–201.

Conover, R. J., 1966. Assimilation of organic matter by zooplankton. Limnol. Oceanogr. 11: 338–345.

Dokulil, M., 1983. Aspects of gut passage of algal cells in *Sarotherodon mossambicus* (Piscus, Cichlidae). In: Schiemer, F. (ed.) Limnology of Parakrama Samudra – Sri Lanka: a case study of an ancient man-made lake in the tropics. Developments in Hydrobiology (this volume). Dr W. Junk, The Hague.

Dowgiallo, A., 1975. Chemical composition of an animal's body and its food. In: Grodzinski, W., Klekowski, R. Z. & Duncan, A. (eds.) Methods for Ecological Bioenergetics. IBP Handbook No. 24: 160–199.

Dubois, M., Gilles, K. A., Hamilton, J. K., Rebers, P. A. & Smith, F., 1956. Colorimetric method for determination of sugars and related substances. Analyt. Chem. 28: 350–356.

Fischer, Z., 1979. Selected problems of fish bioenergetics. Proc. World Symp. Finfish Nutr. Fishfeed Technol. 1: 17–44.

Fischer, Z. & Lyakhnovich, V. P., 1973. Biology and bioenergetics of grass carp (*Ctenopharyngodon idella* Val.). Pol. Arch. Hydrobiol. 20: 521–557.

Fish, G. R., 1951. Digestion in *Tilapia esculenta.* Nature 167: 900–901.

Hofer, R. & Niederholzer, R., 1980. The feeding of roach (*Rutilus rutilus* L.) and rudd (*Scardinius erythrophthalmus* L.). II. Feeding experiments in the laboratory. Ekol. pol. 28: 61–70.

Hofer, R. & Schiemer, F., 1981. Proteolytic activity in the digestive tract of several species of fish with different feeding habits. Oecologia (Berl) 48: 342–345.

Hofer, R., Forstner, H. & Rettenwander, R., 1982. Duration of gut passage and its dependence on temperature and food consumption in roach *(Rutilus rutilus):* Laboratory and field experiments. J. Fish Biol. (in press).

Johnson, M. G. & Brinkhurst, R. O., 1971. Production of benthic macroinvertebrates of Bay of Quinte and Lake Ontario. J. Fish. Res. Board Can. 28: 1699–1714.

Kirilenko, N. S., Mel'nikov, G. B., Grinberg, L. R. & Nazimirova, N. I., 1975. The digestibility of the crude protein and neutral fat of some microalgae by *Tilapia mossambica* Peters. J. Ichthyol. 15: 151–155.

Kraupauer, V., 1967. Food selection of two years old grass carp. Bul. Vyzk. Ust. Ryb. Vodnany 3: 7–17.

Lowry, O. H., Rosebrough, N. J., Farr, A. L. & Randall, R. J., 1951. Protein measurement with the Folin Phenol reagent. J. biol. Chem. 193: 265.

Meakins, R. H., 1976. Variations in the energy content of freshwater fish. J. Fish Biol. 8: 221–224.

Mironova, N. V., 1976. Changes in the energy balance of *Tilapia mossambica* in relation to temperature and ration size. J. Ichthyol. 16: 120–129.

Moriarty, D. J. W., 1973. The physiology of digestion of blue-green algae in cichlid fish, *Tilapia nilotica.* J. Zool. Lond. 171: 25–39.

Moriarty, D. J. W. & Moriarty, C. M., 1973. The assimilation of carbon from phytoplankton by two herbivorous fishes: *Tilapia nilotica* and *Haplochromis nigripinnis.* J. Zool. Lond. 171: 41–55.

Niederholzer, R. & Hofer, R., 1979. The adaptation of digestive enzymes to temperature, season and diet in roach *Rutilus rutilus* L. and rudd *Scardinius erythrophthalmus* L.: Cellulase. J. Fish Biol. 15: 411–416.

Pandian, T. J. & Raghuraman, R., 1972. Effects of feeding rate on conversion efficiency and chemical composition of the fish *Tilapia mossambica.* Marine Biol. 12: 129–136.

Payne, A. I., 1978. Gut pH and digestive strategies in estuarine grey mullet (Mugilidae) and tilapia (Cichlidae). J. Fish Biol. 13: 627–629.

Prejs, A., 1976. The role of fish in the detritus formation. In: Pieczynska, E. (ed.) Selected problems of lake littoral ecology, pp. 173–179. Warszawa.

Prejs, A. & Blaszczyk, M., 1977. Relationship between food and cellulase activity in freshwater fishes. J. Fish Biol. 11: 447–452.

Prus, T., 1975. Measurements of calorific value using Phillipson microbomb calorimeter. In: Grodzinski, W., Klekowski, R. Z. & Duncan, A. (eds.) Methods for Ecological Bioenergetics, IBP Handbook No. 24: 149–160.

Schiemer, F., 1981. Parakrama Samudra (Sri Lanka) Project, a study of a tropical lake ecosystem I. An interim review. Verh. Internat. Verein Limnol. 21: 993–999.

Schiemer, F. & Hofer, R., 1983. A contribution to the ecology of the fish fauna of the Parakrama Samudra reservoir. In: Schiemer, F. (ed.) Limnology of Parakrama Samudra – Sri Lanka: A case study of an ancient man-made lake in the tropics). Developments in Hydrobiology (this volume). Dr W. Junk, The Hague.

Stanley, J. G. & Jones, J. B., 1976. Feeding algae to fish. Aquaculture 7: 219–223.

Author's addresses:
R. Hofer
Institute of Zoophysiology
University of Innsbruck
Peter Maerstr. 1a
A-6020 Innsbruck
Austria

F. Schiemer
Institute of Zoology
University of Vienna
Althanstr. 14
A-1090 Vienna
Austria

14. Aspects of gut passage of algal cells in *Sarotherodon mossambicus* (Peters), (Pisces, cichlide)

M. Dokulil

Keywords: cichlids, *Sarotherodon*, food selectivity, digestion

Abstract

Microscopic examination of algal material from the gut contents of *Sarotherodon mossambicus* reveals live algal cells unaffected by passage through the intestine. The degree of digestion of the cells and filaments taken from different parts of the intestine is assessed by measuring pigment degradation. Experimental evidence for the viability of the algal material from the hind gut is demonstrated by photosynthethic oxygen evolution. Quantification of the relative abundance of algal species in the water and the stomach contents suggests selective feeding.

1. Introduction

Since its introduction into Sri Lanka in 1952 the cichlid fish *Sarotherodon mossambicus* has become one of the dominant fish in Parakrama Samudra (Fernando & Indrasena 1969). This species feeds as a herbivore and detritivore at the base of the food chain in tropical and subtropical freshwaters (Lowe-McConnell 1975). Because of their potential importance as a basic protein resource the tilapias have attracted considerable attention. Food and feeding habits of *Tilapia nilotica* have been studied in detail in Lake George, Uganda, by Moriarty & Moriarty (1973a, b) and Moriarty et al. (1973). Fish (1951, 1955) concluded from his observation of undigested cells of blue-green algae in the rectum that these algae, the dominant elements of the phytoplankton, could not be digested. Later, Moriarty (1973) demonstrated that blue-green algae are lysed by acid in a diurnal cycle related to feeding. Thus digestion may vary from zero to maximum each day.

In the present study the viability of undigested algal material from the rectum of the herbivorous fish *Sarotherodon mossambicus* is demonstrated by visual inspection, pigment estimation and photosynthetic experiments. The food selection of the species is described and explained.

2. Materials and methods

Sarotherodon mossambicus, size class 8–11 cm, was caught by gill net in the northern basin of Parakrama Samudra (PSN). The fish were killed immediately, brought to the laboratory and dissected within half an hour of capture.

Samples for pigment estimation were taken from the stomach, the first and second half of the intestine and the rectum. The fresh weight of the material taken and the length of the intestine were recorded. Pigments were extracted from the gut contents by homogenizing the debris in cold methanol. Concentrations of chlorophyll-a and degradation products were measured spectrophotometrically, according to the procedure of Holm-Hansen & Riemann (1978). Separate samples from the stomach and the hind-gut were fixed with formalin for later microscopical evaluation. The fresh intestinal contents were examined several times under the microscope for living algal material.

Schiemer, F. (ed.), Limnology of Parakrama Samudra – Sri Lanka
 ISBN 90 6193 763 9

To demonstrate the viability of the algal material from the rectum of *S. mossambicus* six photosynthetic experiments were performed. Depending on the size of the fish four to ten individuals were dissected for each experiment and the contents of the last 4–5 cm of the intestine used. The material, about 500 mg fresh weight per test, was suspended in two liters of lake water previously filtered through glass fibre (Whatman GF/C) and Millipore filters (pore size 0.2 μm). The flask containing the stock preparation was thoroughly shaken to give an almost homogeneous suspension. Subsamples were than siphoned off into Winkler-bottles of 120 ml capacity. All glass-ware was carefully cleaned and sterilized before each run.

Photosynthetic activity was estimated from oxygen changes in light and dark bottles using a modified back-titration technique with amperometric end-point detection (Talling 1973). All experiments were performed in triplicate in a container submerged to about 10 cm depth in the lake. Water temperature (around 30 °C) was measured at the beginning and at the end of each incubation period, 3 °C being the greatest temperature difference noted. Incoming radiation was continuously recorded. Triplicate measurements of the chlorophyll-a concentration of the stock preparation served as an indicator of the amount of viable biomass.

Food selection of this herbivorous species was estimated by comparing counts of several algal species in the content of stomach and intestine of nine individuals with pooled counts from eight water samples equally spaced over the entire water column (water depth 4 m) and taken at the same time and place. For each algal species the coefficient of electivity (Ivlev 1961) as modified by Jacobs (1974) was calculated:

$$D = \frac{r - p}{r + p - 2\,rp},$$

where r is the relative abundance of the species in the gut contents, and p is the relative abundance of that species in the environment.

3. Results

Inspection of the gut contents under the microscope revealed that various algal species were healthy and alive in the rectum of *S. mossambicus*. Colonies of blue-green and green algae appeared undamaged and bright green. Species such as *Euglena* or *Peridinium* were seen to be mobile and some of the filamentuous blue-green algae exhibited typical gliding movements. Upon treatment with hypertonic solutions many algal cells plasmolyzed, thus indicating that they were physiologically intact. In contrast, almost all diatom frustules and most cells of *Mougeotia* contained no cytoplasm. The chains of *Melosira* were usually broken into short fragments not longer than three cells.

Analysis of photosynthetic pigments made clear that the relative proportion of chlorophyll-a to phaeopigment varies in different parts of the intestine (Fig. 1). More than 80% of the samples from the stomach had a chlorophyll content of less than 50%, which means that at least half of the pigments were degraded to phaeopigment. In the hind-gut the picture was reversed, giving relative chlorophyll-a concentrations of between 50 and 100% in 80% of the cases. Figures in the first and second half of the intestine were somewhere inbetween. These data suggest that a considerable part of the algal material passes through the intestine undamaged.

The physiological viability of this algal material was demonstrated by photosynthetic experiments (Fig. 2). Results are expressed as gross-oxygen release per unit biomass (mg O_2 mg Chl-a^{-1} h^{-1}) and as efficiency, to allow for the variable light conditions during in situ incubations (mg O_2 mg Chl-a^{-1} J^{-1} cm^2). Positive photosynthesis was observed in all six experiments, but recovery was significantly better in the afternoon. Except for the first experiment in the afternoon specific rates per unit biomass were 6.4, 6.0, and 9.7 mg O_2 mg Chl-a^{-1} h^{-1} compared to 3.4 and 2.4 in the morning. Results expressed as photosynthesis per unit chlorophyll-a per unit light do not alter the general pattern.

Specific rates are on an average about one tenth in the morning and more than one third in the afternoon of the mean value obtained with natural phytoplankton (Dokulil et al. 1983).

A considerable difference was found between the frequency distribution of cell counts from the lake water and from the stomach contents of *Sarotherodon mossambicus* (Table 1). Several species, e.g. *Melosira* and *Microcystis*, are relatively more abundant in the stomach, whereas others, like *Mo-*

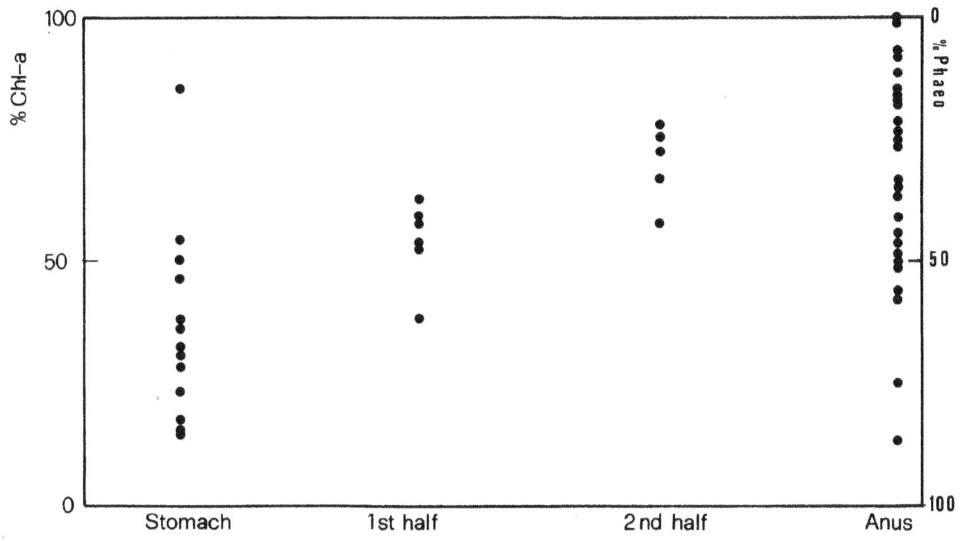

Fig. 1. Relative concentration of chlorophyll-a and phaeopigment (%) in the stomach, first and second half of the gut and near the anus of *S. mossambicus*.

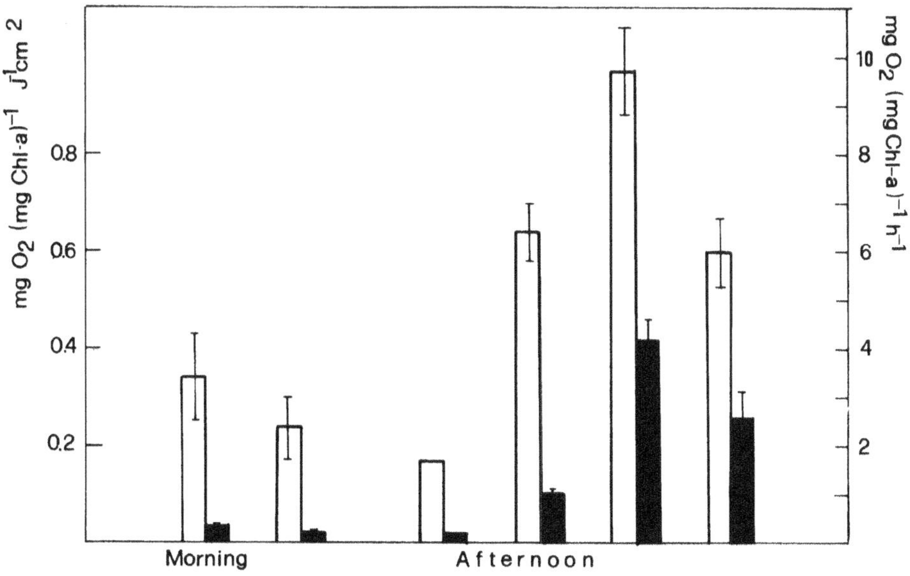

Fig. 2. Specific gross-photosynthetic rate (mg O_2 Chl-a^{-1} h^{-1}) and photosynthetic efficiency (mg O_2 mg Chl-a^{-1} J^{-1} cm^2) of algal material from the hind gut of *S. mossambicus*.

Table 1. Relative abundance of algal species in the lake waters (rel. No.) and modified Ivlev-index (D) according to Jacobs (1974) for the stomach contents ($n = 9$) of *Sarotherodon mossambicus* (size class 8.3–10.5 cm). Mean dimensions, length and width, in μm for colonies, filaments or single cells.

	Rel. No.	D	l	w
Microcystis spp.	0.001	+0.95	40	35
Melosira granulata	0.003	+0.79	60	10
Staurastrum sp.	0.005	+0.38	26	24
Lyngbya circumcreta	0.016	+0.35	35	20
Mougetia sp.	0.032	+0.33	100	4
Synedra acus	0.007	+0.27	45	3
Anabaenopsis raciborskii	0.250	+0.23	75	3
Scenedesmus spp.	0.016	-0.03	24	12
Pediastrum simplex	0.004	-0.09	70	15
Merismopedia spp.	0.177	-0.10	55	10
Peridinium inconspicuum	0.007	-0.35	24	18
Monoraphidium irregulare	0.437	-0.92	20	1

noraphidium, are certainly underrepresented. These observations suggest that the 'gulping' feeding action of the herbivore effects a certain food selection. For each algal species the coefficient of electivity was therefore calculated. The results (Table 1) indicate a high positive selection of *Microcystis, Melosira, Lyngbya,* and two green algae.

Other blue-green and green algal species were neither definitely positively nor negatively selected. The only species with a strictly negative electivity is *Monoraphidium irregulare.*

Selection seems to be based upon size, and perhaps the shape, of the food particles, because measurements of length and width (Table 1) of the algal cells or colonies explain 52% of the variability of the electivity coefficients.

4. Discussion

The condition of the algal material passing through the intestine of *Sarotherodon mossambicus* as reported in the present study closely resembles observations by Fish (1951) on algal digestion in *Tilapia esculenta.* Blue-green and green algae appear to be largely unaffected whereas the protoplasm of the diatoms is completely digested and filaments of *Melosira* are broken into shorter fragments. Algae found in the gut are almost exclusively phytoplankton species, suggesting that food is taken by filtering in the free water or by sucking the upper sediment layer containing phytoplankton, as proposed by Hofer & Schiemer (1983).

The proportion of chlorophyll-a to phaeopigment found in the hind gut near the anus is similar to the ratios reported by Moriarty (1973) for green and green/brown faeces of *Tilapia nilotica.* Brown faeces containing 100% phaeopigment were never observed, because the lowest pH-value recorded by Hofer & Schiemer (1983) was 1.6. For complete lysis of blue-green algae a pH of 1.4 is necessary (Moriarty, 1973).

The 24-h cyclic change of pH, was modest, 1.99 ± 0.37, because of almost continuous feeding activity (Hofer & Schiemer 1983). The variation in the pigment ratio found along the intestinal tract (Fig. 1) reflects the gut passage time of about 13 h (Hofer & Schiemer 1983) and the variability of the pH in the stomach which was frequently above 1.8, a value designated by Moriarty (1973) as the limit for significant lysis of algal cells.

The incomplete digestion of the algal material means that a high proportion of cells passes undamaged through the gut and may resume metabolic activity. The viability of the algal cells from the rectum of *S. mossambicus* was demonstrated by their potential photosynthesis (Fig. 2); this seems to be the first time that physiological methods have been used for this purpose.

The significantly higher photosynthetic rates in the afternoon are indicative of rapid passage through the stomach or of higher pH values at the time of food intake and hence of relatively small cell lysis. The gut passage time suggests that feeding activity and therefore acid secretion must have been lower during the previous night. The apparently lower specific photosynthetic rate in the first afternoon run might reflect the shorter incubation time of 2 h as compared to 4 h in all other experiments, and perhaps indicates the involvement of a time factor in the recovery process. The low rates measured during the morning must be a result of severe stress on algal cells due to acidification during gut passage on the previous late afternoon.

It remains unsettled whether a significant amount of algal material is thus fully recovered and recycled to the natural algal population. However, a large proportion of the feaces, which usually sediment was found floating at the lake surface. This was caused by gas bubbles trapped in the mucuous envelope of the faeces and possibly generated by photosynthesis.

The marked positive selection of the colonial blue-green algae *Microcystis*, the colonial diatom *Melosira*, and of the filamentuous blue-green algae *Lyngbya* confirms the results of Moriarty et al. (1973), although the negative selection which they reported for the blue-green alga *Anabaenopsis* has not been observed in the present study. Instead, the green alga *Monoraphidium* had a highly negative coefficient of electivity. These results, and observations by Fish (1960) on the digestion of *Anabaenopsis*, suggest that the two species studied, *S. mossambicus* and *T. nilotica* differ as to food selection. However, data on negative selection should be interpreted with the utmost caution, especially when based on the microscopic examination of stomach or gut contents. Rapid lysis of small cells of certain algal species may bias the result of such observations. The apparent positive selection for the blue-green algae may be due to better preservation and the ready digestion of *Monoraphidium* may erroneously suggest negative selection. Feeding of *Sarotherodon* on sedimented algal material, as indicated by the ash-content in the stomach (Hofer & Schiemer 1983), may also affect the Ivlev-index.

For juvenile *S. mossambicus* (19 g fresh weight) daily consumption was estimated at 6 g fresh material (820 mg dry substance) by Hofer & Schiemer (1983). The average relative pigment content in the stomach amounts to 0.14%. The daily food consumption is thus calculated to be equivalent to 8.4 mg chlorophyll-a. Using data on net-chlorophyll-a increase per day from Table 6 of Dokulil et al. (1983) the productivity of 1–2 m^3 of lake water have a carrying capacity for one individual.

The biomass of *S. mossambicus* per square meter of the open lake should therefore range from 38–76 g fresh weight (380–760 kg ha^{-1} yr^{-1}). The actual fish catch for the year 1975 of 597 kg ha^{-1} yr^{-1} (Wijeyaratne & Costa 1981) is within this range. Using a relationship between the morphoedaphic index, based on conductivity and mean depth, and the annual fish catch Wijeyaratne & Costa (1981) estimated 392 kg ha^{-1} yr^{-1}.

Allmazan & Boyd (1978) developed curvilinear regressions for predicting net-production of *Tilapia* from different measurements of phytoplankton abundance. Applying data on average secchi-depth, chlorophyll-a and primary productivity in Parakrama Samudra (Dokulil et al. 1983) to these equations gave 254, 300 and 620 kg ha^{-1} yr^{-1} respectively.

All these estimates of potential fish yield suggest a range of 300–600 kg ha^{-1} yr^{-1} of fish production, in agreement with the actual fish catch.

Acknowledgements

I wish to thank Dr. R. Hofer for his cooperation during the field work and for his critical reading of the manuscript, Dr. H. Winkler for computations and helpful comments and to J. Wieser for correcting the English text.

References

Almazan, G. & Boyd, C. E., 1978. Plankton production and *Tilapia* yield in ponds. Aquaculture 15: 75–77.

Dokulil, M., Bauer, K. & Silva, I., 1983. An assessment of the phytoplankton biomass and primary productivity of Parakrama Samudra, a shallow man-made lake in Sri Lanka. In: Schiemer, F. (ed.) Limnology of Parakrama Samudra – Sri Lanka: a case study of an ancient man-made lake in the tropics. Developments in Hydrobiology (this volume). Dr W. Junk, The Hague.

Fernando, C. H., 1973. Man-made lakes of Cylon: A biological resource. In: Ackermann, W. C., White, G. F. & Worthington, E. B. (eds.) Man-made lakes: Their problems and environmental effects. Geophysical Monograph Series 17: 664–671.

Fernando, C. H. & Indrasena, H. H. A., 1969. The freshwater fisheries of Ceylon. Bull. Fish. Stn., Ceylon 20: 101–134.

Fish, G. R., 1951. Digestion in *Tilapia esculenta*. Nature 167: 900–901.

Fish, G. R., 1955. The food of *Tilapia* in East Africa. Uganda J. 19: 85–89.

Fish, G. R., 1960. The comparative activity of some digestive enzymes in the alimentary canal of *Tilapia* and *Perch*. Hydrobiologia 15: 161–178.

Hofer, R. & Schiemer, F., 1983. Feeding ecology, assimilation efficiencies and energetics of two herbivorous fish: *Sarotherodon (Tilapia) mossambicus* (Peters) and *Puntius filamentosus* (Cuv. et Val.) In: Schiemer, F. (ed.) Limnology of Parakrama Samudra – Sri Lanka: a case study of an ancient man-made lake in the tropics. Developments in Hydrobiology (this volume). Dr W. Junk, The Hague.

Holm-Hansen, O. & Riemann, B., 1978. Chlorophyll-a determination: improvement in methodology. Oikos 30: 438–447.

Ivlev, V. S., 1961. Experimental ecology of the feeding of fishes. Yale University Press.

Jacobs, J., 1974. Quantitative measurements of food selection. A modification of the forage ratio and Ivlev's electivity index. Oecologia (Berlin) 14: 413–417.

Lowe-McConnell, R. H., 1975. Fish communities in tropical freshwaters. Longman, New York, USA.

Moriarty, D. J. W., 1973. The physiology of digestion of blue-green algae in the cichlid fish, *Tilapia nilotica*, J. Zool., Lond. 171: 25–39.

Moriarty, C. M. & Moriarty, D. J. W., 1973a. Quantitative estimation of the daily ingestion by *Tilapia nilotica* and *Haplochromis nigripinnis* in Lake George, Uganda. J. Zool., Lond. 171: 15–23.

Moriarty, D. J. W. & Moriarty, C. M., 1973b. The assimilation of carbon from phytoplankton by two herbivorous fishes: *Tilapia nilotica* and *Haplochromis nigripinnis*. J. Zool., Lond. 171: 41–55.

Moriarty, D. J. W., Darlington, J. P. E. C., Dunn, I. G., Moriarty, C. M. & Tevlin, M. P., 1973. Feeding and grazing in Lake George, Uganda. Proc. R. Soc. Lond. B. 184: 299–319.

Talling, J. F., 1973. The application of some electrochemical methods to the measurements of photosynthesis and respiration in freshwaters. Freshwat. Biol. 3: 335–362.

Wijeyaratne, M. J. S. & Costa, H. H., 1981. Stocking rate estimations of *Tilapia mossambica* fingerlings for some inland reservoirs of Sri Lanka. Int. Revue ges. Hydrobiol. 66: 327–333.

Author's address:
M. D. Dokulil
Institute of Limnology
Austrian Academy of Science
Gaisberg 116
A-5310 Mondsee
Austria

15. Contributions to the functional anatomy of the feeding apparatus of five cyprinids of Parakrama Samudra (Sri Lanka)

P. Adamicka

Keywords: functional anatomy, feeding, Puntius, Rasbora

Abstract

The results presented concern the functional anatomy of the head of *Puntius filamentosus, P. chola, P. dorsalis, P. sarana,* and *Rasbora daniconius* from the Parakrama Samudra reservoir, Sri Lanka, with respect to feeding. These species represent a spectrum of interdependent forms and functions.

1. Introduction

Twenty-three species of fish have been encountered in the Parakrama Samudra Reservoir, most of which originate from the catchment area of the tributary rivers. Of the systematic groups represented, the cyprinids comprises the largest number of species, all of which show a remarkable degree of feeding opportunism (Schiemer & Hofer 1983). This chapter presents the results of the head anatomy of five common species in Parakrama Samudra with special consideration given to feeding adaptations.

2. Results

2.1. *Puntius filamentosus* (Cuv. et Val.)

Of the five species investigated *P. filamentosus* appears to have the most primitive feeding apparatus. The eyes are very large (although feeding occurs during the day (Hofer & Schiemer 1983)). The mouth is small and terminal; the adductor mandibulae muscle (A) is not fully differentiated into three parts. The premaxillary is of relatively low protrusibility. The tendons of $A_1\alpha$ and $A_1\beta$ are not crossed. The latter is very weak and, as usual in cyprinids, also arises from the lower jaw (articular). Suspensory and operculum are short, but the lev. arc. pal. muscle is strong enough to facilitate suction snapping. Surprisingly, $A\omega$ and the intermandibularis muscle (which are, respectively, usually and often missing in cyprinids) are both present, although weak.

The branchial spines are short and sturdy (10 on the first arch). The pharyngeal teeth are of the tearing-grinding type (Fig. 4a), antagonistic to the horny pharyngeal plate (on the processus basioccipitalis – Fig. 4b).

The structure of this feeding apparatus suggests a benthos feeder. In Parakrama Samudra the species exhibits a broad food spectrum (Hofer & Schiemer 1983), consisting of benthos, Aufwuchs, wind-driven material and macrophytes. It probably devours *Ceratophyllum,* for example, in such large quantities in order to obtain animal Aufwuchs-organisms. A comparable method of nutrition is seen in *Neoceratodus* which feeds on snails obtained by ingesting large amounts of macrophytes.

2.2. *Puntius chola* (H. B.)

The eyes of *P. chola* are smaller than those of *P. filamentosus.* The olfactory rosette is almost as well developed as that of *P. filamentosus.* The

Schiemer, F. (ed.), Limnology of Parakrama Samudra – Sri Lanka
 ISBN 90 6193 763 9

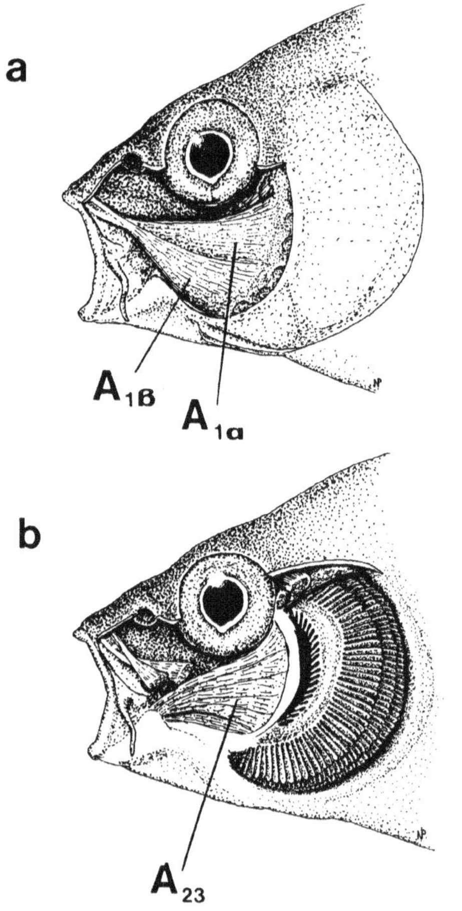

Fig. 1. Puntius chola. (a) superficial add. mand. muscle. (b) The band of the add. mand. to the lower jaw, and the branchial spinae. Levator arcus palatine, dilatator operculi, levator and adductor operculi removed.

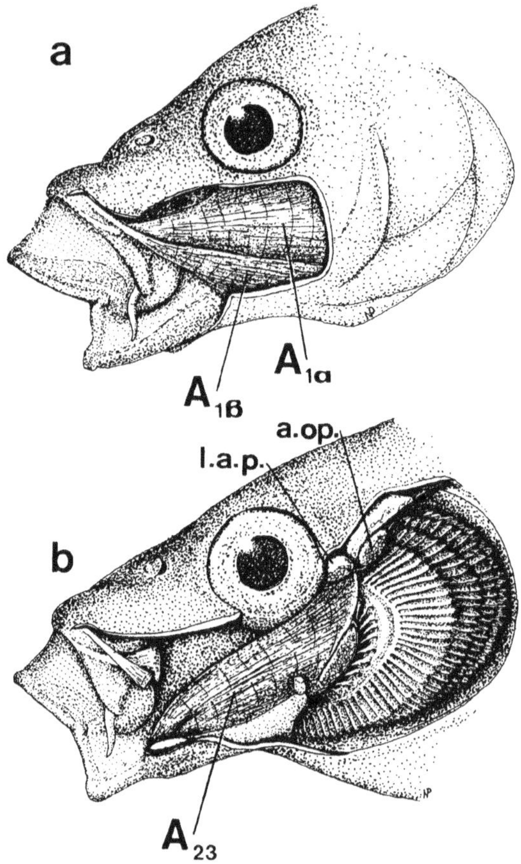

Fig. 2. Puntius dorsalis, as in Figure 1.

mouth is practically terminal, the premaxillary is fairly well protrusible (more so than in *P. filamentosus*). $A_1\alpha$ and $A_1\beta$ are not separate. A_{23} is in principle similar to that of *P. filamentosus*. Here, too, there is a distinct m. intermandibularis, but $A\omega$ is missing (Fig. 1a). The gill rakers or branchial spines on the first branchial arch are finger-like (about 16) and are longer than those of *P. filamentosus,* corresponding to the greater lengths of the dorsal and ventral portions of the arches and of the suspensory and operculum. This must result in better mobility of all these elements in suction snapping. The spines on the other arches are shorter but are also finger-like (Fig. 1b). The pharyngeal teeth (Fig. 5) are strongly curved, and the pharyngeal process bears a kind of grinding plate. These features point to a preference for benthic organisms, caught by means of powerful suction snapping. In P.S. the species feeds on benthos (chironomids, cyclopids, cladocerans) from soft mud habitats (Schiemer & Hofer 1983).

2.3. *Puntius dorsalis* (Jerdon)

As regards functional anatomy *P. dorsalis* is the most interesting of the species investigated. Its eyes

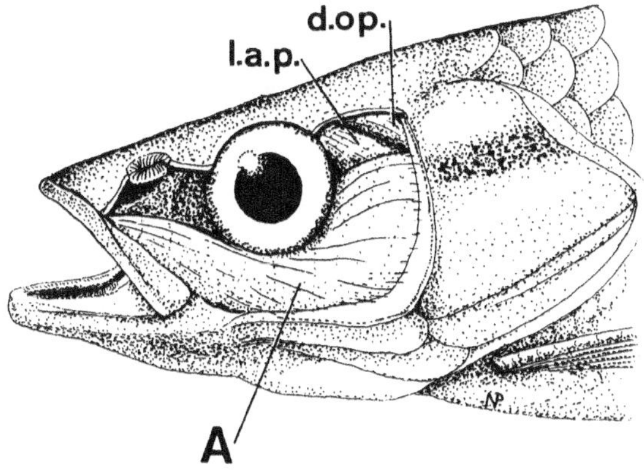

Fig. 3. Rasbora daniconius, head from the left side.

are large (nocturnal activity (Schiemer & Hofer 1983)), and the barbels are well developed. The most salient feature is the size of the mouth compared with its great protrusibility, in which, to the best of our knowledge, it excels all other cyprinids. The mouth is practically terminal and, when protruded, is tubular, pointing forward and slightly downwards.

As is usual, the ventral $A_1\beta$ is weaker than $A_1\alpha$, its broad tendon crossing the narrower one of the $A_1\alpha$ externally (concerning the functional significance of this type of build, cf. Alexander 1966; for cyprinids of similar build, cf. for example, Takahasi 1925; Matthes 1963). $A_1\beta$ is connected with A_{23} in the depth, rostrally it is to some extent pinnate (Fig. 2a). Dorsally, A_{23} extends far up but is rather weak.

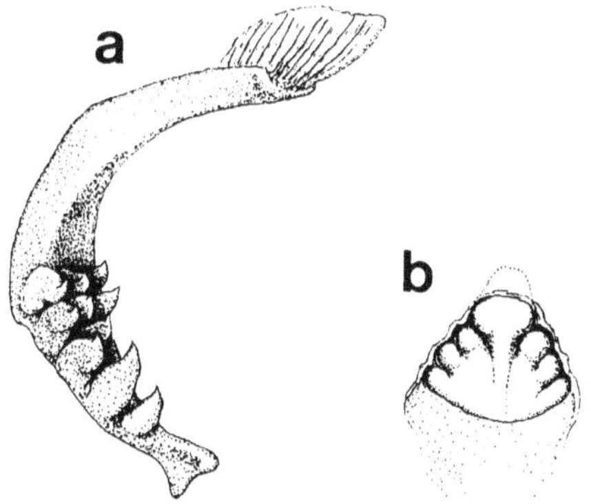

Fig. 4. Puntius filamentosus. (a) pharyngeal teeth. (b) The corresponding pharyngeal (or masticatory) plate. Orientation: top is caudal.

The rostral bone is exceptionally long and lies in a deep groove of the mesethmoid. $A\omega$ and the intermandibularis are lacking.

The gill rakers of the first arch are long, fingerlike, bent inwards at the tip, about 18 in number (somewhat denser than in the previous two species), those of the second arch are shorter, those of its outer row are forked at the tip, those of the inner row are medially denticulated; the same holds for the other arches. The pharyngeal teeth are powerful, for crushing and bruising. The dental formula is 4–5/3/2; the information on this point in De Silva et al. (1980, p. 65) is misleading (Fig. 6). The feeding apparatus of this species seems to be eminently suited to catching elusive, hard prey and is reminiscent of *Epibulus* (labrid of the Indian Ocean), a species that uses its highly protrusible and exceptionally large mouth for snapping swimming decapod crustaceans, i.e. an elusive prey with a light, protective armour. De Silva et al. (1980) give information concerning the nutrition of *P. dorsalis:* crustaceans constitute the larger part of their diet (at least in subadults), supplemented by large quantities of algae and macrophytes. In Parakrama Samudra, the species prefers benthic chironomids, cladocera and copepods (Schiemer & Hofer 1983).

2.4. *Puntius sarana* (H. B.)

This species is morphologically interesting. The mouth is almost terminal, the barbels are conspicuously long; the nose is well developed (ca. 56 lamellae), suggesting that the species seeks its food under conditions where visibility is poor. A_1 is barely divided; A_{23} extends as a broad muscle to the lower jaw. The whole m. add. mand. is fairly powerful. The limited and mainly downward protrusibility of the premaxillary, combined with the blunt form of the snout, indicate that the food organisms preferred cannot be capable of escape. The branchial spines are short and plump on the first branchial arch, 8 on the ceratobranchial, 4 small ones on the epibranchial, the tips of the larger ones are branched, resembling small trees. The pharyngeal teeth (Fig. 7) are arranged in three rows, of which only the innermost contains powerful bi- or tricuspid (bluntly rounded) teeth. On the left side of the fish inspected, the strongest of these teeth were worn down to two oval cyclindrical stumps.

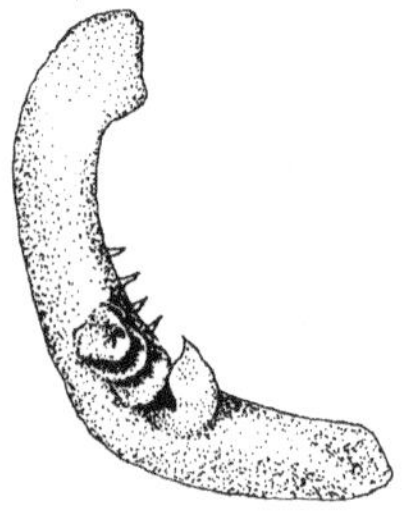

Fig. 5. Puntius chola, pharyngeal teeth.

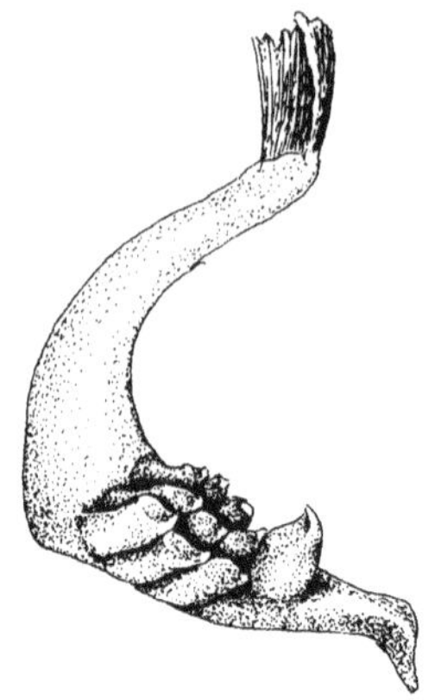

Fig. 6. Puntius dorsalis, pharyngeal teeth.

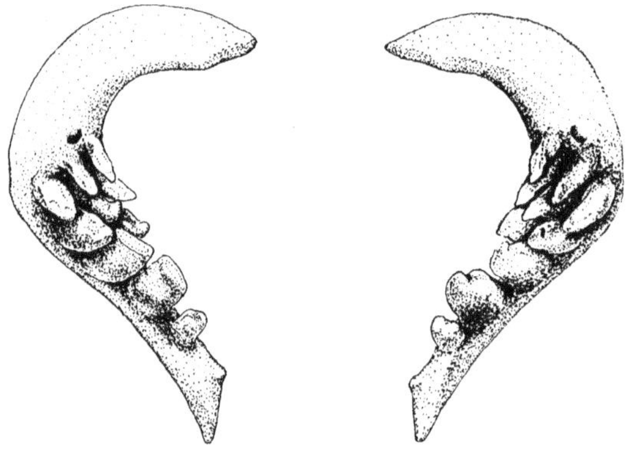

Fig. 7. Puntius sarana, pharyngeal teeth.

Fig. 8. Rasbora daniconius, pharyngeal teeth.

The pharyngeal dentition of cyprinids is hardly ever completely symmetrical, but we have not previously encountered such a degree of asymmetry (probably a purely individual phenomenon). It was matched by a marked asymmetry in the broad, triangular pharyngeal plate: two humps on the left (the surface of the worn-down teeth was correspondingly slightly concave), 3 or 4 grooves on the right. This individual clearly chewed with the left-hand side. The structure suggests that most probably the species feeds upon sessile benthic organisms that require through crushing and grinding; e.g. diatoms, green algae, insect larvae. Of all the species investigated here, *Puntius sarana* has the most powerful pharyngeal musculature.

2.5. Rasbora daniconius (H. B.)

This is a small, agile, predatory fish, mainly feeding upon insects and young fish (Schiemer & Hofer 1983). Its orientation is chiefly visual and thus the nose is considerably less conspicuous than, for example, in *Puntius sarana* (20 lamellae); the barbels are reduced. However, the pores of the sensory canals on the head are very large. The snout is long, and the mouth situated somewhat dorsally, thus indicating that *R. daniconius* lives near the surface. For this reason, Deraniyagala's (1952) statements regarding food (worms, gnat larvae, plant fragments) cannot be considered as characteristic. The maxillary apparatus is well developed, the premaxillary exhibiting a well-developed processus ascendens (on the other hand, the rostral bone is very small and largely replaced by ligamentous tissue). Retraction of the maxillary apparatus is as in many other snapping predatory fishes; and therefore the A_1 is weak and practically not subdivided. The curvature of the mouth cleft is also typical for predatory fish (Fig. 3). The horny edges of the jaws - smooth in the Puntius species investigated here (but not so in many other cyprinids) - have a toothed appearance in *Rasbora* due to the development of sharp-edged furrows (ca. 60 in each half of the jaws). Lev. arc. pal. and dil. op. are powerful (important in suction snapping). The first ceratobranchial bears 14 long, finger-like, outer, and an equal number of pointed, conical, inner spines; the second bears the same number, but shorter, and so on (cf. De Silva et al. 1980). The pharyngeal teeth are suited for tearing, are conical and arranged in two rows (6/4) and serve to lacerate the prey (Fig. 8). The broad pharyngeal plate, triangular in shape, is slightly concave and bears a large number of horny thorns for holding the prey. In the African genus *Barilius,* which greatly resembles *Rasbora* in its structure and behaviour, the horny ridges on the jaw edges are set more irregularly than the furrows of the latter (Matthes 1963). The pharyngeal teeth are very similar in these two genera.

3. Discussion

The anatomical features of an animal permit us to make inferences as to its feeding type. In five cyprinids of a Sri Lanka reservoir, Parakrama Samudra, conformity with known food items was fairly good, although the fish must often behave opportunistically with respect to the available food. Thus *P. filamentosus* can pass for a typical zoobenthos feeder, in spite of the fact that in Parakrama Samudra it devours large quantities of macrophytes which it can only partially digest. *Puntius chola* shows adaptations for snapping benthic insect larvae. *Puntius dorsalis* is characterized as a powerful suction snapper, while *P. sarana* feeds by crushing and grinding food items. Finally, *Rasbora daniconius* is unequivocally predacious, preying on insect larvae and small fish. Matthes (1963), working on African cyprinids, found a comparable feeding opportunism in *Barbus* species (*Barbus* and *Puntius* are very closely related and not clearly distinguishable).

Acknowledgements

Thanks are due to Dr. F. Schiemer and Dr. R. Hacker for the material, Dr. P. Newrkla for the drawings, and Mrs Joy Wieser for the translation.

References

Alexander, R. McN., 1966. The function and mechanisms of the protrusile upper jaws in two species of cyprinid fish. J. Zool. 148: 288-296.

Deraniyagala, P. E. P., 1952. A colored atlas of some vertebrates from Ceylon. Vol. 1: Fishes, Ceylon Natl. Mus. Publ. 196 pp.

De Silva, S. S., Cumaranatunga, P. R. T. & De Silva, C. D., 1980. Food, feeding ecology and morphological features associated with feeding of four co-occurring cyprinids (Pisces: Cyprinidae). Neth. J. Zool. 30: 54-73.

Matthes, H., 1963. A comparative study of the feeding mechanism of some African Cyprinidae (Pisces, Cypriniformes). Bijdr. Dierk. 33: 1-35.

Schiemer, F. & Hofer, R., 1983. A contribution to the ecology of the fish fauna of the Parakrama Samudra reservoir. In: Schiemer, F. (ed.) Limnology of Parakrama Samudra - Sri Lanka: (a case study of an ancient man-made lake in the tropics. Developments in Hydrobiology (this volume) Dr W. Junk, The Hague.

Takahasi, N., 1925. On the homology of the cranial muscles of the Cypriniform fishes. J. Morph. 40: 1-109.

Author's address:
P. Adamicka
Biological Station
A-3293 Lunz
Austria

16. Changes in body condition and proximate composition with maturation in *Puntius sarana* and *Sarotherodon mossambicus*

S. S. De Silva, C. D. De Silva & W. M. K. Perera

Keywords: condition factor, gonadosomatic index, maturity, protein, ash, lipid, *Puntius sarana, Sarotheroden mossambicus*

Abstract

Changes in body condition and moisture 'protein' total lipid and ash content of the musculature in *Puntius sarana* and *Sarotherodon mossambicus* with maturation was studied in individuals obtained from the commercial fishery of the Parakrama Samudra, an ancient man-made lake in Sri Lanka. In *S. mossambicus* the body condition declined, in both sexes, with increasing maturity whilst in *P. sarana* the body condition improved with maturation until Stages IV/V and declined after spawning.

The protein and total lipid content also showed marked changes with maturation, and at times differences were noted between the sexes of any one species. Generally, all changes in proximate composition with maturation were less well defined in the males in comparison to the females. In *S. mossambicus* (both sexes) the lipid content increased with maturation whilst in *P. sarana* the lipid content increased with the onset of maturity and declined from Stage III onwards, when deposition of yolk takes place, until spawning. Changes in protein content with maturity were also found to be different in the two species.

The observations are discussed in the light of known knowledge on the breeding habits of the two species as well as on the different reproductive strategies opted for by *P. sarana* and *S. mossambicus.*

1. Introduction

The proximate composition is known to be affected by a number of factors; with the internal state of the fish as well as with environmental changes (Love 1970). Specific studies on the changes of proximate composition in relation to the maturity have been mainly carried out on temperate species; for example, Medford & Mackay (1978) on the pike (*Esox lucius*), Craig (1977) on perch (*Perca fluviatilis*) and Beamish et al. (1979) on sea lamprey. Similar studies on tropical species are scarce (Caulton & Bursell 1977).

Puntius sarana (Ham. Buch.) and *Sarotherodon mossambicus* (Peters) are important constituent species of the inland fisheries, which are primarily confined to the multitude of man-made reservoirs, in Sri Lanka (De Silva & Fernando 1980; De Silva 1982). The reproductive cycle of these two species have been investigated by De Silva & Chandrasoma (1980) and Chandrasoma & De Silva (1981) and it has been shown that *P. sarana* shows a seasonality in its reproductive activities, reaching a peak during the rainy period and that it undertakes a migration out of the reservoirs for breeding, in contrast to the exotic *S. mossambicus.*

In this chapter changes in the condition factor, gonadosomatic index (GSI) and proximate composition in relation to the maturity stages of both species are presented.

2. Materials and methods

All samples were obtained from the commercial fishery in the Parakrama Samudra, an ancient

Schiemer, F. (ed.), Limnology of Parakrama Samudra - Sri Lanka
 ISBN 90 6193 763 9

man-made lake of 2662 ha at full supply level. The collection of samples and other relevant details of the fishery have been described elsewhere (De Silva & Chandrasoma 1980). For proximate analyses 8–30 fish of each species of known total length, body weight and gonadal maturity were randomly selected, each month, and a portion of the muscle from the widest part of the right fillet, devoid of skin and bone was taken, weighed and dried overnight at 80 °C to constant weight.

The maturity stage of each individual was determined according to Chandrasoma (1981), for both species.

The dried muscle was finely ground and aliquots were used to determine the total lipid (200–300 mg), protein (80–100 mg) as described earlier by Ehrlich (1975), Perera & De Silva (1978) and De Silva & Rangoda (1979). Briefly, total lipid was determined gravimetrically by extracting in 2:1 methanol: chloroform mixture, adding 20% water, and homogenising in a Potter-Elvehejim homogeniser. The filtered homogenate was made up to the mark and allowed to separate at 4 °C, overnight, when the lower chloroform layer containing the lipid was siphoned out; 3 sub-samples of this homogenate was taken, vacuum dried and the total lipid determined (Bligh & Dyer 1959). Protein was determined by digesting aliquots at 100 °C for 30 min in 1N NaOH and adding 10 ml of biuret to 4 ml of the filtered digest and reading the intensity of the colour developed after 40 min at 555 nm in a Perkin-Elmer (S-35) spectrophotometer. Bovine-serum albumen (BDH, Poole, U.K.) was used as the standard. Ash content was determined by igniting at 520 °C overnight, in a muffle furnace. In addition the condition factor (k) was determined for each individual fish according to the formula (LeCren 1951)

$$K = \frac{W}{L^3} \times 100,$$

where W and L are weight in g and the total length in mm. Duncan's multiple range test (Alder & Roessler 1972) was used to determine whether the changes in the mean content of any one component at different stages of maturity in each group of fish was statistically significant ($p < 0.05$).

3. Results

The changes in the gonadosomatic index (GSI), and condition factor in relation to the six maturity stages are shown in Figures 1 and 2, respectively. G.S.I. of *P. sarana* is found to be much higher than in *S. mossambicus*, ranging from 0.22 to 3.71 and 0.01 to 21.21 in male and female, *sarana* respectively. In *S. mossambicus* the values range from 0.007 to 0.32 in males and from 0.071–7.97 in females. As expected the changes in the GSI in both sexes of both species, increase from stages I onwards, the most noticeable increase occurring after stage III, and shows a marked decline in stage VI. The trends in the changes of the condition factor appear to differ in the two species but is similar in the two sexes of any one species. In *P. sarana* the condition factor shows an increase with maturation, particularly between stages I & II (♂ & ♀ s) and a decrease in IV and V, in males and from stage V onwards in females. In *S. mossambicus* the condition factor tends to decline with increasing maturity in both sexes. Table 1 gives the statistical relationship in the condition factor, for any one group, at different stages of maturity and it is evident that in all groups the differences between stages II and III are not significant whereas the 'condition' at stage VI is significantly different from all the others ($P < 0.05$).

Percent moisture, total lipid, protein and ash, the last three parameters expressed on a dry weight basis, at each stage of maturity in *P. sarana* and *S. mossambicus* are given in Tables 2 and 3, respectively, whilst the trends of changes of the above parameters in both sexes of the two species are shown in Figures 3 and 4.

In *S. mossambicus* males the moisture content of the muscle increased gradually from stage I to stage IV but shows a sudden decrease at spawning – in stage-VI males, while in the females there is a reverse trend where the moisture content decreases up to the spawning period followed by a subsequent increase in the post-spawning period.

In *P. sarana* females the moisture content shows a trend similar to *S. mossambicus* females but the increase in moisture content takes place during the spawning period and not afterwards as in female *mossambicus*. In *sarana*, in both sexes, a decrease in the moisture content is seen after stage II till stage IV followed by an increase thereafter.

The percentage of total lipid content in the two

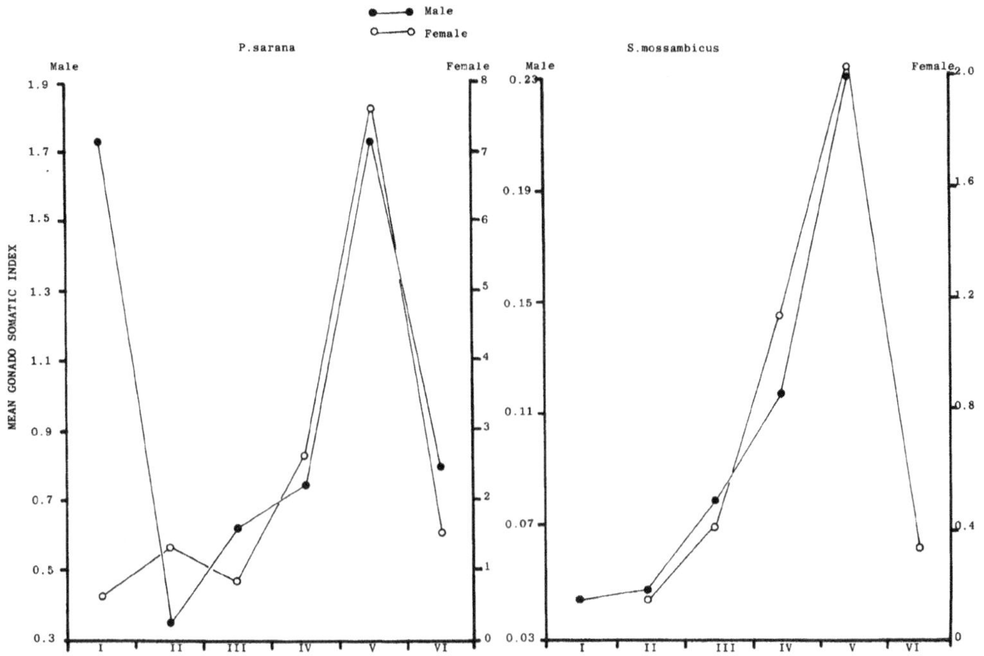

Fig. 1. Changes in the mean GSI in male and female *P. sarana* and *S. mossambicus* with maturity.

species show different trends (the differences between maturity stages), being more pronounced in the females than in the males. In the cichlid the lipid content increases significantly with increasing maturity, in both sexes. In contrast the total lipid content increases significantly in the cyprinid with the onset of maturity (stage II) and begins to decline in the females from stages III to V and then increases markedly after spawning, minimum lipid values of 6.76 being reached in stage V. In the males

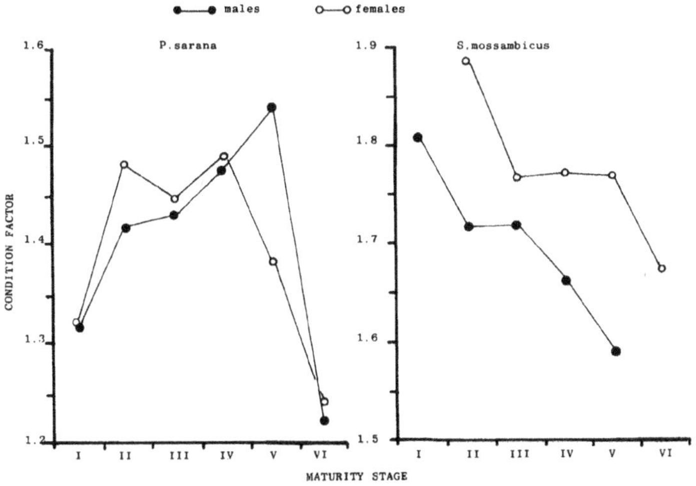

Fig. 2. The mean condition of males and females of *P. sarana* and *S. mossambicus* with maturity.

Table 1. Details of the fish used in the present study (the range of each parameter is given in paranthesis where appropriate; CF – Condition factor; N – Number analysed; GSI – Gonadosomatic index; SE – Standard error; Values with the same super script are not statistically significant).

Stage	*P. sarana* ♂						*P. sarana* ♀					
	N	Length (cm) $\bar{X}$	Weight (g) $\bar{X}$	CF $\bar{X}$	GSI $\bar{X}$	SE	N	Length (cm) $\bar{X}$	Weight (g) $\bar{X}$	CF $\bar{X}$	GSI $\bar{X}$	SE
I	2	24.4 (23.2–25.6)	192 (170–214)	1.31[a] (1.27–1.36)	1.74 (0.35–3.13)	1.39	3	26.5 (24.4–30.0)	252 (185–360)	1.31[a] (1.27–1.34)	0.67 (0.37– 1.01)	0.19
II	7	25.5 (21.8–30.6)	242 (159–415)	1.42[a] (1.32–1.64)	0.35 (0.17–0.69)	0.07	4	23.8 (20.5–25.5)	209 (110–245)	1.48[a] (1.27–1.66)	1.36 (0.49– 3.27)	0.65
III	14	25.6 (22.8–31.5)	250 (165–462)	1.43[a] (1.21–1.77)	0.63 (0.29–1.95)	0.11	20	27.14 (24.7–34.5)	311 (201–725)	1.45[a] (1.18–1.77)	0.87 (0.01– 1.79)	0.11
IV	19	25.8 (20.5–28.4)	246 (123–355)	1.47[a] (1.20–2.37)	0.75 (0.22–2.25)	0.12	14	28.1 (25.2–33.5)	337 (236–515)	1.49[a] (1.36–1.87)	2.68 (1.00– 6.15)	0.43
V	24	27.0 (22.2–30.9)	279 (187–359)	1.54[a] (1.14–2.66)	1.74 (0.39–3.71)	0.21	30	28.2 (20.5–34.0)	321 (105–610)	1.38[a] (0.76–1.79)	7.68 (0.49–21.21)	0.90
VI	1	26.2	221	1.22[b]	0.80	–	8	29.2 (25.2–34.7)	312 (205–474)	1.24[b] (0.86–1.41)	1.56 (1.01– 3.37)	0.28
	S. mossambicus ♂						*S. mossambicus* ♀					
I	22	21.3 (17.2–27.4)	180 (118–342)	1.80[a] (1.29–2.65)	0.05 (0.01–0.31)	0.01	–	–	–	–	–	–
II	31	22.8 (18.0–27.9)	219 (103–369)	1.72[a] (1.14–2.50)	0.05 (0.01–0.11)	0.01	3	20.2 (18.2–21.6)	158 (118–190)	1.89[a] (1.83–1.95)	0.15 (0.13– 0.18)	0.01
III	17	24.8 (19.5–32.8)	283 (140–601)	1.72[a] (0.89–2.08)	0.08 (0.04–0.15)	0.01	22	21.5 (18.2–27.0)	182 (105–348)	1.76[a] (1.40–2.04)	0.41 (0.07– 0.73)	0.03
IV	4	26.5 (22.4–32.8)	324 (197–533)	1.67[a] (1.51–1.77)	0.12 (0.06–0.17)	0.02	29	21.6 (18.2–26.0)	180 (103–290)	1.77[a] (0.85–2.60)	1.17 (0.34– 7.97)	0.26
V	4	23.3 (21.5–24.5)	208 (178–230)	1.59[b] (1.32–1.79)	0.23 (0.04–0.32)	0.06	29	22.71 (18.0–28.1)	206 (135–320)	1.77[a] (0.58–3.77)	2.08 (0.56– 5.58)	0.21
VI	–	–	–	–	–		7	23.1 (20.0–25.5)	210 (140–265)	1.67[b] (1.51–1.91)	0.34 (01.5– 0.48)	0.05

Table 2. Changes in the proximate composition of the muscle of *Puntius sarana* with maturity (ash, protein and lipid are expressed as percentage of dry weight; mean values with the same superscript for anyone component is not significantly different from each other at different stages of maturity at 5% level).

Stage	*P. sarana* ♂				*P. sarana* ♀			
	Moisture	Ash	Lipid	Protein	Moisture	Ash	Lipid	Protein
I	$79.84^{a} \pm 0.16$	5.47^{a}	5.78^{a}	83.95^{a}	$78.94^{a} \pm 1.82$	4.96^{a}	$7.17^{a} \pm 1.75$	$75.57^{a} \pm 3.89$
	(79.67–80.00)				(76.36–82.46)	(4.68–5.38)	(4.58–10.51)	(67.80–79.63)
II	$80.09^{a} \pm 0.54$	$5.27^{a} \pm 0.18$	$10.49^{b} \pm 2.26$	$82.89^{a} \pm 2.80$	$78.68^{a} \pm 1.10$	$6.05^{a} \pm 0.51$	$12.48^{b} \pm 3.27$	$73.19^{a} \pm 1.39$
	(78.35–82.80)	(4.35–5.77)	(4.25–19.44)	(71.53–88.47)	(75.56–80.48)	(4.53–6.62)	(8.16–22.23)	(69.30–75.42)
III	$79.08^{a} \pm 0.59$	$5.63^{a} \pm 0.24$	$10.18^{b} \pm 1.52$	$77.37^{b} \pm 1.96$	$77.59^{b} \pm 0.44$	$4.76^{a} \pm 0.39$	$12.33^{b} \pm 1.23$	$76.35^{a} \pm 11.40$
	(75.59–82.42)	(4.28–6.98)	(1.96–21.31)	(66.96–87.74)	(72.91–81.38)	(2.10–8.59)	(3.27–18.49)	(61.37–86.88)
IV	$78.72^{a} \pm 0.39$	$5.82^{a} \pm 0.23$	$9.45^{b} \pm 0.89$	$77.04^{b} \pm 1.15$	$77.23^{a} \pm 0.92$	$5.44^{a} \pm 0.44$	$9.89^{b} \pm 1.89$	$77.16^{a} \pm 2.44$
	(75.81–81.85)	(4.24–7.50)	(3.41–14.72)	(67.52–83.22)	(67.35–81.50)	(1.85–7.83)	(3.16–26.96)	(57.67–87.76)
V	$79.81^{a} \pm 0.43$	$6.23^{a} \pm 0.40$	$9.62^{b} \pm 0.74$	$77.03^{b} \pm 0.96$	$78.18^{a} \pm 0.50$	$5.31^{a} \pm 0.32$	$6.77^{a} \pm 0.85$	$80.30^{b} \pm 1.11$
	(76.72–84.00)	(4.24–10.10)	(6.01–18.35)	(70.61–84.43)	(72.60–83.48)	(2.04–8.92)	(3.16–17.75)	(70.03–89.06)
VI	–	5.72^{a}	8.34^{b}	–	$78.91^{a} \pm 1.39$	$5.17^{a} \pm 0.33$	$12.92^{b} \pm 2.23$	$75.30^{a} \pm 1.72$
					(75.58–82.54)	(3.78–6.85)	(5.17–24.04)	(68.85–81.13)

Table 3. Changes in the proximate composition of the muscle of *Sarotherodon mossambicus* with maturity (ash, protein and lipid are expressed as percentage of dry weight; mean values with the same superscript for anyone component is not significantly different from each other at different stages of maturity at 5% level).

Stage	*S. mossambicus* ♂				*S. mossambicus* ♀			
	Moisture	Ash	Lipid	Protein	Moisture	Ash	Lipid	Protein
I	$79.44^{a} \pm 0.65$	$6.73^{a} \pm 0.31$	$11.75^{a} \pm 0.97$	$79.30^{a} \pm 0.97$	–	–	–	–
	(74.35–85.74)	(5.02–9.57)	(4.62–18.67)	(73.52–84.23)				
II	$79.48^{a} \pm 0.33$	$6.22^{a} \pm 0.25$	$11.91^{a} \pm 0.82$	$80.64^{a} \pm 1.06$	$77.82^{a} \pm 1.60$	$6.22^{a} \pm 0.51$	$6.94^{a} \pm 1.17$	$80.85^{a} \pm 0.17$
	(75.60–82.00)	(3.47–9.30)	(7.47–17.84)	(68.36–84.84)	(75.00–80.53)	(5.49–7.21)	(5.19–9.16)	(80.51–81.08)
III	$79.58^{a} \pm 0.50$	$6.35^{a} \pm 0.41$	$10.83^{a} \pm 1.88$	$79.36^{a} \pm 1.10$	$79.95^{a} \pm 0.41$	$6.71^{a} \pm 0.32$	$10.53^{b} \pm 1.25$	$79.88^{a} \pm 1.05$
	(75.00–83.85)	(3.64–8.67)	(5.01–23.02)	(73.76–82.74)	(75.51–83.08)	(4.65–9.21)	(3.92–20.07)	(73.75–85.78)
IV	$81.12^{a} \pm 1.60$	$5.26^{a} \pm 0.02$	$13.09^{b} \pm 3.12$	$80.96^{a} \pm 3.16$	$79.02^{a} \pm 0.46$	$6.79^{a} \pm 0.25$	$10.86^{b} \pm 1.18$	$79.45^{a} \pm 1.41$
	(76.91–84.43)	(5.24–5.28)	(9.98–16.19)	(77.80–84.13)	(76.23–83.02)	(5.17–9.54)	(4.69–19.23)	(69.97–89.28)
V	$78.50^{a} \pm 1.20$	$7.88^{a} \pm 1.22$	$21.95^{c} \pm 4.50$	$76.36^{b} \pm 1.91$	$78.87^{a} \pm 0.59$	$6.71^{a} \pm 0.21$	$11.16^{b} \pm 1.31$	$77.58^{b} \pm 0.99$
	(76.29–81.90)	(6.67–9.09)	(17.45–26.46)	(74.38–78.34)	(74.26–86.62)	(5.29–9.21)	(2.34–22.12)	(70.90–83.53)
VI	–	–	–	–	$81.68^{a} \pm 2.23$	$6.96^{a} \pm 0.99$	$16.24^{c} \pm 1.11$	$79.90^{a} \pm 0.71$
					(72.72–87.10)	(5.89–8.93)	(14.98–18.37)	(79.20–80.61)

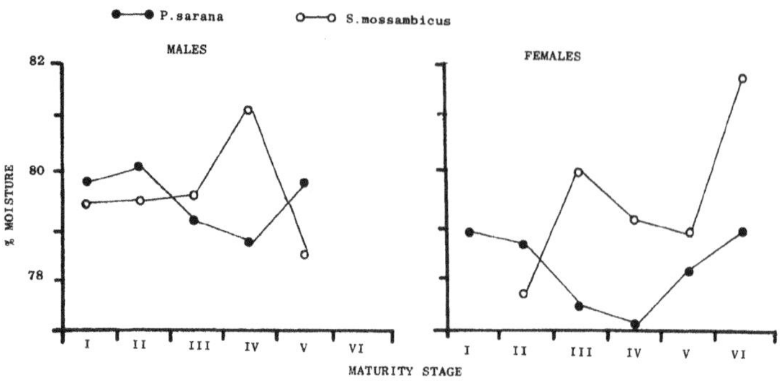

Fig. 3. Percent change in moisture content with maturity in *P. sarana* and *S. mossambicus*.

the decline from stage II onwards is not significantly different from stage to stage (Tables 2 and 3).

The changes in the percentage of protein in the muscle differ in the two sexes in both species. In *sarana* the protein content in the male declines with increasing maturity, the most distinct change occurring between stages II and III. In the female, the percentage protein content increases from stage II until spawning (stage V) and decreased thereafter almost to stage-I levels. In *mossambicus* the minimal protein content is observed in stage-V fish (both sexes), the changes occurring in the males being less clearly defined, unlike in the females, prior to stage V. After spawning female *mossambicus* shows an increase in muscle protein levels but this is not as high as in stage-I females.

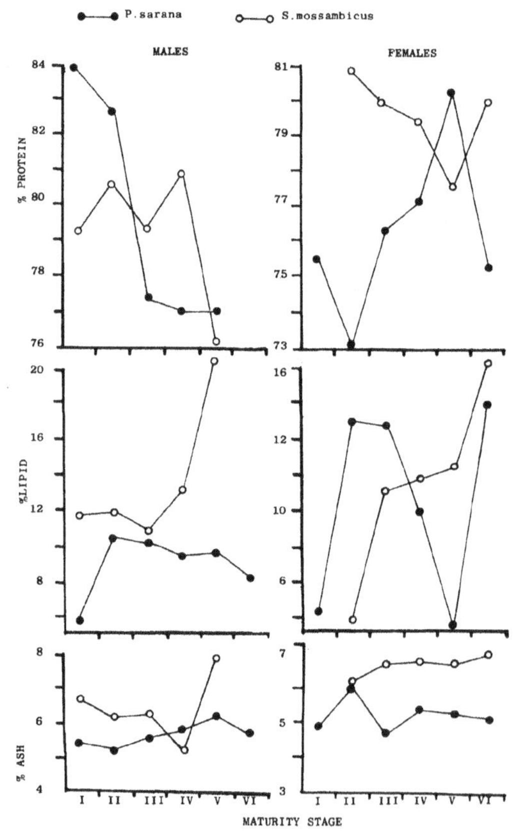

Fig. 4. Percent change in the protein, total lipid and ash content of *P. sarana* and *S. mossambicus* at different stages of maturity.

It could be seen from Figure 4 that in female *mossambicus* an increase in lipid coincides with an increase in protein from stage IV onward. A reverse trend is seen in the males where protein decreases with an increase in lipid from stages IV to V.

In *P. sarana* the females show a similar increase of lipid with a decrease in protein but from stage V to VI, i.e., in the spawning and post spawning period. In male *sarana* protein and lipid show decreases from stage I for protein and stage II for lipid right up to spawning.

4. Discussion

The process of maturation or the ripening of sex cells primarily involves morphological changes in the gonads (Barrington & Jorgensen 1968; Balinsky 1970). These are accompanied by changes in the physiological status of the individual which are reflected in the chemical composition of the body musculature, liver etc.

The decrease in condition observed in *S. mossambicus* with maturation has hitherto not been observed in fishes, whereas the trends observed in

P. sarana is to be expected, viz. a decrease in condition after spawning, especially in view of the supposed spawning migration it has to undergo (Chandrasoma & De Silva 1981).

The proximate composition of adult *Puntius gonionotus* and *S. mossambicus* was investigated by Tan (1971) and De Silva & Rangoda (1979), respectively. The difference observed in the changes in the proximate composition with maturity of the two species presently studied, is probably indicative of their different breeding strategies. *S. mossambicus* is a biparental mouth brooder (Fryer & Iles 1972; Trewavas 1982). Also the cichlid males differ from the female in that the testes weight is very much lower (Peters 1971). It is possible that in *S. mossambicus* during maturation a considerable amount of fat is laid down at the expense of proteins in preparation for its mouth brooding habit, when feeding will be minimal. However, lipid is generally regarded as one of the most important food reserves contributing to condition (Caulton & Bursell 1977) and therefore, the decrease in condition with maturity is contrary to expectations. This may, on the other hand, indicate the importance of protein in determining condition as well as its dominance in the metabolism of *S. mossambicus*.

In the case of *P. sarana* the relative 'investments' in gonads are considerably higher than in cichlid species as reflected by the GSI (Fig. 2). This higher 'investment' may be due to the serial spawning habit of *sarana* in contrast to the cichlid which is a single spawner with parental care. As such it is conceivable that the fat deposits get used up during maturation in this species when the lowest levels, similar to levels prior to spawning, are reached when the gonads are fully developed particularly in the female. The significant increase observed in the total lipid, in spent (stage VI) females could indicate that the 'return' to the reservoirs after spawning in the rivers is achieved by only those individuals having an 'excess' of reserve material.

The decrease in muscle protein during maturation as seen in *S. mossambicus* males and females and subsequent post-spawning increase has also been observed in other species (Bano 1975; Sathyashree 1981). Such changes are supposed to be indicative of muscle protein utilization for gonadal development (Abraham & Balakrishna Nair 1977) and also indirectly supports the earlier views on accumulation of lipids during maturation in preparation of its mouth breeding habit.

The reverse takes place in female *sarana* where increasing amounts of fat are utilized during the initial stages of ovarian maturation, which in a serial spawner such as *sarana* (Chandrasoma & De Silva 1981) takes place synchronously (Forberg 1982). The comparatively high fecundity in *sarana* and the high 'investment' on the ovaries during maturation could possibly make a demand on both muscle proteins and lipids unlike in the cichlid.

Thus it is evident from the present study that changes in proximate composition, as well as that in the traditional condition factor with maturation gives an indirect indication of the different strategies adopted by species with contrasting reproductive strategies.

Acknowledgements

We are most grateful to Mr. J. Chandrasoma for collection of the samples. This work was partially supported by a grant from the National Science Council of Sri Lanka, to the first author.

References

Abraham, M. & Balakrishnan, N., 1977. Studies on the seasonal variations in the major biochemical constituents of *Psettodeserumei* (Bl. Sch.): *Decapierus* dayi (wakiya) and *Stolephorus heierolobus* (Ruppell). of the south west coast of India in relation to breeding cycle and feeding intensity. 'Biochemistry of fish', 'National Colloquium', Annamalainagar, pp. 17–19.

Alder, H. L. & Roessler, E. B., 1972. Introduction to probability and statistics 5th ed. W. H. Freeman & Co., San Francisco.

Balinsky, B. I., 1970. An introduction to embryology. W. B. Saunders, London.

Bano, Y., 1975. Seasonal variations in the biochemical composition of Clarius batrachus L. Proc. Indian Acad. Sci. 85B: 147–155.

Barrington, E. J. W. & Jorgensen, C. B., 1968. Perspectives in endocrinology. Academic Press, New York.

Beamish, F. W. H., Potter, I. C. & Thomes, E., 1979. Proximate composition of the adult anadromus sea lamprey, *Petromyzon marinus*, in relation to feeding, migration and reproduction. Animal Ecology 48: 1–19.

Bligh, E. G. & Dyer, W. G., 1959. A rapid method of total lipid extraction and purification. Comp. J. Bioch. Phys. 37: 911–17.

Caulton, M. S. & Bursell, E., 1977. The relationship between changes in condition and body composition in young *Tilapia rendalli* Boulenger. J. Fish. Biol. 11: 143–150.

Chandrasoma, J., 1981. Aspects of the reproductive biology of *Puntius sarana, Sarotherodon mossambicus* and *Tilapis rendalli*. Parakrama Samudra, Sri Lanka. M. Phil. Thesis, Kelaniya University, Sri Lanka, 180 pp.

Chandrasoma, J. & De Silva, S. S., 1981. Reproductive biology of *Puntius sarana*, an indigenous species, and *Tilapia rendalli* (*melanopleura*), an exotic, in an ancient man-made lake in Sri Lanka. Fish. Mgmt. 12: 17–28.

Craig, J. P., 1977. The body composition of adult perch, *Perch, Perca fluviatilis* in Windermere, with reference to seasonal changes and reproduction. Anim. Ecol. 46: 617–632.

De Silva, S. S. (in press) Freshwater fisheries of Sri Lanka: A case study and strategies for a man-made fishery.

De Silva, S. S. & Chandrasoma, J., 1980. Reproductive biology of *Sarotherodon mossambicus*, an introduced species in a man-made lake in Sri Lanka. Env. Biol. Fish 5: 253–259.

De Silva, S. S. & Fernando, C. H., 1980. Recent trends in the fishery of Parakrama Samudra, an ancient man-made lake in Sri Lanka. In: Furtado J. I. (ed.), Tropical ecology and development, pp. 927–938. University of Malayasia Press, Kuala Lumpur.

De Silva, S. S. & Rangoda, M., 1979. Some chemical characteristics of Fresh and salt-dried *Tilapia mossambica* Peters. J. Natn. Sci. Coun. Sri Lanka 7: 19–27.

Ehrlich, K. F., 1975. A preliminary study of the chemical composition of sea-caught larval hering and plaice. Comp. Biochem. Physiol. 51(B): 25–28.

Forberg, K. G., 1982. A histological study of development of oocytes in capelia, *Mallotus villosus villosus* (Muller). J. Fish. Biol. 20: 143–154.

Fryer, G. & Iles, T. D., 1972. The cichlid fishes of the great lakes of Africa. Published in Great Britain by Oliver & Boyd, Edinburgh.

LeCren, E.D., 1951. The length-weight relationship and seasonal cycle in gonad weight and condition in the perch. (*Perca fluviatilis*). J. Anim. Ecol. 20: 201–219.

Love, R. M., 1970. The chemical biology of fishes, Vol. II. Academic Press, London.

Medford, B. A. & Mackay, W. C., 1978. Protein and lipid content of gonads, liver and muscle of Northern Pike (*Esox lucius*) in relation to gonad growth. J. Fish. Res. Bd. Can., 35: 213–19.

Perera, P. A. B. & De Silva, S. S., 1978. Studies on the chemical biology of young grey mullet, *Mugil cephalus* L. J. Fish. Biol. 13: 297–304.

Peters, H. M., 1971. Testes weights in *Tilapia* (Pisces: Cichlidae). Copeia 33: 13–17.

Sathyashree, P. K., 1981. Biological and biochemical studies on grey mullets (Family Mugilidae) of Porto Novo, S. India. Ph.D. Thesis, Anamalai University, Tamil Nadu, India, 311 pp.

Tan, Y. T., 1971. Proximate composition of freshwater fish, grasscarp, *Puntius goniotus* and *Tilapia*. Hydrobiologia 37: 361–366.

Trewavas, E., 1982. Genetic groupings of Tilapiini used in Aquaculture. Aquaculture 27: 79–81.

Authors' address:
S. S. De Silva
C. D. De Silva
W. M. K. Perera
Department of Zoology
Ruhuna University College
Matara
Sri Lanka

17. Reproductive strategies of some major fish species in Parakrama Samudra Reservoir and their possible impact on the ecosystem – a theoretical consideration

S. S. De Silva

Keywords: tropics, man-made lake, reproduction, fish, *Sarotherodon mossambicus*, *Puntius* spp.

Abstract

Available information and fresh evidence is presented to show that presently, commercially important indigenous cyprinid species such as *Puntius sarana* and *Labeo dussumieri* of the Parakrama Samudra reservoir as well as other species like *P. dorsalis* etc. which are known to constitute a significant fish biomass of the reservoir but commercially yet unexploited, are not likely to breed in the reservoir. Possible reasons for an apparent lack of breeding success in the reservoir are briefly discussed. In addition the possible effects of water level fluctuations in the reservoir on the nest building capability of the exotic cichlid. *Sarotherodon mossambicus,* which is the mainstay of the commercial fishery are discussed. It is postulated that a certain degree of fluctuation in the water level is essential for *S. mossambicus* to maintain its breeding success.

Possible impact of the likely reproductive migrations of some of the indigenous cyprinids, which are shown to be seasonal, and the variation in the breeding success of *S. mossambicus* in relation to water level fluctuations on the ecosystem, as a whole, is briefly evaluated, qualitatively.

1. Introduction

Fifty three species of fresh water fish are now known to occur in Sri Lanka (Deraniyagala 1952; Munro 1955; De Silva *et al.* 1981). Of these 24 species are found in the Parakrama Samudra (Schiemer & Hofer 1983). Like its contiguous Indian mainland fauna (Jhingran 1975), the indigenous Sri Lankan fish fauna lacks truly lacustrine species. The indirect consequences of this on the reservoir fisheries have been aptly documented (Fernando 1973; Fernando & Indrasena 1969).

The freshwater fishery in Sri Lanka is almost entirely dependent on its multitude of reservoirs, of which the fishery of Parakrama Samudra is not only the best documented (Fernando & Indrasena 1969; De Silva & Fernando 1980) in the Island but possibly that of a reservoir in the whole of South East Asia (Fernando 1977). This fishery is dominated by an exotic species, viz. *Sarotherodon mossambicus* (Peters).

Parakrama Samudra is characterised by a fast flow through (Schiemer 1983) and by a widely fluctuating water level throughout the year. Both features are not unique to this reservoir but is characteristic of almost all the perennial reservoirs in the country. However, since 1975 and more regularly and uniformly since 1978/79 Parakrama Samudra receives water from the diversion of the river Mahaweli which should consequently result in a less drastic change in the water level during the year (Fig. 1). The changes would be even more less pronounced when the diversion is expected to be completed in 1985.

In this chapter an attempt is made to speculate on the possible consequences that the recent physical changes may have on the breeding success of *S. mossambicus*. In addition an attempt is also made

Schiemer, F. (ed.), Limnology of Parakrama Samudra – Sri Lanka
 ISBN 90 6193 763 9

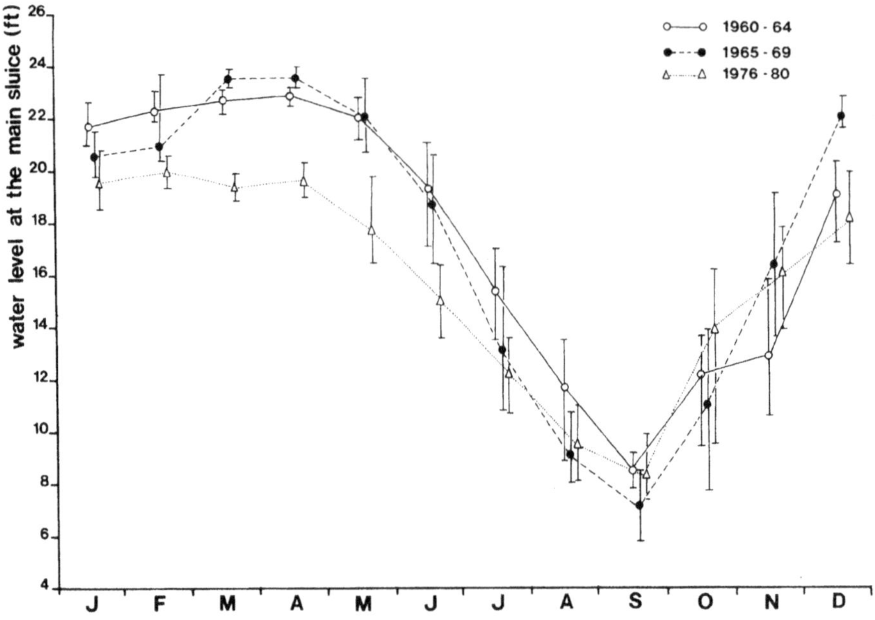

Fig. 1. Changes in the mean water level at the main sluice of the Parakrama Samudra over the periods 1960–1964, 1965–1967, and 1976–1980. The data between 1967 and 1976 were not available. The vertical lines indicate the range for each month. (These data have been supplied by Irrigation Engineer, Polonnaruwa.)

to evaluate the possible effects of the reproductive strategies of some of the major non-lacustrine indigenous species on fish populations in the reservoir and hence the ecosystem as a whole. This chapter as such is based primarily on published work together with some data from the Parakrama Samudra project and studies on the cyprinid biology from two river systems in the island.

2. Results and discussion

2.1. Indigenous species

The most important indigenous species of the fishery are cyprinids, *Puntius sarana* (Ham. Buch.) and *Labeo dussumieri* (Val.) and the estuarine cichlid *Etroplus suratensis* (Bloch) which has been transplanted from its estuarine habitat (De Silva & Fernando 1980). Apart from these a number of species of cyprinids, *Puntius dorsalis* (Jerdon), *Puntius filamentous* (Val.) and *R. daniconius* (Ham. Buch) constitute a significant proportion of the total fish in the reservoir (see Schiemer & Hofer 1983).

Studies on *P. sarana* populations of Parakrama Samudra has shown that this species is most unlikely to reproduce within the reservoir and that it undertakes migrations prior to the onset of the rains (Chandrasoma & De Silva 1981). This view is further supported by the absence of larvae and or fry, as well as indirectly from catch statistics.

The total landings from the reservoir were found to be higher just prior to the rains and during the rains (De Silva & Fernando 1980). These changes are thought to be possibly a result of aggregation of either immigrating or emigrating populations, as the case may be. Statistics of landings of *P. sarana* and *L. dussumieri* from the reservoir exemplifies this further (Fig. 2). Such trends have also been reported for these two species from other major reservoirs in the Island (De Silva & Chandrasoma, in prep.). Fish fry have been reported trapped in rivulets, leading to the reservoir, after the rainy season (and predated upon by macacques – Dr W. Dittus, personal communication), which were iden-

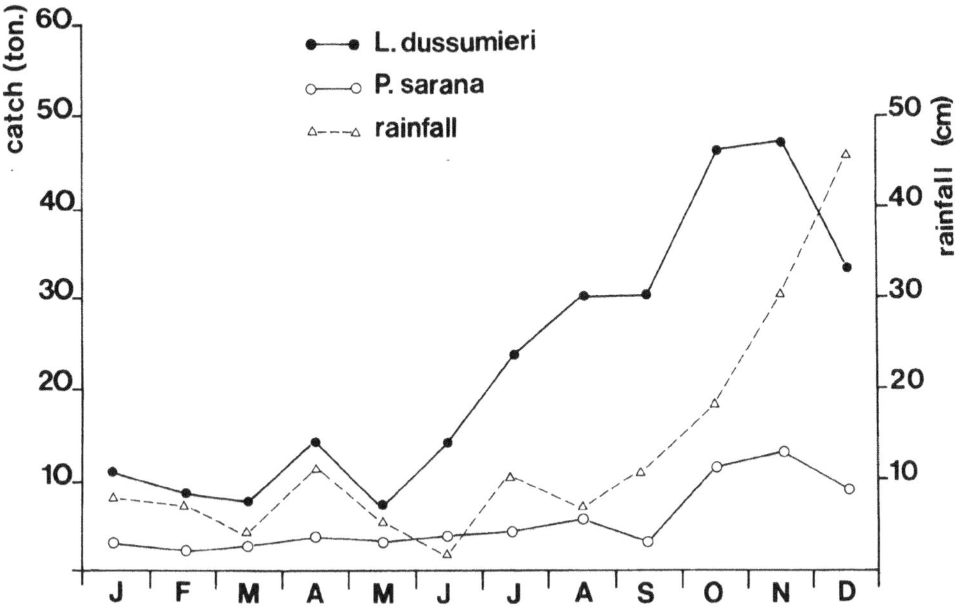

Fig. 2. The mean catches of *P. sarana* and *L. dussumieri* for the period 1975–1979 in the reservoir. Also indicated is the mean rainfall in the catchment for the same period.

tified as those of *P. sarana* and *L. dussumieri*. Gwahaba (1973) also reported increased catches from Lake George, Uganda during the rainy period but the causes responsible were not discussed.

Apart from studies on the reproductive biology of *P. sarana* there have been no similar studies on any indigenous fishes in reservoirs in Sri Lanka. However, comparison of data from a 2-year study on the reproductive biology of *Puntius* species in their natural habitat of two river systems in the Island (De Silva *et al.*, in press) with the information obtained on three species during the Parakrama Samudra project sheds further light on the possible reproductive strategies of the two species in question in their alien environment; the reservoirs. Changes in the gonosomatic index (the ratio of the gonad weight to the body weight, expressed as a percentage), presence of 'running' fish in the population and the occurrence of eggs, larvae and fry are either collectively or individually taken as criteria to determine the reproductive seasonality in fish populations (Weatherly 1972). Table 1 summarizes the monthly mean gonosomatic index (GSI) (together with range) of *P. dorsalis* females from the Ginganga.

Also incorporated into the Table are the mean and range in the GSI of *P. dorsalis*, *P. filamentous* and *P. chola* from the periods September/October 1979 and March/April 1980. Examinations of the gonads of the latter three species indicated that without exception all individuals in the reservoir were either in stage IV or V maturity (De Silva 1973) with mature yolked laden ova almost completely filling the peritoneal cavity, reflecting a higher GSI or in the very early stages of gonadal development (stage II). Individuals which have partially spent their ovaries were not caught. In contrast, during the two years of sampling no more than 10 females of any of the species, with mature ova filling the peritoneal cavity was obtained from either of the two river systems, as reflected in the low mean GSI as well as in the narrow range (De Silva *et al.* in press). De Silva *et al.* also found that of the minor carps investigated *P. dorsalis,* which is of relevance to this chapter, tends to spawn throughout the year shedding its eggs in batches with peaks of reproductive activity related to the rainy periods. Furthermore, each female goes through a short recovery phase before respawning and thus a female with ovaries filling the peritoneal

Table 1. Mean monthly gonosomatic index (GSI) and the range (in parenthesis) in *P. dorsalis* females from the Ginganga and that of *P. dorsalis, P. chola* and *P. filamentosus* from the Parakrama Samudra for the periods of the project (Gg – Ginganga; Ps – Parakrama Samudra; n = number sampled).

	P. dorsalis				P. chola		P. filamentosus			
	Gg		Ps		Ps		Gg		Ps	
Month	n	x	n	x	n	x	n	x	n	x
January	36	2.06 (1.28–3.39)								
February	40	2.40 (0.58–5.80)						1.6 1.2 (0.62–3.2)		
March	35	1.91 (0.89–3.64)	36	7.64 (0.76–14.59)	32	14.4 (6.6–21.7)	18		14	8.61 (0.85–16.72)
April	38	3.13 (0.86–6.23)								
May	32	2.11 (0.46–5.70)								
June	27	1.73 (0.69–2.50)								
July	26	2.20 (0.97–5.78)								
August	43	2.68 (1.23–5.78)								
September	40	1.90 (0.96–5.50)	41	17.9 (5.4–22.4)	18	14.4 (9.8–21.8)				
October	46	1.45 (0.56–3.46)								
November	40	1.89 (0.94–3.60)								
December	44	1.43 (0.44–3.29)								

cavity (therefore of high GSI) rarely occurs in the riverine populations.

During the Parakrama Samudra project, Clark-Bumpus surveys and sporadic fry surveys, using a lift net, were carried out. Cyprinid larvae or very young fry were not recorded in these surveys, confirming the findings of Chandrasoma & De Silva (1981). Thus it is most probable that *P. dorsalis*, which is not commercially fished now but which is an important constituent species of the fish populations in the reservoir (Schiemer & Hofer 1982), also does not spawn in the reservoir.*

Major carps have been used in fish culture, dating back to many hundreds of years (Bardach *et al.* 1972). Both in intensive and extensive culture systems major carps such as the Chinese grass, bighead, silver and Indian rohu, catla etc. are not known to spawn naturally. Sinha *et al.* (1979) reviewed the conditions necessary for spawning of Indian major carps in their natural habitat. There is however, little known about the reproductive biology of minor carps. Sobana & Balakrishnan Nair (1974) found that the subspecies *P. sarana subanastus* (Val.) does not breed in the reservoirs in the mainland of India. On the other hand, a number of minor barbs have been bred in captivity by the aquarium trade and for behavioural studies (Kortmulder 1972).

Changes in the reproductive strategies in fin fish species with changes in the climatic conditions or in the environment in their geographical range are documented (Siddiqui 1979). It is thus likely that at least the indigenous species considered in this paper migrate from the reservoir for spawning, at a particular time of the year – coinciding with the most favourable time of the year for breeding of the

* In the Zoologisch Laboratorium, University of Leiden, The Netherlands, where a number of Asiatic minor barbs have been bred for a number of years, attempts to breed *P. dorsalis* has been unsuccessful (Dr. K. Kortmulder, personal communication).

related populations in the rivers. This of course is a situation not peculiar to Sri Lankan conditions but is known to prevail in most tropical regions where secondary colonisation of lakes have taken place by riverine species (Lowe-McConnell 1975; Welcomme 1970).

2.2. Exotic species

The reproductive biology of two exotics viz. *S. mossambicus* (Peters) and *T. rendalli* (Boulenger) (= *melanopleura*) in the Parakrama Samudra, introduced in 1952 and 1974, respectively, have been investigated by De Silva and Chandrasoma (1980) and Chandrasoma & De Silva (1981). Both species are reported to breed throughout the year, with slightly higher intensities during the rainy months. Similar observations have also been made in other cichlid species (Fryer & Iles 1972). De Silva & Chandrasoma (1980) showed that *S. mossambicus* usually builds nests at depths of about 15 cm, to about 100 cm. The necessity of shallow areas either with soft substrata or vegetation has been observed for other *Sarotherodon* and *Tilapia* species investigated (Welcomme 1967, 1970; Lowe-McConnell 1975). It was stated earlier that Parakrama Samudra was characterised by wide fluctuation in the mean level throughout the year. The breeding areas of *S. mossambicus* in the Parakrama Samudra were mapped by Fernando & Indrasena (1969). They found that the nesting sites were almost totally confined to the west bank(s) and in the area between the water levels in the dry season and at the full supply level (Fig. 3). Observations of the author also indicate that the hard soil on the east bank of the reservoir is not suitable for nest building. It is also, likely that the receding water levels in the reservoir, during the dry season, which almost exposes the soft, muddy bottom of the littoral and the sub-littoral is beneficial for *S. mossambicus* in that it enables it to exploit these areas for nest building by providing a suitable water depth.

Optimum water depth for nest building in *S. mossambicus* is shown to be at 60–80 cm (De Silva & Chandrasoma 1980). The intensity of fry production is bound to be affected if the area between the water levels during the dry season and at full supply level (prior to 1974) continues to have a water depth of over 1 m during most of the year; that is when a

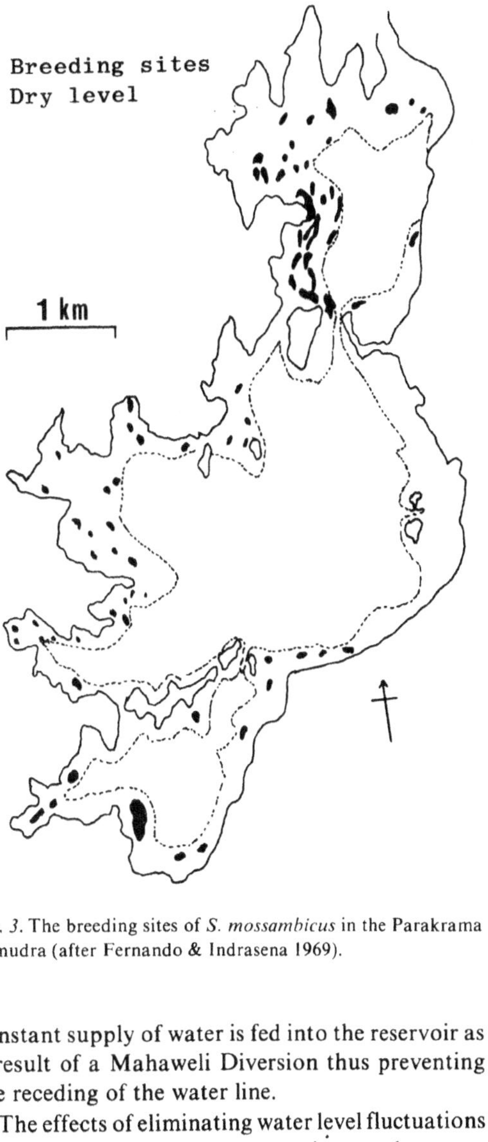

Fig. 3. The breeding sites of *S. mossambicus* in the Parakrama Samudra (after Fernando & Indrasena 1969).

constant supply of water is fed into the reservoir as a result of a Mahaweli Diversion thus preventing the receding of the water line.

The effects of eliminating water level fluctuations such as by building dams across rives on the reproductive capability of cichlid species have also been noted earlier by Babiker & Ibrahim (1979) on *S. niloticus*. Similarly the closure of the Kaufe George Dam, which brought about high post-impoundment water levels in Kaufe flood plains in Zambia is thought to have brought about reductions in reproductive intensities and year-class sizes in *S. andersoni* and *S. macrochir* (Dudley 1979).

There is evidence accumulating that the diversion of the river Mahaweli waters into some of the ancient reservoirs, which were dependent primarily on the rainfall in their respective catchments, and as such were hitherto characterised by wide fluctuations in the water level through the year, is affecting their fish populations. Kalawewa, an ancient reservoir, situated in the Dry Zone of Sri Lanka, with very steep banks, has been supplied with water from the diversion since 1974 and there is evidence to indicate that the catches of cichlids are on the decline, probably as a result of reduced breeding success. Effects of the speculated reduction in the breeding success of cichlids in the Parakrama Samudra is not likely to be visible for a few years yet: the reservoir underwent major repairs in the years 1978/1979 and during this period water level was maintained at about the dry level prior to the diversion (see Fig. 1) and as such in effect pre-diversion conditions were prevalent in the reservoir until 1980.

The migration(s) in and out of the reservoirs of the main riverine species is bound to have an impact on the ecosystem of the reservoir. These reproductive migrations would effect, initially, the standing crop of the fish biomass in the reservoir which will have consequences on their immediate food resources thus triggering a chain reaction in the ecosystem, as a whole. Presently, the data available is insufficient to quantify any of these changes that could result from these behaviour patterns.

The situation seen in Parakrama Samudra is not thought to be unique to this reservoir, but is probably of general occurrence in similar reservoirs in Sri lanka. It is, therefore, suggested that careful evaluations of the fish populations together with their behaviour patterns, particularly those pertaining to the reproductive biology, have to be taken into consideration in the final analysis of these types of lakes.

3. Conclusions

In the foregoing sections an attempt has been made to show from available data and observations that in the Parakrama Samudra at least two and possibly three important constituent, indigenous, species of the fish fauna do not reproduce in the reservoir proper. Further, the recent changes in the water intake regime and the accompanying morphometric changes could limit the available breeding grounds for the exotic species, *S. mossambicus*, which is still by far the mainstay of the fishery (De Silva & Fernando 1980).

Most reservoirs in Sri Lanka are shallow (Fernando 1973). Even in shallow reservoirs the morphometry would, principally, determine the area available for nest building for *S. mossambicus*. At the appropriate water depth details of the bottom substratum would finally determine the suitability of a nesting area (Fryer & Iles 1972). Investigations of these aspects in the multitude of reservoirs of Sri Lanka would be most rewarding in many ways. Such studies would not only shed light on the hypotheses presented here but may provide, clues for the success of 'tilapias' in South-East Asia. These may also reveal whether *S. mossambicus* populations react, to water level fluctuations, nature of substratum etc. of the habitats; and if so how, thus possibly providing further understanding to basic concepts in fish population dynamics.

Acknowledgements

I wish to thank Professor C. H. Fernando of the University of Waterloo, Canada, for reading the draft and also the 'Austrian Group', in particular, Drs. F. Schiemer and R. Hacker for providing data on the gonadal weights and for their valuable criticisms during the preparation of this manuscript.

References

Babiker, M. M. & Ibrahim, H., 1979. Studies on the biology of reproduction in the cichlid, *Tilapia nilotica* (L.): gonadal maturation and fecundity. J. Fish. Biol. 14: 437–448.

Bardach, J. E., Ryther, J. H. & McLarney, W. O., 1972. Aquaculture: the farming and husbandry of freshwater and marine organisms. Wiley Interscience, New York, p. 868.

Chandrasoma, J. & De Silva, S. S., 1981. Reproductive biology of *Puntius sarana*, an indigenous species, and *Tilapia rendalli* (*melanopeura*), an exotic in an ancient man-made lake in Sri Lanka. Fish. Mgmt 12: 17–28.

Deraniyagala, P. E. P., 1952. A colored atlas of some vertebrates in Ceylon. Volume I. Ceylon Govt. Press.

De Silva, S. S., 1973. Aspects of the reproductive biology of the sprat, *Sprattus sprattus* L., in inshore waters of the west coast of Scotland. J. Fish Biol. 6: 689–705.

De Silva, S. S. & Chandrasoma, J., 1980. Reproductive biology of *Sarotherodon mossambicus*, an introduced species, in an ancient man-made lake in Sri Lanka. Env. Bio. Fish 5: 53–60.

De Silva, S. S. & Chandrasoma, C., in prep. The fishery of eight man-made lakes in Sri Lanka.

De Silva, S. S. & Fernando, C. H., 1980. Recent trends in the fishery of Parakrama Samudra, an ancient man-made lake in Sri Lanka. In: J. Fernando (ed.) Tropical Ecology and Development. pp. 927–937. Society for Tropical Ecology, Kuala Lumpur, Malysia.

De Silva, S. S., Schut, J. & Kortmulder, K., (in press). Reproductive biology of 6 *Puntius (= Barbus)* species indigenous to Sri Lanka (Pisces; Cyprinidae). Env. Biol. Fish.

De Silva, S. S., Kortmulder, K. & Maitipe, P., 1981. The identity of *Puntius melanampyx sinhala* (Duncker, 1911) (Pisces, Cyprinidae). Neth. J. Zool. 32: 777–785.

Dunn, I. G., 1973. The commercial fishery of Lake George, Uganda, (East Aprica). Afr. J. Trop. Hydrobiol. Fish. 2: 109–120.

Dudley, R. G., 1979. Changes in the growth and size distribution of *Sarotherodon marochir* and *Sarotherodon andersoni* from the Kaufe Flood Plains, Zambia, since construction of the Kaufe George Dam. J. Fish. Biol. 14: 205–223.

Fernando, C. H., 1973. Man-made lakes in Ceylon. Geophysical Monographs 17: 664–671.

Fernando, C. H., 1977. Reservoir fisheries in Southeast Asia: past, present and future. Proc. Indo-Pacific Fish. Coun. 17th Session, Sec. III, 475–489.

Fernando, C. H. & Indrasena, H. H. A., 1969. The freshwater fisheries of Ceylon. Bull. Fish. Res. Stn., Ceylon, 20: 101–134.

Fryer, G. & Iles, T. D., 1972. Cichlids of the African great lakes. Oliver & Boyd, Edinburgh, pp. 532.

Gwahaba, J. J., 1973. Effects of fishing on the *Tilapia nilotica* (Linne, 1957) population in Lake George, Uganda over the past 20 years. E. Afr. Wild. J. 11: 317–328.

Jhingran, V. G., 1975. Fish and fisheries of India. Hindustan Publishing Corporation Delhi. 954 pp.

Kortmulder, K., 1972. A comparative study in colour patterns and behaviour in seven Asiatic barbus species (Cyprinidae, Ostariophysi, Osteichthyes). Supplement XIX, Behaviour, pp. 328.

Lowe-McConnell, R. H., 1975. Fish communities in tropical freshwaters. Longmans, London, 337 pp.

Munro, I. S. R., 1955. The marine and freshwater fishes of Ceylon. Dept. of Ext. Affairs. Canberra, Australia, 351 pp.

Schiemer, F., 1983. The Parakrama Samudra Project – scope and objectives. In: Schiemer, F. (ed.) Limnology of Parakrama Samudra – Sri Lanka: a case study of an ancient man-made lake in the tropics. Developments in Hydrobiology (this volume). Dr W. Junk, The Hague.

Schiemer, F. & Hofer, R., 1983. A contribution to the ecology of the fish fauna of the Parakrama Samudra reservoir. In: Schiemer, F. (ed.) Limnology of Parakrama Samudra – Sri Lanka: a case study of an ancient man-made lake in the tropics. Developments in Hydrobiology (this volume). Dr W. Junk, The Hague.

Siddiqui, A. O., 1979. Reproductive biology of *Tilapia zilli* (Gervais) in lake Naivasha, Kenya. Env. Biol. Fish. 4: 257–262.

Sinha, V. R. P., Jhingran, V. G. & Ganapati, S. V., 1974. A review on spawning of the Indian major carps. Arch. Hydrobiol. 73: 518–536.

Sobhana, B. & Balakrishna Nair, N., 1974. Observations on the maturation and spawning of *Puntius sarana subanastus* (Valenciennes). Indian J. Fish. 21: 357–368.

Weatherly, A. H., 1972. Growth and ecology of fish populations. Academic Press, New York & London.

Welcomme, R. L., 1967. Observations on the biology of the introduced species of Tilapia in Lake Victoria. Rev. Zool. BA Afr. 76: 249–279.

Welcomme, R. L., 1970. Studies on the effects of abnormally high water levels on the ecology of fish in certain shallow regions in Lake Victoria. J. Zool. Lond. 160: 405-436.

Author's address:
S. S. De Silva
Department of Zoology
Ruhuna University College
Matara
Sri Lanka

18. The ecology of cormorants (genus *Phalacrocorax*)

H. Winkler

Keywords: Chichlids, birds, fish, nutrient-cycling, *Phalacrocorax*, predation

Abstract

Three species of cormorants, *Phalacrocorax carbo, Ph. fuscicollis,* and *Ph. niger,* exploit the fish populations of Parakrama Samudra. Peak numbers were 102, 13700, and 1850 individuals, respectively. The estimated mean daily fish consumption amounts to 696 kg fish fresh weight d^{-1}. Nutrient export by these birds can reach almost one third of the nutrient loss through the outflow. These three species are well separated ecologically, but cichlids are the predominant prey for all of them.

1. Introduction

Probably ever since man included fish in his diet cormorants have been regarded as powerful competitors. These birds have always been ill reputed for their voracious appetite for fish. These prejudices led to the diminishing or even extinction of many populations of cormorants especially in European countries (cf. Bowmaker 1963). The Indian subcontinent, including Sri Lanka, is inhabited by three species of cormorants, namely the Southern or Large Cormorant *Phalacrocorax carbo sinensis,* the Indian Shag *Ph. fuscicollis,* and the Little Cormorant *Ph. niger* Ali & Ripley 1968; Henry 1971). All these species occur at Parakrama Samudra. Studies on the ecology and potential impact of cormorants in tropical and subtropical countries have mainly been carried out in Africa and Australia (Bowmaker 1963; Donelly 1967; Birkhead 1978; Whitfield & Blaber 1978; Miller 1979). Arguments concerning the possible harm or value of these birds concentrate mainly on the diet (composition and amount of food eaten) and possible beneficial effects through short-circuiting the nutrient flow.

With increasing inland fisheries and especially with the introduction of *Tilapia mossambica* cormorant predation became a problem in Sri Lanka as well. The cormorants were therefore studied during the PS-project and some additional information was collected in January/February 1982.

The methods used are referred to in the respective sections of this chapter. For reasons of easy reference the generic name *Tilapia* is used throughout including, especially, the genus *Sarotherodon* recognized by some taxonomists.

1. Foraging behaviour and habitat requirements

Even in the early days of niche theory cormorants served as an example for the ecological separation of closely related and overtly similar species (Lack 1945). Ecological separation of cormorants is expressed as differences not only in the size and quality of their food but also in feeding habits and habitat preferences (e.g. Lack 1945; Stonehouse 1967; Miller 1979; Whitefield & Blaber 1978).

The diet of the cormorants at Parakrama Samudra, not surprisingly, also follows this general pattern (Winkler in preparation): Little Cormorants have the most diverse diet (cf. also Mukherjee 1969). They eat not only many species of fish but

Schiemer, F. (ed.), Limnology of Parakrama Samudra – Sri Lanka
© 1983, Dr W. Junk Publishers, The Hague. ISBN 90 6193 763 9

also insect larvae and larger crustaceans. Indian Shags and even more so Southern Cormorants have a less variable diet. The size range of the food species is correlated with the size of the three bird species. Southern Cormorants hunt fish of approximately 130 mm (standard length), Indian Shags prefer fishes of between 60 and 120 mm and, finally, Little Cormorants prey upon fishes 30 to 70 mm in length (Winkler, loc. cit.). Cichlids seem to form the main prey of all three species. Southern Cormorants have been observed to catch mainly large *Tilapia* (approximately two thirds of the fish caught), *Heteropneustes* being the main alternative food. Cichlids comprise two thirds of the food of Indian Shags also, but *Etroplus* and *Tilapia* are taken equally. *Puntius* spec. play a minor role (about 8% of the individuals taken). Cichlids (*Etroplus: Tilapia* roughly 2:1) form about half of the diet of Little Cormorants. *Puntius* (predominantly juv. *P. filamentosus*) may comprise another 40% of the fish eaten.

The many differences in feeding behaviour are reflected in the composition of the diet. Southern Cormorants usually terminate a hunting sortie after having caught one fish whereas the average meal size of Indian Shags at Parakrama Samudra was 2.4 fishes and of Little Cormorants 8 fishes. The resulting relationship between meal size and size of fish eaten is in good accordance with a similar relationship found by Whitfield & Blaber (1978) among *Ph.* (*carbo*) *lucidus*. Various differences can be observed in the diving methods. Here, I restrict myself to the readily quantifiable aspects of diving. Stonehouse (1967) tried to evaluate the diving efficiency of the comorants using diving and resting periods of four species of cormorants (including *Ph. carbo novaehollandiae*). Similar data were collected for the three species present at Parakrama Samudra. To gain sufficient data (especially for *Ph. fuscicollis*) observations made at neighbouring tanks were included in the analysis presented in Table 1. Diving periods are defined as the time spent under water and resting periods as the time at the surface between two successive dives. For estimating diving/resting ratios diving periods and the immediately ensuing resting period were measured using an electronic stopwatch. The ratios given in Table 1 are the means of all the individually computed ratios and are not the ratios of the means as in Stonehouse (1967). During diving heart rates do not change much, but resting is associated with tachycardia of 50% (Kanwisher et al. 1981) which may be part of the physiological mechanism for post-diving recovery. Southern Cormorants at Parakrama Samudra never seemed to work under physiological strain. The correlation ($r = 0.02$) between diving and subsequent resting is not significant. Significant correlations do exist in Indian Shags ($r = 0.586, p < 0.1\%$) and Little Cormorants ($r = 0.646, p < 0.5\%$). Diving periods do not differ significantly between Little Cormorants and Indian Shags but the birds do differ significantly ($p < 5\%$) in their resting periods. Thus, in a given unit of time, Indian Shags spend more time under water than Little Cormorants. Comparing the ratios of the means with the data presented by Stonehouse (1967) the ratios of 3.5 and 3.6 for Southern Cormorants and Indian Shag, respectively, lie well above the ratio of 3.0 reached by *Ph. carbo novaehollandiae*, the most efficient of the New Zealand shags investigated by Stonehouse. The Little Cormorant's ratio of 2.3 equals that of *Ph. melanoleucos* (weight about 775 g, Serventy 1939) the least efficient cormorant in Stonehouse's list. The social

Table 1. Mean diving and resting periods of three species of cormorants. 95% confidence limits of the mean are given. See text for further comments.

Species	Diving		Resting		Ratio
	No. of observations	mean (sec)	No. of observations	mean (sec)	Diving/resting
Ph. carbo	134	29.8 ± 1.43	26	8.5 ± 1.13	3.7 ± 1.07
Ph. fuscicollis	42	12.9 ± 1.91	38	3.6 ± 0.61	3.4 ± 0.64
Ph. niger	103	12.4 ± 1.33	19	5.4 ± 2.15	2.5 ± 0.57

structure of the birds also has a considerable influence on their feeding ecology. Southern Cormorants breeding in the lake do seem not to defend any feeding territory and hunt mostly alone. They do not sleep in large communal roosts. Indian Shags, on the other hand, are nomadic visitors to the lake, do not defend feeding territories, hunt preferably with other individuals of their own species and form enormous communal roosts. Even two individuals of thus species synchronized their diving when close together. The shags may form very large feeding flocks which immediately attract the other two species as well as pelicans, herons and terns. Adult Little Cormorants defend feeding territories. Only immature birds may form feeding flocks which, among other things, may serve to overcome the aggression of the adults. Little Cormorants also form large communal roosts. These findings are somewhat at variance with the information compiled by Gadgil & Salim Ali (1975) which record the three species as all forming small to medium sized (five to twenty and several tens of individuals, respectively) roosts only. The main purpose of the large feeding flocks obviously is the enhancement of hunting success (cf. Curio 1976; Hoffman et al., 1981). Efficiencies, and even the type of prey eaten, may be quite different in large flocks as compared with birds feeding by themselves. The above data concerning the feeding habits of the cormorants at Parakrama Samudra, obtained exclusively from single birds, should thus be used cautiously when assessing the ecology of these species.

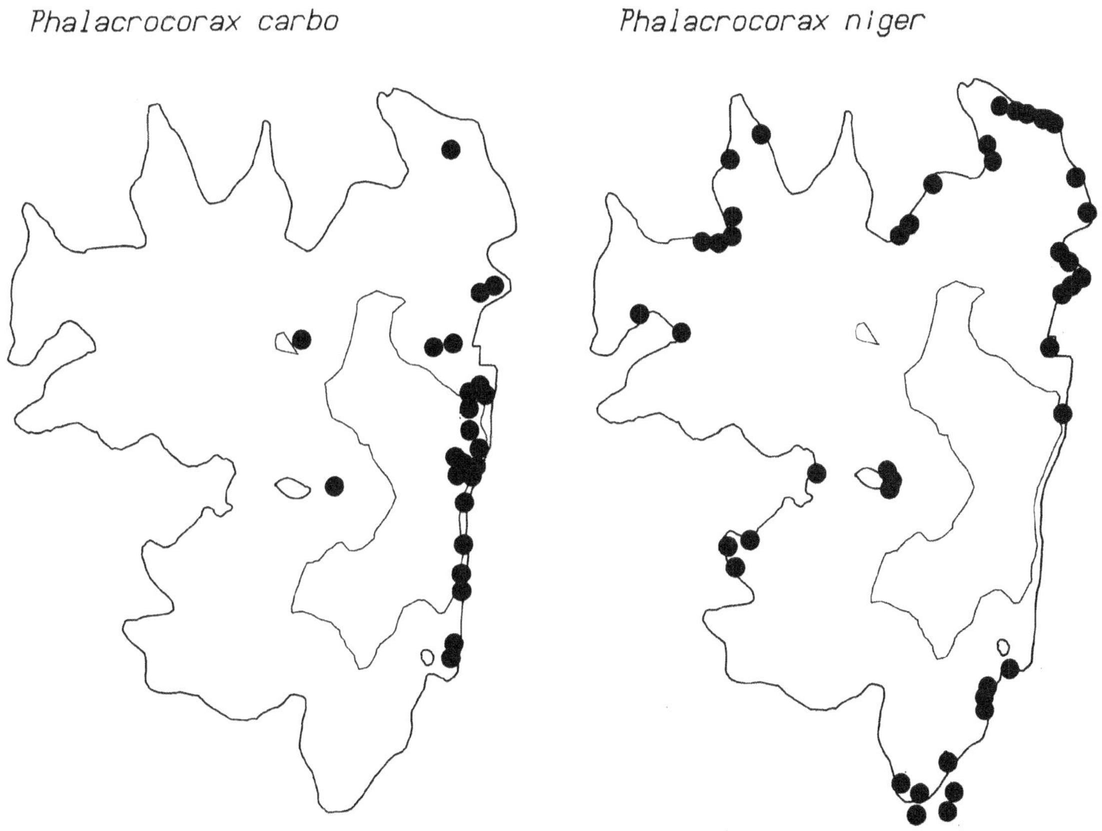

Fig. 1. Feeding sites of Southern Cormorants and Little Cormorants in 1980 in PSN. Dots, marking the feeding stations, may be based on more than one observation. The outer boundary indicates the water line and the inner one the depth of 4.3 m. Dots outside the lake boundary refer to observations of Little Cormorants in the channel connecting the northern and central parts of the lake and in pools cut off from the main water body.

Diving abilities are closely related to the depth reached and consequently to the habitat selected for hunting (cf. Stonehouse 1967; Whitfield & Blaber 1978). Southern Cormorants may dive in very shallow water hardly covering their backs. Indian Shags are somewhere between those extremes. The different preferences with respect to depth are also reflected in the distribution of the cormorants over the lake. This is illustrated in Figure 1. Furthermore, Little Cormorants manage to dive among the flooded vegetation, giving them access to those parts of the lake which presumably act as nurseries for the young of *Tilapia* (cf. Cambray et al. 1978) and *Puntius*. Indian Shags hunt along the edge of stands of macrophytes or flooded vegetation only and Southern Cormorants hunt the open water.

Summarizing, the data so far presented show clearly that the three species of cormorants at Parakrama Samudra are ecologically well separated. Therefore, any detailed analysis of their impact and any possible attempts at managing them have to take into account their different food preferences, diving habits, social behaviour and habitat requirements.

2. The ecological impact of the cormorants

Basically two methods might be employed for counting the numbers of cormorants present at the lake: firstly, direct counts obtained either by surveys from the shore using a telescope, or by scanning the lake by boat, or secondly, counts of the birds flying to their roosts. The first method was the only feasible one for the Southern Cormorant since this species practically never left the lake to visit a roost (single birds did leave the lake in February 1982). Indian Shags and Little Cormorants always left the lake to visit either a roost near the lake or a more distant one, using in either case the same flight routes. Only in 1982, with the lake at almost full supply level, did Little Cormorants and even Indian Shags form a roost at the northern end of the lake. Furthermore, birds of the southern lobe of the lake not only left along the usual route but some flew directly to the more distant roost. Thus, the count from 29.1.82 for *Ph. niger* is far from complete for the first (and second) reason, and that from 18.2.82 may be somewhat too low for the second reason. For all the other counts the assumption was made that the counts covered all (and only) the birds from the lake. This is a safe assumption although birds invading the lake may have come from other directions. Immigration of this sort has been observed several times, but took place in the morning hours only, so that the assumption made above is still valid with respect to the assessment of the ecological role of the cormorants.

The results are presented in Table 2 which also gives 'weighted means' which were obtained by stipulating that the counts are representative for a particular day of a year, regardless of the actual year studied, and by integrating these values using the trapezoidal rule. These means can be taken as a first approximation to the daily numbers of birds present during a year and are the basis for later calculations.

There is insufficient room here to discuss the causes of the large variations in numbers. It should, however, be mentioned that wind is one of the more important factors making the lake unaccessible to birds roosting some distance away but, it is as important that the roost is accessible to the lake.

Southern Cormorants were estimated to have a breeding population of 51 pairs in 1979. A survey by boat in 1980 gave an estimate of 66 birds.

Food consumption was not measured directly and has therefore to be estimated by some other means. The bad reputation of cormorants is largely based on over-estimates of their fish consumption. Even recent attempts at evaluating the daily food intake on the basis of stomach content analysis give widely differing results. Miller (1979) estimated the daily consumption of *Ph. sulcirostris* (778–960 g; Serventy 1939) to be around 60 g per bird (maximum 188.2 g) and of *Ph. melanoleucos* (730–820 g, Serventy 1939) to be roughly 40 g per bird. He took the stomach content of birds shot when returning to their communal roost as representative for their daily intake. Mukherjee (1969) based his statement that *Ph. niger* consumes 300 g fish per day on the supposition that the gut contents (birds were collected between 6 and 9 h) are representative for one meal only and that these cormorants ingest five meals per day. Other methods are usually based on equations using relationships between body weight and basic metabolism. In this study a regression derived from data compiled by Drent et al. (1981) was used. Only fish-eating birds larger than 50 g

Table 2. Numbers of Little Cormorants, Indian Shags and Indian Darters at Parakrama Samudra and the amount of fish eaten by them. Equation (1) was used for assessing fish consumption.

Date	*Ph. niger*		*Ph. fuscicollis*		*Anhinga rufa*	
	Number	fish eaten kg fr. wt	number	fish eaten kg fr. wt.	number	fish eaten kg fr. wt.
12.9.79	66	6	120	21	117	27
14.9.79	84	8	13 089	2 238	–	–
18.9.79	529	51	13 699	2 343	–	–
4.3.80	1 026	98	81	14	17	4
17.3.80	1 851	178	241	41	22	5
24.3.80	1 470	141	212	36	–	–
28.3.80	1 770	170	346	59	86	20
1.4.80	1 350	130	280	48	–	–
8.4.80	1 631	157	388	66	103	23
29.1.82	41[a]	4	3 251	556	–	–
3.2.82	296	28	1 594	273	–	–
18.2.82	1 264	121	214	37	–	–
weighted means	747.2	71.7	3 479.3	595.0	–	–

[a] Count not complete; see text for further explanation.

were considered, the 15 data pairs yield the following equation:

(1) $\log(y) = 0.861188 \cdot \log(x) - 0.3233879$

where y is fish fresh weight in grams and x is bird weight in grams. Equation (1) gives the amount of fish needed to cover the daily energy expenditure of a non-breeding bird. Based on this equation the fish consumption of *Ph.* carbo (2 100 g) would be 345 g fish^{-1} d^{-1} bird^{-1}, of *Ph. fuscicollis* (930 g) 171 g fish/d/bird, and of *Ph. niger* (475 g) 96 g fish d^{-1} bird^{-1}. These values were used for calculating the corresponding entries in Table 2. Assuming a mean daily number of 84 Southern Cormorants the mean daily fish consumption of the three species together can be estimated as 696 kg fish d^{-1}.

The data above may be used for evaluating the nutrient flow also. Daily uptake rates may be computed (cf. Vareschi 1979) by assuming with Bull & Mackay (1976) that 0.04% of the body weight of fish is phosphorus and 2.7% nitrogen and stipulating a balanced nutrient budget. Supposing that Little Cormorants and Indian Shags drop half of their feces outside the lake (cf. Vareschi 1979), they export 2.32 mg P m^{-2} a^{-1} (15.63 mg N m^{-2} a^{-1}) and 19.20 mg P m^{-2} a^{-1} (129.11 mg N m^{-2} a^{-1}) when the corresponding values are related to the area covered at full supply level (2 262 ha). This is about 2–27% of the P lost through the outflow (data from 1980, Gunatilaka, personal communication) or 8–32% of the N, if both species are considered together. Bédard et al. (1980) measured the amount of soluble nutrients in the feces of fish-eating birds (gulls). Using their data and adjusting for the weight of the cormorants, an Indian Shag excretes 18.2 mg N d^{-1} and 6.2 mg P d^{-1} in soluble form. The corresponding figures for Little Cormorants are 14.7 mg P d^{-1} and 4.9 mg N d^{-1}. However, the rest of the P and N in the feces is eventually transformed into soluble fractions (Bédard et al. 1980), certainly rather a fast process under tropical conditions.

3. Discussion

The impact of the cormorants in terms of amount of fish eaten can be estimated at 254 metric tons per year. Expressed as fish eaten per ha per year the estimate can be put at 112–161 kg fish fresh weight ha^{-1} a^{-1}, based on the lake's area at full supply level and at the dry season (1979), respectively.

For Polish temperate zone lakes Dobrowolski et al. (1976) estimated the total fish consumption by birds to be between 10–20 kg ha^{-1} a^{-1}. *Ph. lucidus* was reputed to eat 810 kg fish ha^{-1} a^{-1} in a South

African lake (Donelly 1967). The data of Bowmaker (1963) on *Ph. africanus* are much more difficult to compare because of the heterogeneity of the 388 500 ha – 777 000 ha of the Bangweula swamps (Zimbabwe) which he investigated. There cormorants would only consume 0.33–0.63 kg fish ha^{-1} a^{-1}. At Lake Kariba, Zimbabwe, *Ph. africanus* would remove as much as 9 180 kg fish ha^{-1} a^{-1}, based on figures from the zone where they usually hunt a figure certainly too high for the whole lake. At any rate, the data from Parakrama Samudra are by no means excessively high and might even be lower than in areas with a well established lake-fish fauna.

Another important point in assessing impact is the amount of commercially valuable fish consumed by the cormorants. Here cichlids, especially *Tilapia*, are of most interest. The diet of *Ph. africanus* in the Bangweula swamp (Bowmaker 1963) contained (by numbers) 16% *Tilapia* (3 species); at Lake Kariba *Tilapia rendalli* and *T. mortimeri* accounted, in numbers, for 9.7% of the food of this species and for 52.6% by weight (Birkhead 1978). Whitfield & Blaber (1978) fond that guts of *Ph africanus* at Lake St. Lucia, South Africa contained 33% (by number) *T. mossambica* and 9‰ *Ph (carbo) lucidus*. Is is worth mentioning that *Ph. sulcirostris* in New South Wales preferred (95% by weight) to eat introduced species, especially Common Carp. This suggests that newly introduced species may be responsible for an increase of native bird predators. For instance, the introduction of *T. grahami* into Lake Nakuru, Kenya, in 1960 led to an increase of *Pelecanus onocrotalus*, which now probably harvests about 1 640 kg fish ha^{-1} a^{-1} (Vareschi 1979). Considering the data given above, the fact that cormorants at Parakrama Samudra eat 50–67% (by numbers) cichlids is not surprising. The size class of the prey harvested by predators is important for the effects on the population dynamics of the former. Riedel (1965) concludes from his study of *Tilapia mossambica* in Nicaragua that eradication of birds would necessitate intensification of netting activities in order to harvest the lower size classes. Otherwise fishing should continue to be restricted to the upper size classes.

The extension of fishing from the larger size classes to the smaller ones may cause dwarfing (cf. Gwahaba 1973) whereas moderate natural predation on the small size classes may help to achieve an economic, sustained yield of the larger ones.

Bowmaker (1963) argues that the presence of *Ph. africanus* is advantageous to the general productivity of his study area chiefly by making available the nutrients locked away within living non-commercial species. Fish-eating caimans in South America are thought to help to introduce alochthonous nutrients into aquatic systems, leading to the statement 'more predators on fish, the more fish' (Fittkau 1970). Quantitative data or dynamic models supporting these views are not presented in either case. Great White Pelicans at Lake Nakuru have been estimated to export about 330 mg P m^{-2} a^{-1} (2 200 mg N m^{-2} a^{-1}), which is almost 10% (this lake is without outflow) of the lake's total phosphorus (Vareschi 1979). Generally, however, the role of birds in nutrient cycling is probably a minor one. Bédard et al. (1980) suggested that seabirds can

Table 3. Fish-eating birds observed at Parakrama Samudra in the years 1979, 1980, and 1982. Taxonomy follows largely W. W. A. Phillips, Annotated Checklist of the Birds of Ceylon, Colombo 1978.

Indian Little Grebe	*Podiceps ruficollis capensis*
Grey Pelican	*Pelicanus philippensis philippensis*
Southern Cormorant	*Phalacrocorax carbo sinensis*
Indian Shag	*Phalacrocorax fuscicollis*
Little Cormorant	*Phalacrocorax niger*
Darter	*Anhinga rufa*
Eastern Grey Heron	*Ardea cinerea rectirostris*
Eastern Purple Heron	*Ardea purpurea manilensis*
Little Green Heron	*Butorides striatus javanicus*
Indian Pond Heron	*Ardeola grayii grayii*
Great White Heron	*Egretta alba modesta*
Median Egret	*Egretta intermedia intermedia*
Little Egret	*Egretta garzetta garzetta*
Night Heron	*Nycticorax nycticorax nycticorax*
Black Bittern	*Dupetor flavicollis flavicollis*
Painted Stork	*Ibis leucocephalus*
Indian White-necked Stork	*Ciconia episcopus episcopus*
Brahminy Kite	*Haliastur indus indus*
White-bellied Sea-Eagle	*Haliaeetus leucogaster*
Ceylon Grey-headed Fish-Eagle	*Ichthyophaga ichthyaetus plumbeiceps*
Indian Whiskered Tern	*Chlidonias hybrida indica*
Gull-billed Tern	*Gelochelidon nilotica nilotica*
Bridled Tern	*Sterna anaethetus anaethetus*
Little Tern	*Sterna albifrons*
Ceylon Pied Kingfisher	*Ceryle rudis leucomelanura*
Ceylon Common Kingfisher	*Alcedo atthis tabrobana*
Stork-billed Kingfisher	*Pelargopsis capensis capensis*

hardly be viewed as important agents in the dynamic nutrient regeneration process of marine coastal waters. Sturges et al. (1974) reach similar conclusions for a terrestrial system. In Polish Lakes (Dobrowolski et al. 1976) birds (8% fish eating) return up to 29 mg P m^{-2} a^{-1} (47 mg N m^{-2} a^{-1}) as feces. Impact or export is of minor importance. The corresponding values for Parakrama Samudra are comparatively low. The role of piscivorous birds in the nutrient dynamics of this system can only be appreciated by collecting more data and testing the 'impact hypothesis' with a model of the nutrient dynamics. Cormorants are not the only fish-eating birds at Parakrama Samudra. Indian Darter *Anhinga rufa* and Spotted-billed Pelicans *Pelecanus philippensis* are quite common close relatives of the cormorants, and together with the other piscivorus species (Table 3) contribute to the continuous predation upon the fishes of this lake. Although predation is ubiquitous no harmful effects to the fisheries have yet been detected.

Acknowledgement

The author wishes to acknowledge the help of R. Hacker and P. Newrkla in both field and laboratory.

References

Ali, Sálim & Ripley, S. D., 1968. Handbook of the birds of India and Pakistan, vol. 1.

Bédard, J., Therriault, J. C. & Bérubé, J., 1980. Assessment of the importance of nutrient recycling by seabirds in the St. Lawrence Estuary. Can. J. Fish. Aquat. Sci. 37: 583–588.

Birkhead, M. E., 1978. Some aspects of the feeding ecology of the Reed Cormorant and Darter on Lake Kariba, Rhodesia. Ostrich 49: 1–7.

Bowmaker, A. P., 1963. Cormorant predation on two Central African lakes. Ostrich 34: 2–26.

Bull, C. J. & Mackay, W. C., 1976. Nitrogen and phosphorus removal from lakes by fish harvest. J. Fish. Res. Board Can. 33: 1374–1376.

Cambray, J. A., Handiek, S. & Handiek, Q., 1978. The juvenile fish population in the marginal areas of the Hendrik Verwoerd Dam. J. Limnol. Soc. sth. Afr. 4: 21–30.

Curio, E., 1976. The ethology of predation. Springer. Berlin, Heidelberg, New York.

Dobrowolski, K. A., Halba, R. & Nowicki, J., 1976. The role of birds in eutrophication by import and export of trophic substances of various waters. Limnologica 10: 543–549.

Donelly, B. G., 1967. An index of the predation impact by birds on the fish population of North End Lake for December 1966. Limnol. Soc. sth. Africa Newsletter No. 9: 23–27.

Drent, R., Ebbinge, B. & Weijand, B., 1981. Balancing the energy budgets of arctic-breeding geese throughout the annual cycle: a progress report. Verh. orn. Ges. Bayern 23: 239–264.

Fittkav, E.-J., 1970. Role of colimans in the nutrient regime of mouth-lakes of Amazon affluents (an hypothesis). Biotropica 2: 138–142.

Gadgil, M. & Ali, Sálim, 1975. Communal roosting habits of Indian birds. J. Bombay Nat. Hist. Soc. 72: 716–727.

Gwahaba, J. J., 1973. Effects of fishing on the *Tilapia nilotica* (Linné 1957) population in Lake George, Uganda over the past 20 years. E. Afr. Wildl. J. 12: 317–328.

Henry, G. M., 1971. A guide to the birds of Ceylon. 2nd ed. Oxford University Press, Oxford.

Hoffman, W., Heinemann, D. & Wiens, J. A., 1981. The ecology of seabird feeding flocks in Alaska. Auk 98: 437–456.

Kanwisher, J. W., Gabrielsen, G. & Kanwisher, N., 1981. Free and forced diving in birds. Science 211: 717–719.

Lack, D., 1945. The ecology of closely related species with special reference to Cormorant (*Phalacrocorax carbo*) and Shag (*P. aristotelis*). J. Anim. Ecol. 14: 12–16.

Miller, B., 1979. Ecology of the Little Black Cormorant, *Phalacrocorax sulcirostris*, and Little Pied Cormorant, *P. melanoleucos*, in inland New South Wales I. Food and feeding habits. Aust. Wildl. Res. 6: 79–95.

Mukherjee, A. K., 1969. Food-habits of waterbirds of the Sundarban, 24-Parganas district, West Bengal, India. J. Bombay nat. Hist. Soc. 66: 345–360.

Riedel, D., 1965. Some remarks on the fecunditiy of *Tilapia* (*T. mossambica* Peters) and its introduction into Middle Central America (Nicaragua) together with a first contribution toward the limnology of Nicaragua. Hydrobiologia 25: 357–388.

Serventy, D. L., 1939. Notes on cormorants. Emu 38: 357–371.

Stonehouse, B., 1967. Feeding behaviour and diving rhythms of some New Zealand Shags, Phalacrocoracidae. Ibis 109: 600–605.

Sturges, F. W., Holmes, R. T. & Likens, G. E., 1974. The role of birds in nutrient cycling in a northern hardwoods ecosystem. Ecology 55: 149–155.

Vareschi, E., 1979. The ecology of Lake Nakuru (Kenya) II. Biomass and distribution of fish (*Tilapia grahami* = *Sarotherodon alcalicum grahami* Boulenger.) Oecologia 37: 321–335.

Whitfield, A. K. & Blaber, S. J. M., 1978. Feeding ecology of piscivorous birds at Lake St. Lucia, Part 3: Swimming birds. Ostrich 50: 10–20.

Author's address:
H. Winkler
Institute of Limnology
Austrian Academy of Sciences
A-5310 Mondsee
Gaisberg 116
Austria

19. Parakrama Samudra Project – a summary of main results

F. Schiemer & A. Duncan

Keywords: tropics, reservoir, trophic structure, ecosystem

Abstract

The following chapter provides a short synopsis of main results obtained during two research visits (August/September 1979, February/April 1980). It summarizes the major physiographical characteristics of the lake and outlines its trophic structure. Emphasis has been given to factors governing production processes.

1. Physiography

The physiographical and hydrological features of the Parakrama Samudra reservoir (PS) have been discussed in Schiemer (1983). PS consists of three well-separated basins, the northern (PSN) and southern (PSS) being smaller and shallower than the middle part (PSM) (6.5, 3.6 vs 15.4 km^2 at full supply level; maximal depth, $z_m = 8.2$ vs 12.7 m). The lake forms a flow-through system from the artificial inflow channel of the Amban Ganga into PSS to outflow channels situated in PSS (one) and PSN (two). Most water is passing from the inflow channel to the northernmost outflow at PSN.

Due to the monsoonal cycle in the 'dry zone' of Sri Lanka (rains from October–December) and due to the seasonal usage of the reservoir for irrigation of rice fields, the hydrological regime is characterized by high water levels from December to April and low water levels from June to August, with an average annual amplitude of 3.7 m. As a consequence of the seasonality of inflow, outflow and water storage, the water renewal rates exhibits strong seasonal differences with highest relative rates in July and low ones from December to May.

The temperature regime is characterized by a strong diurnal cycle (e.g. from 28–34 °C in September), expressed either over the whole water column during windy periods or in form of strong thermal stratification patterns building up during the day. Under calm conditions, the stability of this stratification prevents mixing of the whole column by even strong afternoon thunderstorms.

The kinetic energy released by the breakdown of thermal gradients in the course of the night, however, results in homeothermy plus an erosional effect on the bottom sediments which lead to resuspension of bottom material during night and in the early morning hours. In the course of the day, differences in heating and cooling between littoral and offshore areas lead to considerable advective currents (Bauer 1983). The underwater light conditions are considered in connection with the primary production processes in the lake (see below).

The ionic composition of the lake water is discussed by Gunatilaka & Senaratne (1981). The ratio of anions (in % of mequ sum), HCO_3^-: Ce^-: SO_4^{--} is 74 : 14 : 13; that of cations, Ca^{++}: Mg^{++}: Na^+: K^+ is 48 : 34 : 15 : 3. pH ranged from 8.7–9.6 in September 1979 and from 8.3–8.9 in March 1980. A higher ionic concentration in March (218–243 μS compared to 130–180 μS in September is likely to

Schiemer, F. (ed.), Limnology of Parakrama Samudra – Sri Lanka
 ISBN 90 6193 763 9

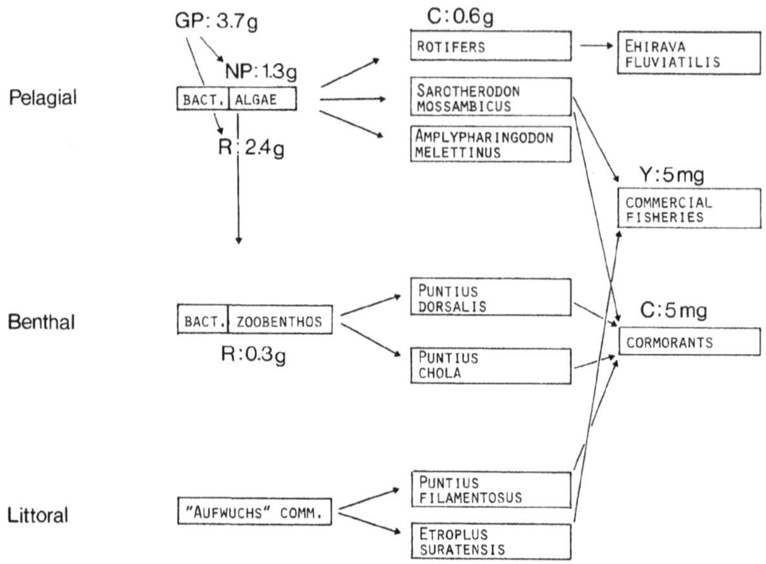

Fig. 1. Main trophic interrelationship of the PS ecosystem with some data on carbon transfer. The data refer to the offshore region of PSN on 1 September 1979, except for the commercial fisheries yield (Y) and food consumption (C) of cormorants, which are based on annual averages for the whole lake area. Data in g (or mg) C per m^2 lake area per day. GP, NP gross, respectively, net primary production, R = community respiration of the limnetic and benthic zone.

be associated with leaching of soluble ions and slow seepage after monsoonal rains.

The lake water is characterized by a high content of nutrients (e.g. phosphorus and nitrogen), present mainly in particulate organic form and thus not readily available to algal growth. The main components of an input-output budget have been discussed in Schiemer (1980, 1981) and Gunatilaka & Senaratne (1981). Dissolved inorganic nutrients other than silica (5.5–11.8 $\mu g\ l^{-1}$) are present only in low concentrations (e.g. soluble reactive phosphate – SRP – 0–6 $\mu g\ l^{-1}$ in September 1979, 2.5–13 $\mu g\ l^{-1}$ in March 1980.

2. Trophic structure of the lake

In PSN where most research work was carried out, three main habitat types can be distinguished (a) a shallow limnetic zone, (b) soft mud sediments in the deeper parts of the lake and (c) a littoral zone mostly with sandy bottoms, to a minor extent with gravel and rocks. Aquatic macrophytes are only at a few locations of periodic importance. Flooded terrestrial vegetation is generally of higher importance as food and substrate of an epiphytic community.

Ad (a) The limnetic zone is characterized by a highly diverse phytoplankton assemblage (83 taxa acc. to Rott 1983). Blue-green algae and diatoms predominated almost everywhere.

In August/September 1979 *Microcystis* spp., *Anabaenopsis raciborskii* Wol. and the diatom *Melosira granulata* (Ehrb.) Ralfs were most abundant in PSN. In March/April 1980 the dinoflagellates *Peridiniopsis* spp. were additionally of importance. Thus, phytoplankton consists to some extent of large colonial and filamentous forms, considered to be unsuitable as food, especially for small-sized zooplankton. However, a high proportion of the limnetic algal biomass (57–94% measured as chlorophyll-a < 94% measured as sestonic carbon) is contained in the filter size-fraction < 33 μm. This suggests that small algal forms contribute considerably to the algal biomass and its production. The concentration of the < 33 μm fraction of sestonic carbon was high in comparison to temperate lakes (e.g. 11 mg.C l^{-1} in 1979; 3 mg C l^{-1} in 1980), (Duncan & Gulati 1983).

The zooplankton consisted of very small sized

rotifers (14 ssp) and protozoans. Crustaceans and large-sized rotifers were virtually absent. A considerable amount of indirect evidence (Duncan 1983) indicate that fish predation is the most important of several factors affecting the zooplankton, accounting both for the very low abundance of crustaceans and the size distribution of rotifers. Amongst the potential zooplanktivorous species are the clupeiform *Ehirava fluviatilis* Deraniyagala, larvae of a hemiramphid species, the smaller size classes of *Rasbora daniconius* (Ham.-Buch.) and *Sarotherodon mossambicus* (Peters). Gut analysis of the small sized *Ehirava fluviatilis* in 1980 revealed the presence of rotifers as well as crustaceans (Duncan 1983) and intensive sampling during the summer 1982 confirmed that this species was the most abundant zooplanktivorous fish in the lake. Potentially, *Rasbora daniconius* could be predatory on large-sized crustacean plankton, but the species has not been found to feed on rotifers. Small *S. mossambicus* (Tl 8–10 mm) capture zooplankton but their distribution in the shallow littoral precludes them to exert a major predatory pressure on the open water zooplankton.

Food limitation is a further factor which could influence body size and density of rotifer populations in the lake. Porriot (1973) suggests that small adults and eggs may arise from poor nutritive conditions during the juvenile phase. Although the concentration of sestonic particles less than 33 μm was quite high compared to temperate lakes (3–11 mg C l^{-1}), there is evidence from the rotifer feeding ecology that concentrations may be suboptimal, i.e. not allowing for maximal performances with respect to body growth and population dynamics.

Two common fish species feed on phytoplankton, either in form of suspended algae or on flocculant material on the sediment surface. These are *Sarotherodon mossambicus* and the small sized cyprinid *Amblypharyngodon melettinus* (Cuv. et Val.). *S. mossambicus* feeds continuously throughout the day diurnal changes in the vertical distribution of the species in the water column suggests a day-time feeding on the sedimented material and a night-time feeding in the open water (Hofer & Schiemer 1983). *A. melettinus*, the second common phytoplanktivorous species in the lake is filtering seston in the water column.

Ad (b) The zoobenthos living in soft mud in PSN is predominated by meiobenthos, especially the cladocerans *Macrothrix* spp. and *Alona* spp., and lower densities of cyclopids, ostracods and nematodes (Schiemer, in preparation). Macrobenthos is sparse, consisting of small-sized chironomid species and few oligochaetes (naididae mainly, *Branchiura sowerbyi* Beddard, very rare).

Heavy predation by zoobenthivorous fish viz. *Puntius chola* (Ham.-Buch.) and *Puntius dorsalis* (Jerdon) and to a lesser degree by *Mystus vittatus* (Bloch) is the cause for the dominance of small-sized zoobenthos. Gut analysis of mature *P. chola* and *P. dorsalis* showed a clear selectivity of larger sized prey. High food overlap values between the two species indicated food competition, which is relaxed by different diurnal feeding patterns (*P. chola* continuous, *P. dorsalis* during the night) and a tendency of spatial segregation by inshore migrations of *P. dorsalis* during the period of its highest feeding intensity (Schiemer & Hofer 1983).

Ad (c) The littoral zone with rocks and flooded terrestrial vegetation develop a dense and diverse Aufwuchs community with sponges, bryozoans, freshwater shrimps (*Caridina* sp.), entomostracans and various aquatic insects of quantitative importance. The nature of the marginal habitats is changing continually due to the water level fluctuations. Molluscs are restricted to a belt between the soft mud sediments and an upper border which appears to be mainly affected by the drawdown regime, and predation by mollusc-eating birds (Bretschko in litt.).

The mean width of the mollusc-belt is approximately 50 m. The following species, ranked according to their quantitative importance, have been recorded: *Parreysia corrugata* (Müll.), *Lamellidens marginalis* (Müll.), *Melanoides tuberculatus* (Müll.), *Bellamya dissimilis* (O.F. Müller) var. *ceylonica* (Dohrn), *Thiara scabra* (Müll.). The very low water level in 1979 accompanied by strong predation, e.g. by storks, lead to a strong reduction in the population density of the mollusc fauna. (In 1982 population densities had again increased.)

Several species of fish, some of them of economical importance, show a predominance of inshore sites and are feeding on animal and plant Aufwuchs organisms and macrophytes. The most important among them are *Etroplus suratensis* (Val.), *Puntius filamentosus* (Val.), *Puntius sarana* (Ham.-Buch.) and *Labeo dussumieri* (Val.). Areas with submerged trees (e.g. PSM) form good conditions for this fish assemblage. The littoral area is the breeding zone of *S. mossambicus* and the favoured habitat of its fry. Short-term water level fluctuations may result in severe disturbance of the breeding activity of the species. In summer 1982 we observed that receeding water level leeds to a considerable loss of *S. mossambicus* fry in marginal pools which are drying out. A strong predation effect on the fry is exerted by several species of fish, e.g. *Rasbora daniconius, Glossogobius giuris* Russell, *Ophiocephalus* sp., *Puntius sarana* and by fish-eating birds.

Cormorants by their high population densities (maximal numbers approximately 15 000 ind.) exert an overall important effect on the fish population of the lake (see below). They are represented by three species, the Large Cormorant *Phalacrocorax carbo sinensis* (Shaw), the Indian Shag *P. fuscicollis* Stephens and the Little Cormorant *P. niger* (Vieillot). These three species differ in size, foraging behaviour, fishing distance from the shore and diving depth. Little Cormorants have the most diverse diet including many species of fish, insect larvae and crustaceans. Indian Shag and even more so Large Cormorants have a less variable diet. The size range of their food species is correlated with the size of the three species. Large Cormorants hunt fishes of an average size of 130 mm (standard length), Indian Shag of 60–120 mm, Little Cormorants of 30–70 mm.

Cichlids (*Sarotherodon mossambicus* and *Etroplus suratensis*) seem to form the main prey of all three species. Little Cormorants hunt within the littoral zone, Indian Shag along the edge of macrophytic vegetation and Large Cormorants at the open water.

Winkler (1983) calculated the mean fish consumption by cormorants as 969 kg fish per day, which roughly converts to 0.5 mg C m^{-2} d^{-1}. It is of significance to note that these values approximately corresponded to the daily catch of the commercial fisheries. In order to assess the overall effects of cormorants on the lake ecosystem this data can also be used to evaluate their role in the nutrient budget and nutrient recycling. Cormorants will cause a nutrient loss from the lake since part of the defecation occurs when birds leave the lake for roosting. Part of the feces with high fractions of soluble nutrients, however, will be deposited in the lake and should be of significance in accelerating nutrient recycling.

3. Factors governing production processes

The phytoplankton biomass of PSN was generally high. The mean areal biomass figures recorded were 95 mg chlorophyll-a m^{-2} (n = 8) for September 1979 and 59 mg chl-a m^{-2} (n = 18) for March 1980 (mean concentrations were 47 and 15 μg chl-a l^{-1} for the two periods, respectively). These seasonal differences in areal algal biomass have to be contrasted with differences in light extinction and water level. In September 79 at low water level the light extinction coefficient for the photosynthetical active radiation (ϵ_{phar} m^{-1}) ranged from 2.9–4.4, which corresponded with a Secchi disc depth (z_{SD}) of 0.4–0.6 m and a depth of the euphotic zone (z_{eu}, taken as the 1% level of the incoming irradiance) of 1.05–1.66 m. In March/April 1980, at higher water level and less wind, light penetration was higher (ϵ_{phar} = 0.9–1.3; z_{SD} = 1.1–1.4 m; z_{eu} = 3.1–4.3 m). The mean compensation depth was coinciding in both seasons roughly with the mean depth of PSN (z = 1.5 m in 1979 and 2.9 m in 1980), i.e. the lake was optically shallow during both visits (Dokulil *et al.* 1983).

Applying Steele's (1975) general considerations of algal productivity in reservoirs and assuming no nutrient limitation, a B_{max} (maximal sustainable areal algal biomass) of 150 and 100 mg chl-a m^{-2} respectively for the two seasons can be predicted, based on extinction coefficients and the depth of the mixed water column. The predicted value of B_{max} is depressed by higher levels of 'r' (respiration per unit chl-a as a fraction of photosynthesis per unit chl-a) (see also Dokulil *et al.* 1983). Since 'r' is based on community respiration the model may have its limitations for eutrophic, tropical lakes with high bacterial and zooplanktic respiration.

The differences between the predicted B_{max} and the realized values can be due to (a) nutrient limitations of photosynthesis (CO_2 and/or nutrients), (b) biomass losses due to outflow, grazing and sedimentation.

Although areal gross primary production values are comparatively high (7.5–14.7 g O_2 m^{-2} d^{-1} in 1979; 3.8–8.6 g O_2 m^{-2} d^{-1} in 1980), there is indication that the productivity of the lake is to some extent nutrient limited. For example, nutrient deficiency indicators such as phosphatase activity and the stored algae phosphorous content showed strong diurnal dynamics, indicating nutrient limitations in the course of the day (Gunatilaka & Seneratna 1981; Gunatilaka 1983). Further, P_{max} values (light saturated chlorophyll specific production) calculated for different periods of the day showed a positive correlation with the diurnally changing SRP levels (Dokulil *et al.* 1983).

Effects of hydrological flushing and dilution on the limnetic biomass were mainly observed during periods of low water level and times of strong changes in water throughput rates (e.g. from 2–40% daily throughput rates of the PSN volume). Periods of decline in phytoplankton biomass were associated with high water inflow from PSM resulting in a dilution effect (lower phytoplankton concentration in this part) and high biomass losses due to the outflow at PSN. When flushing rates decreased, the concentrations increased within a few days. Exactly the same pattern was observed in zooplankton densities (Duncan & Gulati 1981). The importance of flushing became also apparent from the spatial pattern of biomass distribution in the southern and northern part of PSN, which at low water level in 1979 were separated by a peninsula (represented by PSN 8 and PSN 3, see map in Schiemer 1983). Due to the more direct exposure of flushing phytoplankton (Dokulil *et al.* 1983), zooplankton (Duncan & Gulati 1981) and meiobenthos (Schiemer, in preparation) occurred at significantly lower densities in the southern bay. In March/April 1980, at a higher water volume and generally lower flushing rates (%) no similar effects were recognized.

Grazing effects on algal biomass are severe. Studies were conducted on planktonic rotifers (*Brachionus* ssp.), using a C_{14}-technique with 33 μm lake seston as food source. At the food densities prevailing in the PS reservoir, a mean grazing rate of 0.013 $\mu g\ C\ ind^{-1}\ h^{-1}$ (Duncan & Gulati, 1983) was determined. Assuming continuous feeding, the grazing rate of the total *Brachionus* population was calculated as 0.6 $g\ C\ m^{-2}\ d^{-1}$ for September 1979. Under the assumption that all rotifer species encountered in the limnetic zone have similar feeding rates, a total areal grazing rate of 2.4 $g\ C\ m^{-2}\ d^{-1}$ is estimated for the period of high rotifer abundance in September 1979 and 0.6 $g\ C\ m^{-2}\ d^{-1}$ for March 1980 (mean population densities of 7.5×10^6 and 2.4×10^6 ind m^{-2}, respectively). This is a high portion of the areal primary production of the lake. A carbon budget has been calculated for 1 September 1979 and according to above values the net primary production would not even balance the total rotifer grazing rate.

Feeding rates have also been established for one size class of *Sarotherodon mossambicus* (19 g fresh weight) which was abundant in the open lake in 1980. Daily food consumption was estimated as 0.8 g dry weight $ind^{-1}\ d^{-1}$ (or roughly 0.4 g C $ind^{-1}\ d^{-1}$) which roughly equals a third of the net primary production per m^2 lake area.

The role of the herbivorous grazers is controlling primary production processes in the lake will have to be considered in more detail both with regard to its effects on algal biomass and with regard to nutrient recycling in the open lake.

References

Bauer, K., 1983. Thermal stratification, mixis and advective currents in the Parakrama Samudra Reservoir, Sri Lanka. In: Schiemer, F. (ed.) Limnology of the Parakrama Samudra – Sri Lanka: a case study of an ancient man-made lake in the tropics. Developments in Hydrobiology (this volume). Dr W. Junk, The Hague.

Dokulil, M., Bauer, K. & Silva, I., 1983. An assessment of the phytoplankton biomass and primary productivity of Parakrama Samudra, a shallow man-made lake in Sri Lanka. In: Schiemer, F. (ed.) Limnology of the Parakrama Samudra – Sri Lanka: a case study of an ancient man-made lake in the tropics. Developments in Hydrobiology (this volume). Dr W. Junk, The Hague.

Duncan, A., 1983. The composition, density and distribution of the zooplankton in Parakrama Samudra. In: Schiemer, F. (ed.) Limnology of the Parakrama Samudra – Sri Lanka: a case study of an ancient man-made lake in the tropics. Developments in Hydrobiology (this volume). Dr W. Junk, The Hague.

Duncan, A. & Gulati, R., 1981. Parakrama Samudra (Sri Lanka) Project – a study of a tropical lake ecosystem. III. Composition, density and distribution of the zooplankton in 1979. Verh. Internat. Verein. Limnol. 21: 1001–1006.

Duncan, A. & Gulati, R., 1983. Feeding studies with natural food particles on tropical species of planctonic rotifers. In: Schiemer, F. (ed.) Limnology of the Parakrama Samudra – Sri Lanka: a case study of an ancient man-made lake in the tropics. Developments in Hydrobiology (this volume). Dr W. Junk, The Hague.

Gunatilaka, A., 1983. Phosphorus and phosphatase dynamics in Parakrama Samudra based on diurnal observations. In: Schiemer, F. (ed.) Limnology of the Parakrama Samudra – Sri Lanka: a case study of an ancient man-made lake in the tropics. Developments in Hydrobiology (this volume). Dr W. Junk, The Hague.

Gunatilaka, A. & Senaratna, C., 1981. Parakrama Samudra (Sri Lanka) Project, a study of a tropical lake ecosystem. II. Chemical environment with special reference to nutrients. Verh. Internat. Verein. Limnol. 21: 994–1000.

Hofer, R. & Schiemer, F., 1983. Feeding ecology, assimilation efficiencies and energetics of two herbivorous fish: *Sarotherodon (Tilapia) mossambicus* (Peters) and *Puntius filamentosus* (Cuv. et Val.). In: Schiemer, F. (ed.) Limnology of the Parakrama Samudra - Sri Lanka: a case study of an ancient man-made lake in the tropics. Developments in Hydrobiology (this volume). Dr W. Junk, The Hague.

Pourriot, R., 1973. Rapports entre la temperature, la taille des adultes, la longueur des oeufs et le taux de developpment embryonnaire chez *Brachionus calyciflorus* Pallas (Rotifere). Ann. Hvdrobiol. 4: 103–115.

Rott, E., 1983. A contribution to the phytoplankton species composition of Parakrama Samudra, an ancient man-made lake in Sri Lanka. In: Schiemer, F. (ed.) Limnology of the Parakrama Samudra – Sri Lanka: a case study of an ancient man-made lake in the tropics. Developments in Hydrobiology (this volume). Dr W. Junk, The Hague.

Schiemer, F., 1980. Parakrama Samudra (Sri Lanka) Limnology Project. Interim Report. IIZ. Vienna, 112 pp.

Schiemer, F., 1981. Parakrama Samudra (Sri Lanka) Project, a study of a tropical lake ecosystem. I. An interim review. Verh. Internat. Verein. Limnol. 21: 993–999.

Schiemer, F., 1983. Parakrama Samudra Project - Scope and objectives. In: Schiemer, F. (ed.) Limnology of the Parakrama Samudra - Sri Lanka: a case study of an ancient man-made lake in the tropics. Developments in Hydrobiology (this volume). Dr W. Junk, The Hague.

Schiemer, F. & Hofer, R., 1983. A contribution to the ecology of the fish fauna of the Parakrama Samudra reservoir. In: Schiemer, F. (ed.) Limnology of the Parakrama Samudra – Sri Lanka: a case study of an ancient man-made lake in the tropics. Developments in Hydrobiology (this volume). Dr W. Junk, The Hague.

Starkweather, P. L., 1980. Aspects of the feeding behaviour and trophic ecology of suspension-feeding rotifers. Hydrobiologia 73: 63–72.

Winkler, H., 1983. The ecology of cormorants (genus *Phalacrocorax*). In: Schiemer, F. (ed.) Limnology of the Parakrama Samudra – Sri Lanka: a case study of an ancient man-made lake in the tropics. Developments in Hydrobiology (this volume). Dr W. Junk, The Hague.

Authors' addresses:
F. Schiemer
Institute of Zoology
University of Vienna
Althanstrasse 14
A-1090 Vienna
Austria

A. Duncan
Department of Zoology
Royal Holloway College
Englefield Green
Surry TW 20 9TY
United Kingdom

Floristic and faunistic surveys

20. A contribution to the phytoplankton species composition of Parakrama Samudra, an ancient man-made lake in Sri Lanka

Eugen Rott

Keywords: Phytoplankton, tropical reservoir, SE-Asia

Abstract

Phytoplankton species composition of Parakrama Samudra was investigated. On two sampling occasions the algal flora showed a total number of approx. 84 different taxa (Table 1). The taxonomic groups most rich in species were Chlorococcales (Chlorophyceae sensu stricto) with 30, Cyanophyceae with 22 and Zygnemaphyceae with 11 different taxa. The morphology of almost all taxa observed is discussed and the most frequent phytoplankton species are illustrated.

1. Introduction

As compared to the neighbouring tropical regions, few investigations have been published on the planktonic algal flora of the inland waters of Sri Lanka. Of the more recent studies those of Foged (1976) on the freshwater diatoms deserve special mentioning, while an older paper of Holsinger (1955) covers almost all taxonomic groups of algae from three different lakes. As early as 1902 the Desmidiaceae of Ceylon were studied by W. West and G. S. West, and in 1923 W. B. Crow's taxonomic treatment of the blue-green and green algae from a total of 63 lakes, ponds and tanks was published, the material having been collected by F. E. Fritsch. The latter author had already published information on the ecological conditions prevailing in these water bodies, as well as a characterization of the planktonic and littoral algal associations (Fritsch 1907). The first quantitative studies on the phytoplankton of Lake Colombo and Lake Gregory were carried out by Apstein (1907, 1910) and Lemmermann (1907).

The present paper is based on material from Parakrama Samudra, an ancient man-made lake in the 'Dry Zone' of Sri Lanka (7° 55′N; 81° E; 59.1 m above sea level). The lake consists of 3 shallow basins ($z_{max.}$ = 12.7 m), hereafter referred to as PSN, PSM and PSS.

A general description of the limnology of the Parakrama Samudra Reservoir is given by Schiemer (1981).

2. Material and methods

The present study is based on samples taken with water sampler in September 1979 (position PSM, PSN) and with plankton net in April 1980 (mesh size 20 μm, sampling positions PSS, PSM, PSN). Fixation was carried out in September 1979 using Lugol's solution and in April 1980 with neutralized formalin. Rare species of the samples were observed in sedimentation chambers (Reichert-Biovert inverted microscope). For the more frequent species a convenient research microscope (Diapan-Reichert) was employed. Drawings were made with the help of a drawing apparatus and photographs taken with a Reichert automatic. In order to measure the cells a measuring scale was constructed for each magnification, with the aid of an objective micrometer. The dimensions given in this paper are

Schiemer, F. (ed.), Limnology of Parakrama Samudra – Sri Lanka
 ISBN 90 6193 763 9

usually extreme values and/or estimated means. The number of individuals that it was possible to measure was insufficient for a statistical evaluation.

3. Results and discussion

The data contained in previous studies shows that the dominant algal groups in the lowland lakes of Sri Lanka were either blue-green and green algae (Fritsch 1907; Crow 1923b; Svortzow 1928) or blue-green algae and diatoms (Apstein 1907; Holsinger

Table 1. Phytoplankton species list.

CYANOPHYCEAE
Anabaena aphanizomenoides Forti fa.
Anabaena sp. ad *A. solitaria* Kleb.
Anabaenopsis elenkinii fa. *circularis* (G. S. West) Jeeji-Bai
Anabaenopsis raciborskii Wol.
Chroococcus sp. ad *C. dispersus* Lemm.
Chroococcus limneticus Lemm.
Coelosphaerium kuetzingianum Näg.
Dactylococcopsis smithii R. et F. Chodat
Gomphosphaeria naegeliana (Unger.) Lemm.
Gomphosphaeria pusilla (Van Gor) Kom.
Lyngbya circumcreta G. S. West
Lyngbya limnetica Lemm.
Merismopedia punctata Meyen
Merismopedia tenuissima Lemm.
Microcystis aeruginosa Kütz.
Microcystis incerta Lemm.
Microcystis wesenbergii Kom.
Oscillatoria chlorina (Gom.) Kütz.
Oscillatoria raciborskii Wol.
Oscillatoria sp.
Pseudanabaena galeata Böcher
Spirulina sp. ad *S. subsalsa* Oerst.

CRYPTOPHYCEAE
Cryptomonas sp.
Rhodomonas lacustris Pasch. et Ruttn.

DINOPHYCEAE
Cystodinium sp.
Gymnodinium sp.
Peridiniopsis cunningtonii Lemm. *tab. remotum* (Lindem.) Lefèvre
Peridiniopsis pygmaeum (Lindem.) Bourr.
Peridinium inconspicuum Lemm.

DIATOMOPHYCEAE
Cyclotella sp.
Melosira granulata (Ehrenb.) Ralfs et var. *angustissima* Müller
Synedra sp.

XANTHOPHYCEAE
Goniochloris contorta (Bourr.) Ettl
Goniochloris sp. ad *G. smithii* (Bourr.) Fott
Isthmochloron lobulatum (Näg.) Skuja
Pseudostaurastrum limneticum (Borge) Chod.
Tetraplektron sp.

EUGLENOPHYCEAE
Euglena sp.
Lepocinclis sp. ad *L. ovum* (Ehrenb.) Lemm.
Phacus sp.
Trachelomonas abrupta Swir.
Tachelomonas hispida (Perta) Stein fa.
Trachelomonas volvocina Ehrenb.

CHLOROPHYCEAE
Actinastrum hantzschii Lagerh.
Ankistrodesmus bibraianus (Reinsch.) Korš.
Botryococcus braunii Kütz.
Coelastrum astroideum De Notaris var. *astroideum* Sodomk.
Coelastrum microporum Näg.
Coelastrum reticulatum (Dang.) Senn.
Crucigenia tetrapedia (Kirchn.) W. et G. S. West
Crucigeniella apiculata (Lemm.) Kom.
Dictyosphaerium tetrachotomum Printz
Franceia armata (Lemm.) Korš.
Golenkinia radiata Chod.
Monoraphidium irregularis (G. M. Smith) Kom.-Legn.
Monoraphidium sp. ad *M. setiforme* (Nyg.) Kom.-Legn. fa. *brevis* Nyg.
Oocystis sp. ad *O. marssonii* Lemm.
Pediastrum duplex Meyen
Pediastrum simplex Meyen
Pediastrum tetras (Ehrenb.) Ralfs
Quadricoccus sp. ad *Q. verrucosus* Fott
Scenedesmus acuminatus (Lagerh.) Chod.
Scenedesmus sp. ad *S. arcuatus* Lemm.
Scenedesmus sp. ad *S. ellipticus* Corda
Scenedesmus sp. ad *S. opoliensis* P. Richt.
Scenedesmus sp. ad *S. polyglobus* Hortob.
Scenedesmus sp. ad *S. spinosus* Chod.
Tetraedron caudatum (Corda) Hansg.
Tetraedron incus (Teil.) G. M. Smith
Tetraedron minimum (A. Bréb.) Hansg. var. *minimum* Kov. et var. *scrobiculatum* Lagerh.
Tetraedron triangulare Korš.
Tetrastrum heteracanthum (Nordst.) Chod.
Treubaria sp. ad *T. triappendiculata* Bern.

ZYGNEMAPHYCEAE
Cosmarium depressum (Näg.) Lund var. *planctonicum* Rev.
Cosmarium minimum W. et G. S. West
Cosmarium pusillum (Bréb.) Arch.
Cosmarium sp. ad *C. abbreviatum* Racib.
Cosmarium sp.
Mougeotia sp.
Staurastrum sp. ad *St. crenulatum* (Näg.) Delp.
Staurastrum sp. ad *St. brachioprominens* Boerg.
Staurastrum tetracerum Ralfs
Staurastrum sp.
Staurodesmus sp. ad *St. pterosporus* (Lund) Bourr. vel *St. o'mearii* (Arch.) Teil.

1955). Only in Lake Gregory (about 1800 m above sealevel) Chrysophyceae were also well represented (Apstein 1907), although completely absent from the lowland lakes.

The species list from Parakrama Samudra shows (Table 1) that there, too, the taxonomic groups represented by the most species are the blue-green and green algae. The great variability in cell size, colony form and development of gas vacuoles in the *Microcystis* species was particularly striking in all samples. This had already been remarked upon in nearly all material from Sri Lanka, and has led to the description of new species (Crow 1923a; Holsinger 1955). However, these appear to be species that are not easy to differentiate. The complete absence of Chrysophyceae from Parakrama Samudra should also be mentioned (probably as a consequence of the high trophic level). The occurrence of several (5) Xanthophyceae species, some of them morphologically variable, may be characteristic for tropical lakes.

As can be seen from Figures 1A–C, blue-green algae and diatoms were almost everywhere the predominant taxonomic groups, only in April 1980 in

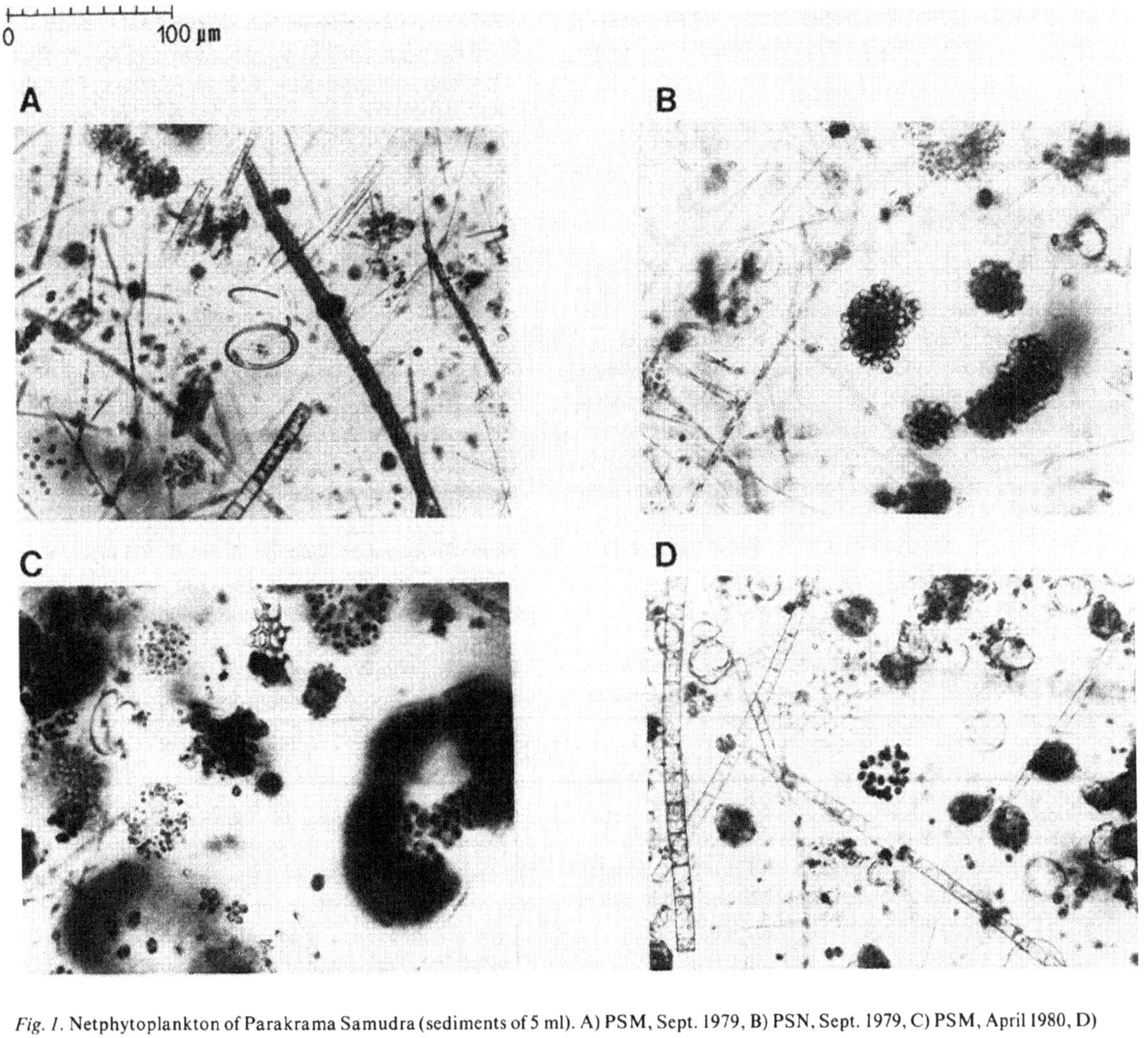

Fig. 1. Netphytoplankton of Parakrama Samudra (sediments of 5 ml). A) PSM, Sept. 1979, B) PSN, Sept. 1979, C) PSM, April 1980, D) PSN, April 1980.

PSN they were joined in any appreciable numbers by dinoflagellates (Fig. 1D). With the exception of a species of *Mougeotia,* green algae were represented by large numbers of species but only in low densities (respective quantitative data can be found in Dokulil et al. (1983)). The comparison of Figures 1A and 1B shows that there are no clear differences between the sampling sites PSN and PSM. The dominant species are *Microcystis* ssp., *Anabaenopsis raciborskii* Wol., *Melosira granulata* (Ehrb.) Ralfs and a *Mougeotia* species. In April 1980 distinct differences between PSS, PSM and PSN were noticeable. In PSS *Oscillatoria raciborskii* Wol. was dominant, in PSM a water bloom of *Microcystis* could be presumed, and in PSN Dinophyceae were well represented by the two *Peridiniopsis* species.

4. Taxonomic notes

4.1. Cyanophyceae

Anabaena aphanizomenoides Forti fa. (Fig. 2A). Cell shape variable, from globose to elongate barrel-shaped, either as long as broad (trichome diameter 3.5 to 4.5 μm) or clearly elongate (more than 11/2 times longer than broad) and slightly truncate; gasvacuoles present. Filaments rounded at one end and gradually narrowing towards the other (from 3 to 3.5 μm), with a larger terminal cell. Heterocysts elongate. In a single filament, two akinetes were seen close to one side of a heterocyst (size of akinetes approx. 7×10 μm).

Anabaena sp. ad *A. solitaria* Kleb. (Fig. 2B). Filaments solitary, straight (diameter 8–9 μm); cells clearly truncate, with gasvacuoles; heterocysts globose; akinetes, observed in one filament, solitary.

Anabaenopsis elenkinii fa. *circularis* (G. S. West) Jeeji-Bai (Fig. 2F). Coiling of filaments (3.5–4 μm in diameter) variable; cell length very variable (6–12 μm). Heterocysts at both ends of filaments, small, globose. Filaments short, often with only one coil, before division with up to 1½–2 coils.

Anabaenopsis raciborskii Wol. (Fig. 2C). Filaments always straight, very similar to those described in Skuja (1949). Filament diameter centrally 2.5–3.5 μm; cell length 6–18 *μm; heterocysts 3*–3.5 $\times$ 5–8 μm; filament length 25–130 μm. In many filaments only one heterocyst could be found, rarely two, one at each end of the filament. As proposed by Jeeji-Bai et al. (1977) *Anabaenopsis raciborskii* should be conserved as a good species of the genus *Anabaenopsis.*

Chroococcus sp. ad *Chr. dispersus* Lemm. (Fig. 2E). Cells of colonies (32, 64 or more) mostly arranged in 3 dimensions, showing irregular planes of division. Cells globose, or oblong, 3.5–4.5 μm in diameter. This alga corresponds with the species recorded by Holsinger (1952) and Desicachari (1959) as *Chroococcus dispersus* Lemm. and also with the alga, which was placed by Crow (1923b) in *Chroococcus minimus* (Keissler) Lemm. This species seems to be connected to the next one by transitional stages.

Chroococcus limneticus Lemm. (Fig. 2D). Cells ovate or globose, 4.5 $\times$ 5–6 μm, without individual cell sheaths often in 8- or 16-celled flat colonies. This species corresponds in colony form to *Chr. limneticus* but its dimensions are smaller.

Coelosphaerium kuetzingianum Naeg. (Fig. 2H). Cells globose, approx. 3–4 μm; colonies small often 30 μm in diameter, with mucilage 50 μm; with a compact inner layer of light-braking mucilage and an outer diffuse layer.

Dactylococcopsis smithii R. et F. Chodat. Cells oblong-ovate, approx. 2×8 μm; colonies small, containing often 8 cells per colony; rare species.

Gomphosphaeria naegeliana (Unger). Lemm. Cells ovate with many gasvacuoles, 2.5–3 $\times$ 5 μm; the single colony observed was 50 $\times$ 70 μm (PSN April 1980).

Lyngbya circumcreta G. S. West (2I). Trichomes coiled, 1.5–2(–2.5) μm in diameter; cells as long as broad or slightly longer; cross walls clearly visible; trichomes frequently with 1 to 3 (at maximum 5) flat coils; diameter of spirals variable (22–44 μm). This is the same *Lyngbya*-species as that of Holsinger (1955). Komarek (1958) already pointed out that *Lyngbya circumcreta* is very similar to *Lyngbya contorta* apart from the cell length (*L. circum-*

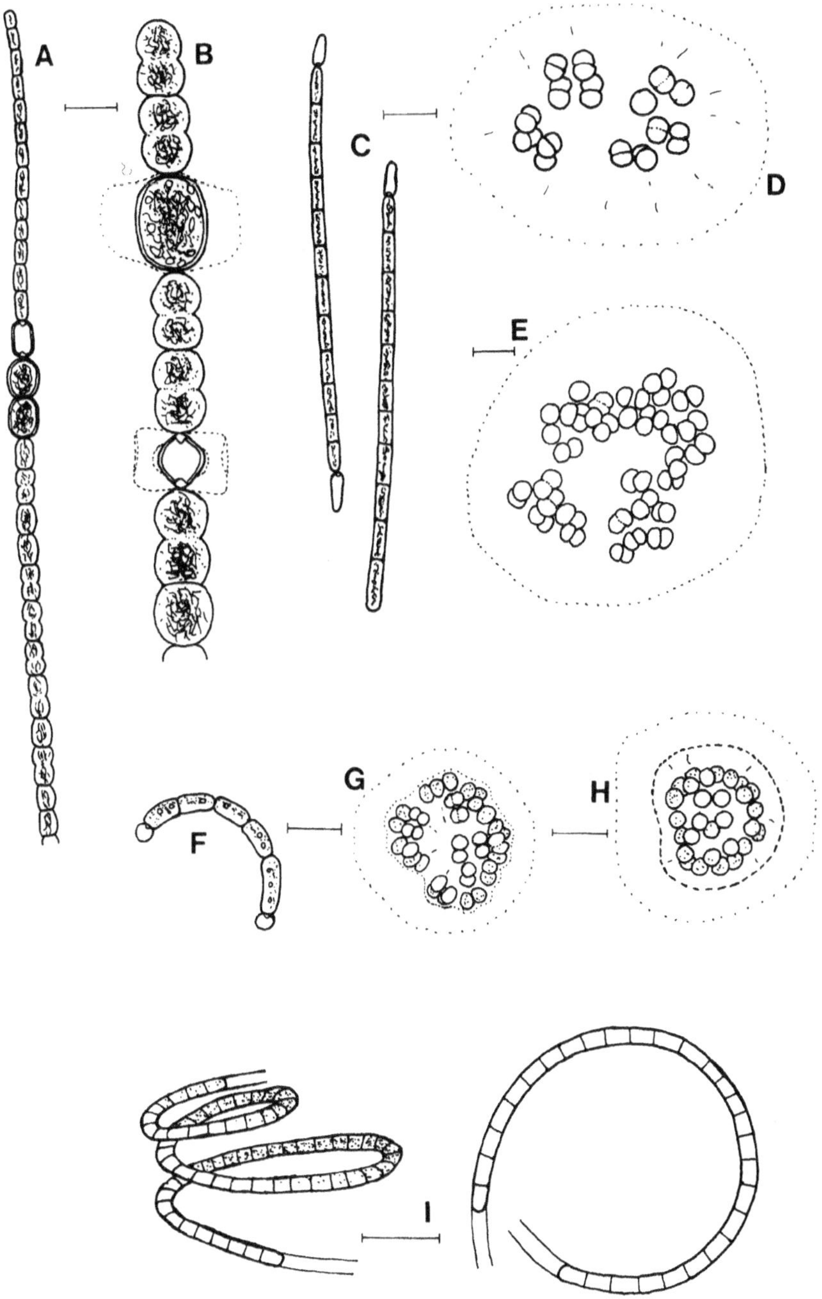

Fig. 2. A) *Anabaena aphanizomenoides* fa., B) Anabaena sp. ad *A. solitaria*, C) *Anabaenopsis raciborskii*, D) *Chroococcus limneticus*, E) *Chroococcus* sp. ad *Chr. dispersus*, F) *Anabaenopsis elenkinii* fa. *circularis*, G) *Gomphosphaeria pusilla*, H) *Coelosphaerium kützingianum*, I) *Lyngbya circumcreta*. Scale = 10 μm.

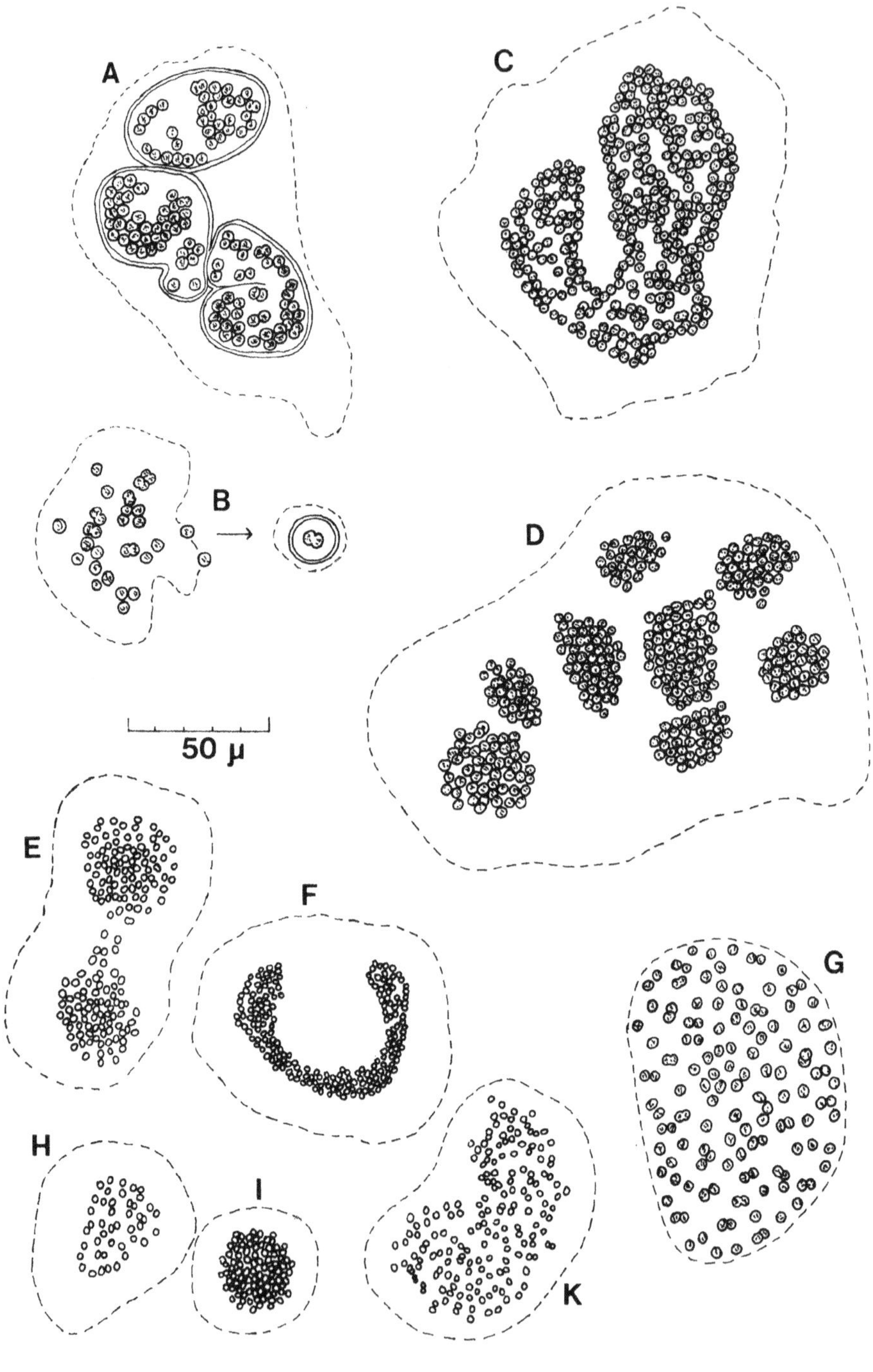

Fig. 3. A) *Microcystis wesenbergii*, B) *Microcystis wesenbergii*, young colony and reproductive stage, C) and G) *Microcystis aeruginosa*, type 1, D) *Microcystis aeruginosa*, type 2, E), F) and I) *Microcystis incerta* type 1, H) and K) *Microcystis incerta* type 2.

creta has square-shaped cells, *L. contorta* elongated cells.).

Merismopedia punctata Meyen (Fig. 4C). 4–128 ovate cells, 1.5–2 × 2–3 μm, in rectangular, flat colonies; the margin of the colonial mucilage is often easily descernible. This species seems to be identical with that described in Komarek (1975).

Merismopedia tenuissima Lemm. Cells ovate, very small, approx. 0.5 × 0.8 μm, in rectangular, 8- to 128-celled colonies. A minute but common species.

In these samples species of the genus *Microcystis* show a specially great variability in respect of cell dimensions, shape of colonies and presence and development of gasvacuoles. On account to this variability several new species have already been described from Sri Lanka (Crow 1923a; Holsinger 1955), but it seems that some of them are not clearly distinct and can be assumed to represent stages or types of common cosmopolitan species. Since the exact identification of the *Microcystis*-species present in the samples is not possible from fixed material, the several forms have been placed in following species:

Microcystis aeruginosa Kütz. The structure of colonies is very variable, all transitional types from compact to very loose can be found. Sometimes the cells are homogeneously distributed in the mucilage up to the margin of the colony (fa. *flos-aquae?*) or they are compactly arranged and surrounded by a transparent layer of marginal mucilage.

Type 1 (Fig. 3C, G): Cells ovate or globose, often 3.5 × 4 or 4 × 5 μm, containing less gasvacuoles than type 2; cells in compact or loose colonies.

Type 2 (Fig. 3D): Cells ovate or globose, often 4.5 × 5 to 6 × 7 μm, with many gasvacuoles; colonies mostly smaller than those of type 1, cells partly arranged in small aggregations.

Microcystis incerta Lemm. Cells spherical, variable in size, always without gasvacuoles; colony shape, density of cells within, and cell size very variable.

Type 1 (Fig. 3E, F, I): Cells spherical (1,5 μm) or slightly elongate (1 × 1.5 μm); colonies very dense, globose or irregular elongate.

Type 2 (Fig. 3H, K): Cells globose, 1.5–2 μm in diameter; cells widely distributed within the colonies (similar to the figures in Komarek (1958)).

Type 3 : Cells spherical, approx. 0.5 μm; in loose colonies of variable size (10 to 50 μm in diameter).

Type 4 : Cells spherical, 2–3(–3.5 μm) in diameter; colonies ± loose, of in part discoid or globose shape (= ? *M. lamelliformis* Holsinger).

Microcystis wesenbergii Komarek (Fig. 3A, B). Cells ovate or globose, 4–5 × 6–7 μm, with prominent gasvacuoles. Colonies few-celled (50–100 μm in diameter) with an inner clear hyaline mucilage layer and outside it a further indistinct layer (visible in indian ink). Young colonies appear to originate from separated pairs of cells, which form soon a refractive mucilage margin (Fig. 3B).

Oscillatoria chlorina (Gom.) Kütz. Trichomes straight, 4 μm in diameter; cells of quadratic shape, showing keritomy. Tychoplanktonic species.

Oscillatoria raciborskii Wol. (Fig. 4A). Trichomes straight, with straight and rounded base and attenuated, uncinate apex. Diameter of trichomes in the basal and central regions approx. 9 μm, narrowing to 7–5 μm at the accute apex. Cells up to the central part of the filaments shorter than broad (approx. 3–5 μm in length), at the apex almost as long as broad, not constricted at the cross-walls. Filaments never calyptrate nor sagitate. Attenuation of filaments and curvature of apices variable.

Oscillatoria sp. Trichomes straight, 5.5–6 μm in diameter. Cells as long as broad. Total length of trichomes up to 900 μm. Apical cell rounded. Cells without gasvacuoles, in fixed stage light blue.

Pseudanabaena galeata Böcher (Fig. 4D). Trichomes straight, 2 μm in diameter; cells 3–7 (–19) μm in length. Apical cell with 2 apical and 1 basal granules (gasvacuoles), other cells with 1 granule at each end. Distinct mucilage connecting pieces jouning cells together. Trichome length 50–80 μm.

Spirulina sp. ad *S. subsalsa* Oerst. (Fig. 4B). Trichomes from closely to loosely coiled, showing all transitional types. Width of spirals often 3 μm, distance of spirals varies from very close to slightly or widely open, up to 4 μm in length at a maximum. Diameter of filaments 2 μm. On account to the constancy in width of filaments and spirals all individuals observed can be referred to the species complex *S. subsalsa* Oerst., *S. meneghiana* (Gom.) Zanard and *S. major* (Gom.) Kütz.

4.2. Cryptophyceae

Cryptomonas sp. Cells ovate. The most frequent species is straight and slightly attenuated at the antapical end. Dimensions approx. 12 × 18 μm. Identification is not possible from fixed samples. Probably some additional species are also present (?).

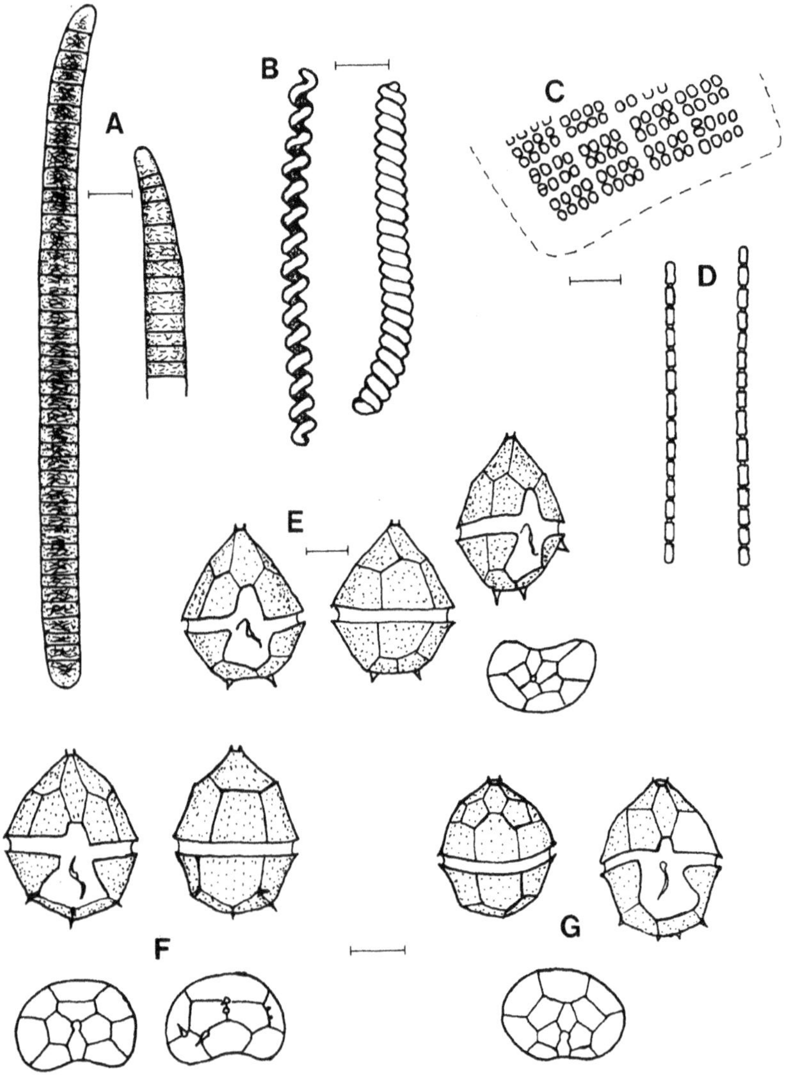

Fig. 4. A) *Oscillatoria raciborskii,* B) *Spirulina* sp. ad *Sp. subsalsa,* C) *Merismopedia punctata,* D) *Pseudanabaena galeata,* E) *Peridiniopsis cunningtonii* tab. *remotum,* F) *Peridiniopsis pygmaeum,* G) *Peridinium inconspicuum.* Scale = 10 μm.

Rhodomonas lacustris Pascher et Ruttner. Cells oblong-ovate, distinctly attenuated at the basal end; dimensions 6–8 × 10–12 μm. Not very frequent species.

4.3. Dinophyceae

Cystodinium sp. Cells elongated spindle-shaped, containing one, before cell division two stellate chloroplasts; dimensions 13 × 30, respectively 18 × 80 μm.

Gymnodinium sp. Dimensions approx. 30 × 35 μm. Identification from fixed material is not possible. Unfrequent species.

Peridinium cunningtonii Lemm. tab. *remotum* (Lindem.) Lef. (Fig. 4E). Size and shape of cells very variable. Dimensions 16–24 × 20–35 × 10–16 μm, sometimes strongly flattened dorsiventrally. The epicone has only six pre-equatorial plates. The left dorsal apical plate is distant from the apex, therefore the plate structure from the dorsal view is asymmetric (distinction from *P. pygmaeum* (Lindem.) Bourr.). Angles of the epicone with in each case well developped median spine (between the antapical plates) or an additional spine on the left antapical plate – this can also be found on the smallest type (16 × 20 × 10 μm). In some cases widely spaced spines are developed on the first and the fifth postequatorial plate. The longitudinal farrow is remarkably widening towards the base.

Peridiniopsis pygmaeum (Lindem.) Bourr. (Fig. 4F). Cells globose, from the ventral side slightly flattened; dimensions 18–24 × 22–32 μm; never as strongly flattened as *P. cunningtonii*. Shape of the epicone pentagonal. Dorsal arrangement of plates symmetric. Seven pre-equatorial plates. Margin of the lower part of the longitudinal farrow (largely widening towards the antapex) armed with two or more spines.

Peridinium inconspicuum Lemm. (Fig. 4G). Cells obovoid, dorsiventrally less flattened than *Peridiniopsis cunningtonii*. Dimensions frequently 21 × 24 × 18 μm. Arrangement of scales strictly symmetrical and corresponding to the type species with two well developed dorsal apical plates. Less frequent than the two *Peridiniopsis* species.

4.4. Diatomophyceae

Cyclotella sp. Margin of frustules arched, with apparently spiral ripps (12 on 10 μm). Comparably rare. It can not be decided, if this species is identical with *Cyclotella meneghiniana* Kütz. var. *laevissima* (Van Goor) Hust. or with *C. ceylonica* Holsinger. However the typical strongly inflated area in the center of the valves of *C. ceylonica* could not be seen.

Melosira granulata (Ehrb.) Ralfs (Fig. 5A) et var. *angustissima* Müller (Fif. 5B). Chaines straight or slightly curved. Cell diameter, in the type form (7–) 8–12 (–14) μm, and in the variety approx. 3 μm; length of cells for both types 18–22 μm. The rows of puncta are very clear and follow a spiral course in the central cells. The structure is well discernible without preparation for the species (10 to 12 points on 10 μm). The terminal cells very often have well developed spines. Transitional stages between the type and the variety occur. As also discussed by Crow (1923 b) and Holsinger (1955) *Melosira*-species are very prominent features of the present samples.

Synedra sp. Since preparation of frustules could not be done, it can not be decided, if there is only one or even two different species of *Synedra* in the lake. Two different size classes at least, one with frustules of 2.5 × 45 μm and the other with frustules of 3–4 × 80 – 100–160 μm were found. The valves are widening continuously against the central part and have slightly capitate ends.

4.5. Xanthophyceae

Goniochloris contorta (Bourr.) Ettl (Fig. 5C). Cell shape triangular, each angle passing into a long process, ornamented by a 5 μm long spine. Dimensions approx. 40 × 45 μm. Six chloroplasts in each cell, arranged stellately. The cell wall is apparently smooth.

Goniochloris sp. ad *G. smithii* (Bourr.) Fott (Fig. 5D). Cell-shape similar to *G. contorta* but more slender. Cell wall distinctly granulate. Dimensions with spines 30 μm, without spines 20 μm.

Isthmochloron lobulatum (Näg.) Skuja (Fig. 5E). Cell-shape rectangular, flat with 4 bifurcated pro-

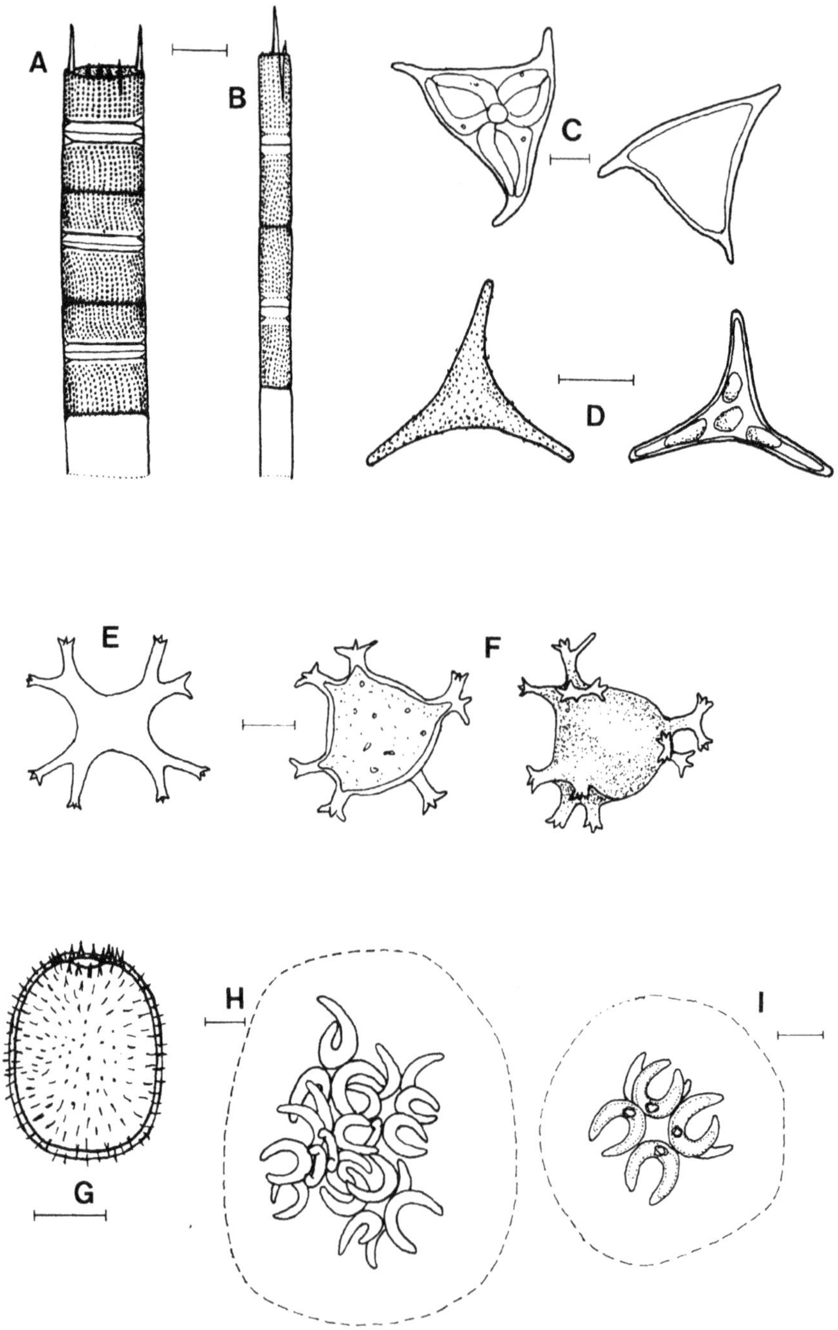

Fig. 5. A) *Melosira granulata*, B) *Melosira granulata* var. *angustissima*, C) *Goniochloris contorta*, D) *Goniochloris* sp. ad *G. smithii*, E) *Isthmochloron lobulatum*, F) *Pseudostaurastrum limneticum*, G) *Trachelomonas hispidafa* fa., H) and I) *Ankistrodesmus bibraianus*. Scale = 10 μm.

cesses, which narrow into three spines. Several parietal chromatophores. Total dimensions of cells 40 × 40 μm.

Pseudostaurastrum limneticum (Borge) Chod. (Fig. 5F). Cell-shape tetrahedral, with four large extensions bifurcating into two arms, each of which ends in two to three short spines. The structure of processes of this species is similar to those developed on *Isthmochloron lobulatum* (Näg.) Skuja, but the processes of *Pseudostaurastrum* are in three-dimensional arrangement; similar to the figures given by Skuja (1949) for *Tetraedron limneticum* Borge from Burma. According to Kovacik (1975) *Tetraedron limneticum* Borge is synonymous with *Pseudostaurastrum limneticum* (Borge) Chod.

Tetraplektron sp. Cell-shape tetrahedral, slightly asymmetric. Each angle composed by one curved lobe with rounded end. Similar to *T. acutum*, which has pointed processes. Dimensions 12 × 14 × 12 μm.

4.6. *Euglenophyceae*

This taxonomic group is represented by a variety of different species, most of which are only present in low cell densities. Therefore only the most frequent ones are discussed.

Euglena sp. Cells oblong-ovate, dimensions approx. 15–30 × 60 μm. Periplast with spiral structure. The species can not be identified from preserved material.

Lepocinclis sp. ad *L. ovum* (Ehrb.) Lemm. Cells obovoid, with a distinctly separated and pointed process. Dimensions of the cells 10 × 10 (-13) × 35 μm, process 8–10 μm. Cell wall with distinct, spiral ribbing. Apex widely rounded.

Phacus sp. Cells obovate, flat, with a short spine (approx. 4–5 μm); dimensions 15 × 28 × 35 μm.

Trachelomonas abrupta Swir. Test oblong, approx. 15 × 23 μm, light brown, with scarce small points. Opening without collar, approx. 3.5–4 μm in diameter.

Trachelomonas hispida (Perty) Stein fa. (Fig. 5G). Test oblong, covered densly with short spines. Aperture of flagellum (approx. 4 μm in diameter) surrounded by an area of stouter approx. 2 μm long spines.

Trachelomonas volvocina Ehrb. Test globose, slightly oblong, most frequently without collar; dimensions 10–14 × 8–12 μm. Frequent species.

4.7. *Chlorophyceae*

Actinastrum hantzschii Lagerh. Colonies identical to the description. Only one single exemplare could be observed.

Ankistrodesmus bibraianus (Reinsch) Korš. (Fig. 5H, I). Cells arcuate, approx. 2.5 × 10 μm, joining together by the convex surface in rather large colonies, surrounded by homogenous mucilage. Rare species.

Botryococcus braunii Kütz. (Fig. 6A). Cells largely oblong-ovate, 4 × 6 to 5 × 8 μm, joint together by dark compact mucilage and embeded in more diffluent mucilage, which had radiated extensions. Diameter of colonies often 30 μm.

Coelastrum astroideum De Not. var. *astroideum* Sod. (Fig. 6B). Cells broadly pyramidal with one widely rounded process. Dimensions 7 × 8 to 11 × 13 μm. Colonies strictly symmetrical of variable size. Cell wall without structures. Colonies frequently composed of 4, 8 or 24 cells; diameter of colonies frequently 18–20 μm.

Coelastrum microporum Näg. Cells spherical, frequently 7 μm in diameter. Colonies compact, frequently 30 μm in diameter. This species was found in lower cell densities than *C. astroideum*.

Coelastrum reticulatum (Dang.) Senn. (Fig. 6D). Cells spherical; each cell with five to seven connecting processes, but joint to a neighbouring cell by one, less frequently two processes; cell diameter variable, 3.5–6 μm. Colonies frequently with 8–16 cells (10–20 μm in diameter). Young daughter colonies are often attached together by remnants of mother colony cell walls (50 μm in diameter).

Crucigenia tetrapedia (Kirchn.) W. et G. S. West. Cells triangular, cruciately arranged in four-celled coenobia; cells frequently 4.5, colonies 10 μm in diameter.

Crucigeniella apiculata (Lemm.) Kom. (Fig. 6C). Cells oblong, straight or slightly reniform, 3.5–4 × 6–7 μm. Apical thickening present, with a more or less well developed tooth; marginal thickening variable. Colonies consisting of four cells arranged around a rhomboid central space. Often two rarely four colonies hold together by remnants of the mother cell wall. According to Hindak (1977) *Crucigeniella pulchra* (W. et G. S. West) Kom. can be included in *C. apiculata* (Lemm.) Kom.

Dictyosphaerium tetrachotomum Printz (Fig. 6E, G). Cells oblong-ovate. 3–10 × 8–12 μm, frequently

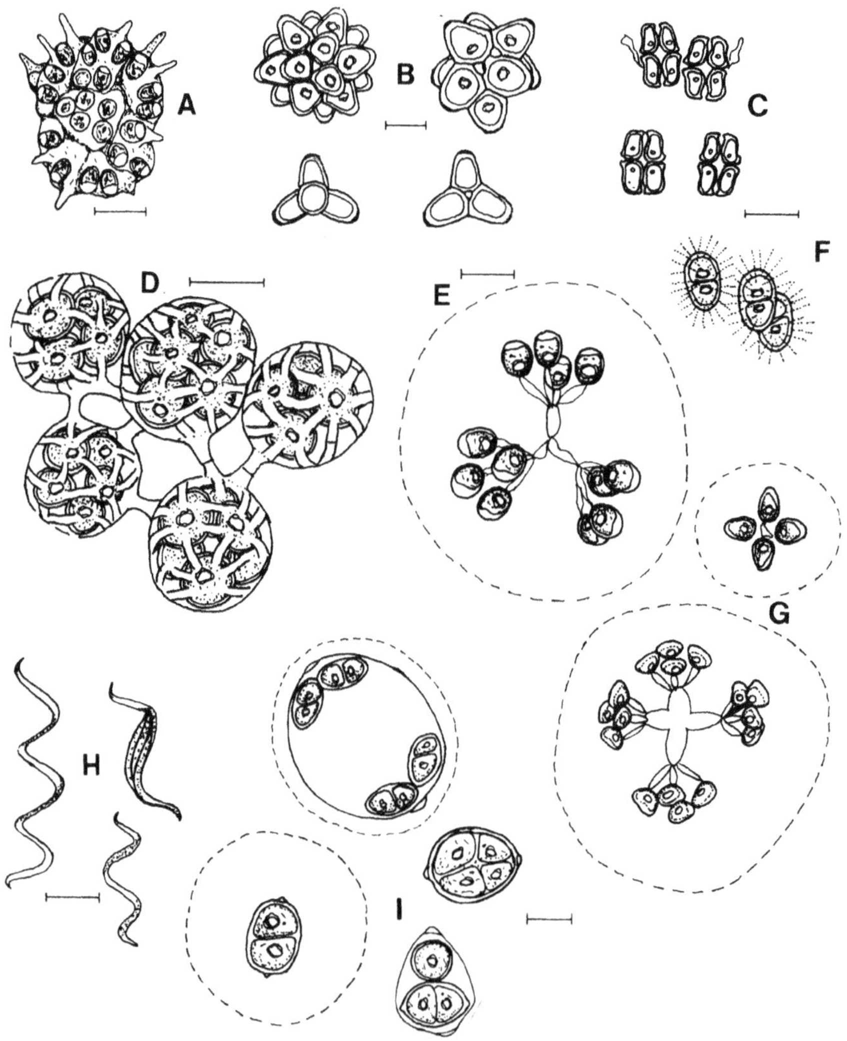

Fig. 6. A) *Botryococcus braunii,* B) *Coelastrum astroideum* var. *astroideum,* C) *Crucigeniella apiculata,* D) *Coelastrum reticulatum,* E) *Dictyosphaerium tetrachotomum,* adult colony, F) *Franceia armata,* G) *Dictyosphaerium tetrachotomum,* joung colony, H) *Monoraphidium irregularis,* I) *Oocystis* sp. ad O. *marssonii.* Scale = 10 μm.

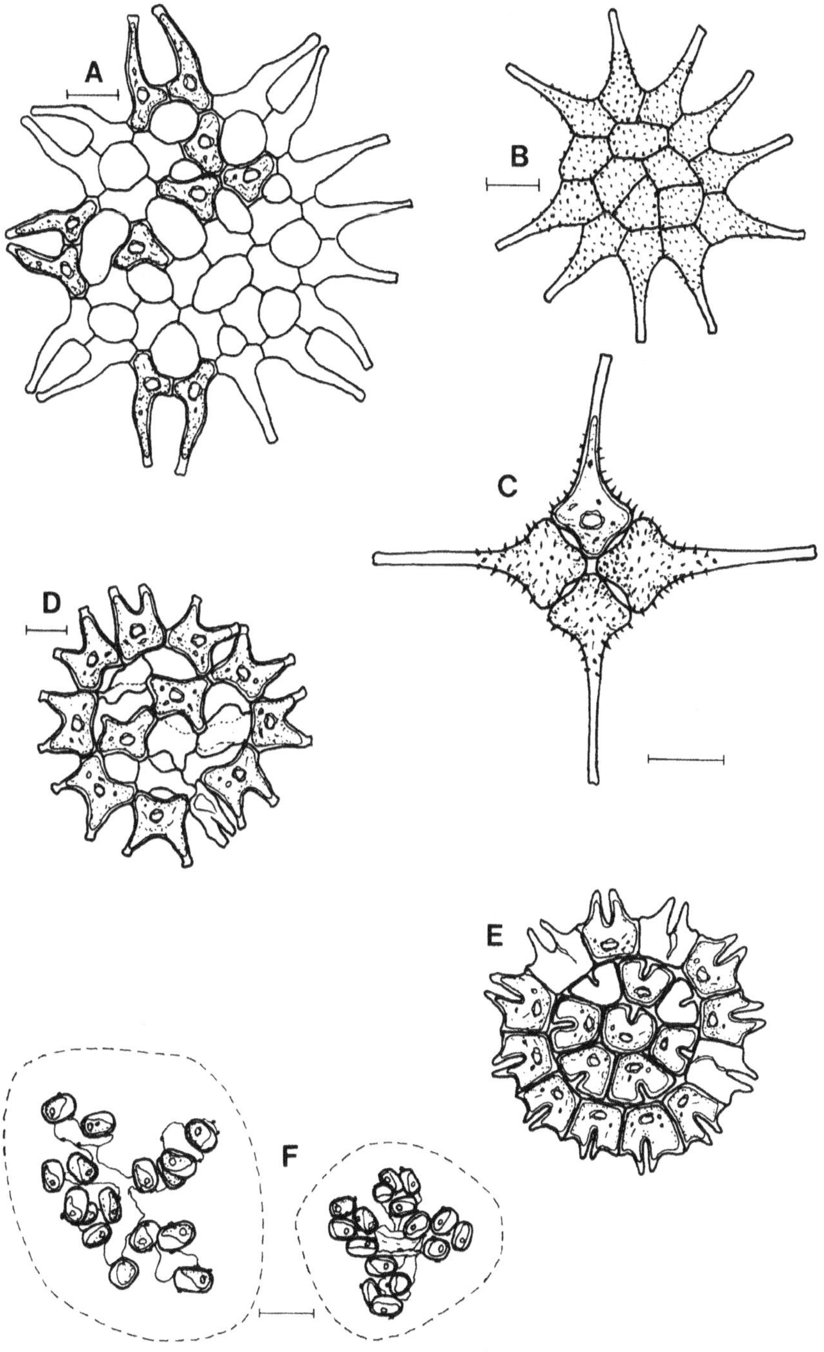

Fig. 7. A) *Pediastrum simplex* var. *simplex*, type 1, B) and C) *Pediastrum simplex* var. *simplex*, type 2, D) *Pediastrum simplex*, E) *Pediastrum tetras*, F) *Quadricoccus* sp. ad *Qu. verrucosus*. Scale = 10 μm.

3.5–7 μm in diameter; with pyrenoid. Mother cells much widened before division. Daughter cells oblong, arranged parallel. Colonies with 4 to 64 cells.

Franceia armata (Lemm.) Korš. (Fig. 6F). Cells broadly ellipsoid, containing two or four chromatophors, each with one pyrenoid. Cell wall covered with thin, regularely dispersed bristles of approx. 5 μm length. Cells solitary or in two-celled coenobia.

Golenkinia radiata Chod. Cells spherical, 5–10–18 μm in diameter. Cell wall furnished with thin bristles. Chromatophor with one reniform pyrenoid. Cell diameter with processes 20–25 (–10) μm. Cells solitary, only a short time after division in four-celled colonies.

Monoraphidium irregularis (G. M. Smith) Kom.-Legn. (Fig. 6H). Cells spindle-shaped, narrow, spirally twisted; mean diameter frequently 1 μm, mother cells not more than 2 μm. Cell ends 10 to 30 μm appart. Division into two or four autospores. Frequent species.

Monoraphidium sp. ad *M. setiforme* (Nyg.) Kom.-Legn. fa. *brevis* Nyg. Cells straight, spindle-shaped, attenuated into long colourless processes, occupying ⅓ to ½ of total cell length. Dimensions 2.5–3.5 × 40–90 μm.

Oocystis sp. ad *O. marssonii* Lemm. (Fig. 6I). Cells broadly ellipsoid, with a very small, often obscure papilla at each end. Young cells containing soon after division four chromatophors. Dimensions 7 × 12 to 8 × 16 μm in two- to eight-celled coenobia, whereas solitary cells are larger (10 × 16 to 17 × 22 μm).

Pediastrum duplex Meyen (Fig. 7D). Marginal cells of colonies extended into two conical blunt-tipped processes. In one case a colony of 16 cells, 50 μm in diameter, was observed (periferal cells 10 × 12 μm, inner cells 10 × 10 μm). Rare species.

Pediastrum simplex Meyen. This species was found in the following two types, which according to Barrientos (1979) can both be included in var. *simplex.*

Type 1 (Fig. 7A): Cell wall smooth; processes of the outer cells not clearly separated from the cell itself and often curved. Dimensions for 70 μm colony diameter: outer cell 12 × 25 μm, inner cells 10 × 10 μm, for 115 μm colony diameter: outer cells 20 × 32 μm, inner cells 18 × 20 μm.

Type 2 (Fig. 7B, C): Cell wall regularly ornamented by 1–2 μm long, minute spines. Processes of the outer cells straight, separated by a slight curvature from the cell itself. Frequent dimensions of the outer cells 7 × 15–8 × 20 μm. Small colonies of this type more frequent than of type 1.

Since no transitional forms between the two types were observed, it can be supposed that there were two genetically separate populations of the same species (or even varieties or forms?) present.

Pediastrum tetras (Ehrb.) Ralfs (Fig. 7E). Marginal cells deeply incised, with two processes near the central incision and often two smaller ones near the cross walls. Dimensions of cells approx. 6 × 8 μm. Colonies of four or eight cells could be observed. Rare species.

Quadricoccus sp. ad *Q. verrucosus* Fott (Fig. 7F). Cells broadly ellipsoid, ornamented with ± well developed granules at the ends; dimensions approx. 3–3.5 × 5–7 μm. Cells joint by remnants of the mother cell wall on their broad side into groups of two or four cells and colonies with 4–8–12 (–16) cells, surrounded by mucilage.

Scenedesmus acuminatus (Lagerh.) Chod. (Fig. 8A, B). Cells spindle-shaped, straight or slightly and often irregularely curved, very variable in shape. For one type the cells are either straight in the whole colony or the marginal cells are curved. Dimensions often 4 × 12 μm. Also more or less dissociating two-celled colonies or four-celled colonies arranged in three dimensions (frequent cell dimensions 2 × 20 μm similar to var. elongatus G. M. Smith) and very large specimens with straight cells (8–10 × 40–50 μm) can be observed.

Scenedesmus sp. ad *S. arcuatus* Lemm. (Fig. 8C). Cells oblong or slightly ovate; marginal cells straight or curved. Colonies often small and in one series of four cells. Dimensions: 3–3.5 × 8–10 μm, sometimes in two raws of alternating cells, up to 7 × 17 μm.

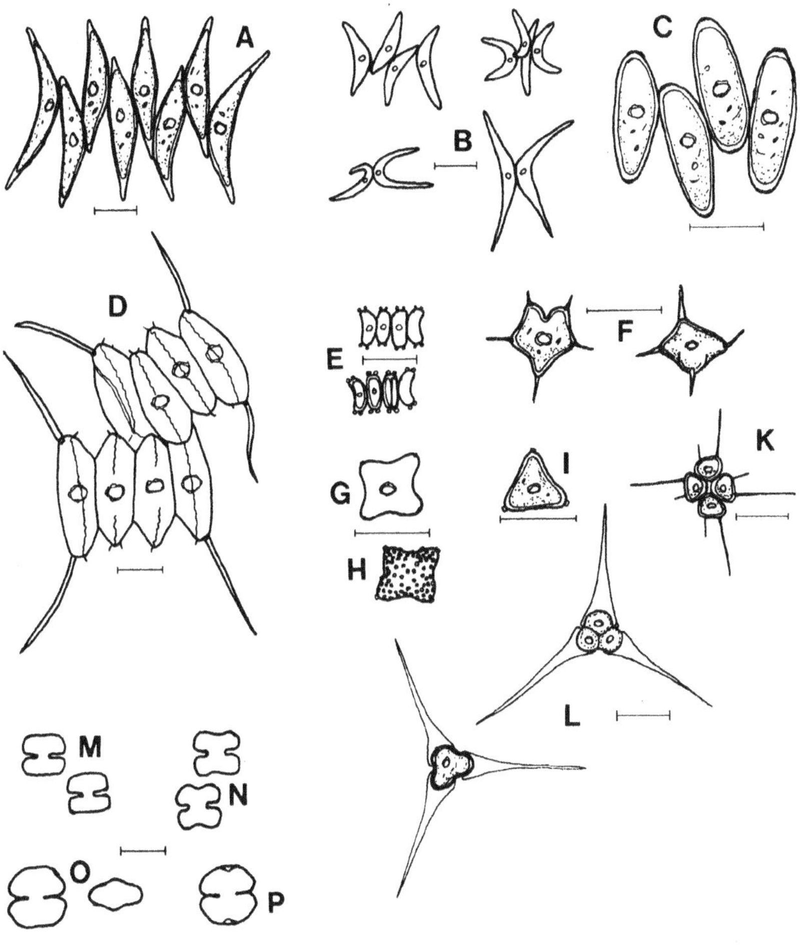

Fig. 8. A) & B) *Scenedesmus acuminatus,* C) *Scenedesmus* sp. ad *Sc. arcuatus,* D) *Scenedesmus* sp. ad *Sc. opoliensis,* E) *Scenedesmus* sp. ad *Sc. polyglobus,* F) *Tetraedron caudatum,* G) *Tetraedron minimum* var. *minimum,* H) *Tetraedron minimum* var. *scrobiculatum,* I) *Tetraedron triangulare,* K) *Tetrastrum heteracanthum,* L) *Treubaria* sp., M) *Cosmarium minimum,* N) *Cosmarium pusillum,* O) *Cosmarium* sp. ad *C. abbreviatum,* P) *Cosmarium* sp.. Scale = 10 μm.

Scenedesmus sp. ad *S. ellipticus* Corda. Cells oblong, structureless, straight, in two- or four-celled colonies. Dimensions 3–3.5 × 5–8 μm.

Scenedesmus sp. ad *S. opoliensis* P. Richt. (Fig. 8D). Cells naviculoid, with truncated ends, joined for ⅓–⅔ of the total length into two- or four-celled coenobia. Marginal cells with a long spine at each end and a small tooth. Normally all cells with median ribbs. Ribbs of inner cells often interrupted. Inner cells ornamented with one or two apical teeth. This species was found in two different size groups with cell dimensions of 3 × 9–5 × 18 μm and 7–8 × 22–28 μm.

Scenedesmus sp. ad *S. polyglobus* Hortob. (Fig. 8E). Cells oblong, with two apparently spherical granules at each end. Marginal cells curved and truncated. Dimensions of cells approx. 3 × 7 μm.

Scenedesmus sp. ad *S. spinosus* Chod. Cells oblong, each with one spine at each end. Marginal cells with one to four additional, marginal spines of variable length. Length of apical spines approx. ½ of total cell length; most frequently coenobia of two cells, rarely four-celled ones. Cell dimensions approx. 4 × 10 μm.

Tetraedron caudatum (Corda) Hansg. (Fig. 8F). Cells five-angle-shaped, flat, with rounded angles and 2–4 μm long spines at each angle. Dimensions of cells without spines 8 × 8 to 12 × 12 μm. In part of somewhat three-dimensional shape, in which the central angle is turned upwards and spines of the other angles are irregularly bent downwards.

Tetraedron incus (Teil.) G. M. Smith. Cells often tetrahedron-shaped, 20 μm in diameter including a 5 μm long spine at each angle (side of tetrahedron 7 μm). Flat cells quadratic, 12 × 12 μm.

Tetraedron minimum (A. Bréb.) G. M. Smith. Cells small, tetragonal, flat, approx. 6 × 7 μm, rarely up to 12 × 14 μm. Angles provided with small verrucae. This species occurs in two different varieties: var. *minimum* Kov. (Fig. 8G) with smooth cell wall and var. *scrobiculatum* Lagerh. (Fig. 8H) with densly granulated walls.

Tetraedron triangulare Korš. (Fig. 8I). Cells small, flat, triangular. Each angle provided with a small verruca. Dimensions often 6 × 7 μm.

Tetrastrum heteracanthum (Nordst.) Chod. (Fig. 8K). Cells triangular, broadly convex at the outer wall. Cells arranged cruciately around a small rectangular opening into four-celled coenobia. Each cell provided with a long (10 μm) and a short (3 μm) spine. Dimensions of cells without spines 4.5 × 5 μm.

Treubaria sp. ad *T. triappendiculata* Bern. (Fig. 8L). Cells triangular, flat; the angles broadly rounded and ornamented with large colourless spines; with one or more pyrenoids. Dimensions of cells approx. 8 × 10 μm, length of processes 20 × 25 μm.

4.8. Zygnemaphyceae

Cosmarium depressum (Näg.) Lund. var. *planctonicum* Rev. Cells with elliptical semicells, deeply constricted, 17 × 20 μm; isthmus 8 μm. Cells surrounded by a mucilage sheeth, which is approx. 40 μm in diameter. Smaller than the type. Rare species.

Cosmarium minimum W. et G. S. West (Fig. 8M). Cells deeply constricted, sinus open, semicells rectangular. The angles of the semicells seem to be slightly more rounded than in the original description, based on material from Sri Lanka (W. et G. S. West, 1904).

Cosmarium pusillum (Bréb.) Arch. (Fig. 8N). Cells deeply constricted, sinus open, semicells rectangular with incurved apices. Dimensions 8 × 8 × 4.5 μm; rarer than the former species.

Cosmarium sp. ad *C. abbreviatum* Racib. (Fig. 8O). Cells deeply constricted, sinus narrow, semicells subhexagonal with rounded angles. Marginal angles nearer to the isthmus than to the apices, not central as for *C. abbreviatum*. In apical view with central inflation. More or less frequent species.

Cosmarium sp. (Fig. 8P). Cells deeply constricted, sinus narrow; semicells subrectangular, with an internal nipple-like prominence. Dimensions 10 × 10 × 3 to 13 × 12 × 4 μm.

Mougeotia sp. Filaments with cylindrical cells. Chloroplast a broad axial plate. Diameter of filaments 3.5–4 μm, length of cells 35–70 μm, length of filaments up to 500 μm. Since no zygotes were found, the species can not be identified. More or less frequent in September 1979.

Staurastrum sp. ad *St. crenulatum* (Näg.) Delp. (Fig. 9A). Cells deeply constricted. Semicells with a ring of granules above the isthmus. Vertical view triangular with three pairs of emarginate verrucae in the central part. Processes straight and extending horizontally from the apex or bent slightly downwards, ornamented with rings of small spines (as compared with *St. crenulatum* the semicells are more slender and the cellshape is more constant!).

Staurastrum sp. ad *St. brachioprominens* Boerg. (Fig. 9B). Cells deeply constricted. Sinus narrow. Semicells campanulate, with two long, divergent, ± straight processes, ornamented with several series of minute spines and terminated by two or three

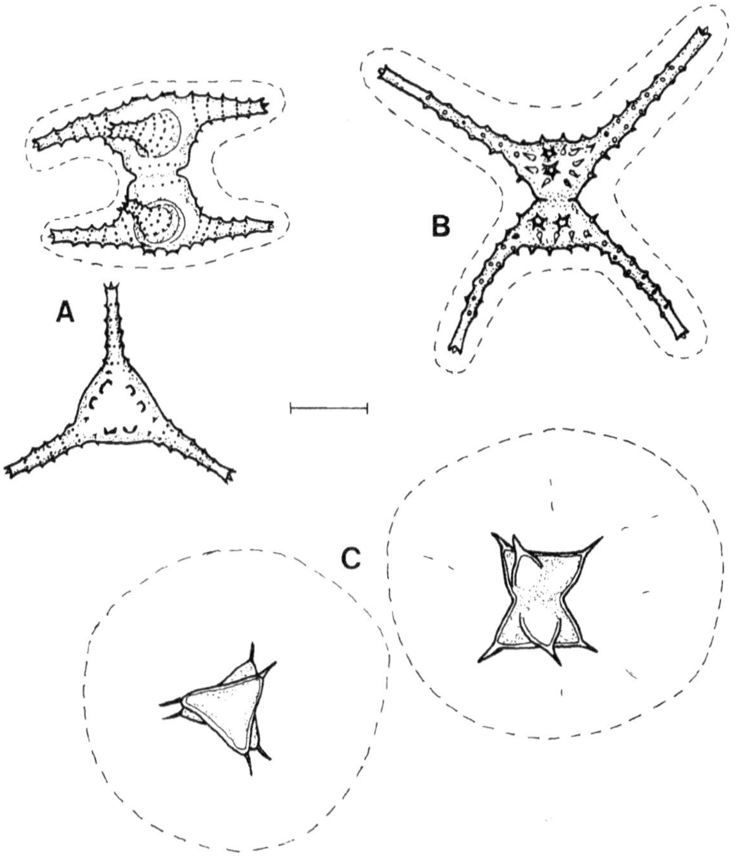

Fig. 9. A) *Staurastrum* sp. ad *St. crenulatum,* B) *Staurastrum* sp. ad *St. brachioprominens,* C) *Staurodesmus* sp.. Scale = 10 μm.

divergent spines. Apices of semicells flat, with two or three small emarginate verrucae. Cells 1½ times longer than broad. Dimensions with processes 30 × 32–38 × 42 μm, isthmus 4–5 μm, processes 15–22 μm. Ventral sides of semicells with two or three large verrucae, ornamented with five spines each (difference to *St. brachioprominens?*). The apical view shows an irregular curvature of the spines of both semicells towards the same side, not a diagonal arrangement found by Holsinger (1955) for *St. brachioprominens* in Sri Lanka.

Staurastrum tetracerum Ralfs. Cells deeply constricted. Semicells campanulate with two long, divergent and straight processes, terminated by three spines. Processes arranged diagonally to one another. Dimensions: with processes 22 × 28 – 25 × 25 μm, isthmus 4–5 μm (smaller and thinner than the former species!).

Staurastrum sp. Cells deeply constricted. Sinus narrow, semicells campanulate, produced into three processes. Processes ornamented with several rings of small spines (6–7), ending in three apical spines. Semicells ornamented with a ring of eight small verrucae near the isthmus. Apices of semicells slightly convex with three pairs of divergent small spines at each side, at the basis of processes one distinct spine more. In apical view the semicells are set at an angle of approx. 30° to another. Dimensions with processes 20 × 30–25 × 40 μm, isthmus 7–8 μm. Processes approx. 15 μm long (minimum dimensions 25 × 20 × 6 μm). This species is similar to *St. boreale* W. et G. S. West.

Staurodesmus sp. ad *St. pterosporus* (Lund) Bourr. vel *St. o'mearii* (Arch.) Teil (Fig. 9C). Cells constricted; sinus open. Cell wall smooth. Semicells triangular, three-radiate with three slightly produced angles, armed with short, divergent spines (3–5 μm long). Dimensions of cells 14 × 13 × 6 to 16 × 16 × 7 μm; mucilage envelope 40–50 μm in diameter. Since no zygospores were found, it is impossible to find out, to which of the two species this alga belongs (as already mentioned by Teiling (1967)).

Acknowledgement

The author is indebted to Dr. F. Schiemer, University of Vienna, for kindly providing the samples for this work. He is also grateful to Professor J. W. G. Lund, Ambleside, for indefatigable help with the translation of the taxonomic notes and to Joy Wieser for the additional translations.

References

Apstein, C., 1907. Das Plankton im Colombo-See auf Ceylon. Zool. Jb., Abt. Syst. 25: 201–244.
Apstein, C., 1910. Das Plankton des Gregory-Sees auf Ceylon. Zool. Jb., Abt. Syst. 29: 661–679.
Barrientos, P. O. O., 1979. Revision der Gattung *Pediastrum* Meyen (Chlorophyta), Bibl. Phycol. J. Cramer, Vaduz, 48, 183 pp.
Crow, W. B., 1923a. The taxonomy and variation of the genus *Microcystis* in Ceylon. New Phytologist 21: 59–68.
Crow, W. B., 1923b. Freshwater plankton algae from Ceylon. J. Botany, London, 61: 110–114, 138–145, 164–171.
Desicachary, T. V., 1959. Cyanophyta. I.C.A.R. Monographs on Algae. New Dehli, 686 pp.
Dokulil, M., Bauer, K. & Silva, I., 1983. An assessment of the phytoplankton biomass and primary productivity of Parakrama Samudra, a shallow man-made lake in Sri Lanka. In: Schiemer, F. (ed.) Limnology of Parakrama Samudra – Sri Lanka: a case study of an ancient man-made lake in the tropics. Developments in Hydrobiology (this volume). Dr W. Junk, The Hague.
Foged, N., 1976. Freshwater diatoms in Sri Lanka (Ceylon). Bibliotheca Phycol. 23, 100 pp.
Fritsch, F. E., 1907. A general consideration of the subaerial and freshwater algal flora of Ceylon. A contribution to the study of tropical algal ecology, Part I. – Subaerial algae and algae of the inland fresh-waters. Proc. Roy. Soc., London, Ser. B. 79: 197–254.
Hindak, F., 1977. Studies on the chlorococcal algae, Chlorophyceae. I. Veda, Slovak Acad. Sc., Bratislava, 190 pp.
Holsinger, E. C. T., 1955. The plankton algae of three Ceylon lakes. Hydrobiol. 7: 8–24.
Jeeji-Bai, N., E. Hegewald & Soeder, C. J., 1977. Revision and taxonomic analysis of the genus *Anabaenopsis*. Arch. Hydrobiol. Suppl. 51, Algol. Stud. 18: 3–24.
Komarek, J., 1958. Die taxonomische Revision der planktischen Blaualgen der Tschechoslowakei. Algol. Stud., Tschechosl. Akad. Wiss., Prag, 358 pp.
Kovacik, L., 1975. Taxonomic review of the genus *Tetraedron* (Chlorococcales). Arch. Hydrobiol. Suppl. 51, Algol. Stud. 13: 354–391.
Lemmermann, E., 1907. Protophyten-Plancton von Ceylon. Zoolog. Jahrb. 25/2: 263–268.
Schiemer, F., 1981. Parakrama Samudra (Sri Lanka) Project, a study of a tropical lake ecosystem I. An interim review. Verh. Internat. Verein. Limnol. 21: 993–999.
Skuja, H., 1949. Zur Süßwasseralgen-Flora Burmas. Nova Acta Reg. Soc. Sc. Upsal. IV, 14,5; 186 pp.
Skvortzow, B. M., 1928. On some freshwater algae from Ceylon, collected by W. P. Lipsky in 1908. Ann. Roy. Bot. Garden, Peradenia: 109ff.
Teiling, E., 1967. The desmid genus *Staurodesmus*. A taxonomic study. Ark. Botanik 6, 11: 437–629.
West, W. & West, G. S., 1902. A contribution to the freshwater algae of Ceylon. Trans. Linn. Soc. 2 Ser. Bot. 6: 123–215.
West, W. & West, G. S., 1904. A monograph of the British Desmidiaceae. Roy. Soc., London, Nr. 82, 84, 88, 92, 108. Johns Repr. Coop. Reprint 1971. 5 Vols.

Author's address:
E. Rott
Botanical Institute
University of Innsbruck
Sternwartestrasse 15
A-6020 Innsbruck
Austria

21. Caddisflies (Trichoptera) from Parakrama Samudra, an ancient man-made lake in Sri Lanka

Hans Malicky

Keywords: Trichoptera, light traps, flight range

Abstract

Light-trap catches at the shore of Parakrama Samudra confirmed the existence of a caddisfly fauna which is typical for paleaotropical lakes, with dominance of Ecnomidae and Leptoceridae. *Ecnomus moselyi* was new for Sri Lanka.

1. Introduction

Within the context of a limnological study of Parakrama Samudra (for a description see Schiemer (1981)) in northeastern Sri Lanka, Dr. F. Schiemer made occasional light-trap catches to obtain information about the aquatic insects of the lake. The small light trap used contained an actinic mercury lamp, and was operated by a car battery. One sample was taken in September 1979, another in March 1980. The results are shown in Table 1.

Table 1. Light trap catches from Parakrama Samudra.

	September 1979		March 1980	
	Males/females		Males/females	
Orthotrichia indica Mart.			10	7
Orthotrichia sp.				6
Dipseudopsis horni Ulmer				2
Paduniella pandya Schmid			3	23
Ecnomus moselyi Mart.	16	24	279	}
E. dutthagamani Schmid			1	} 329
E. vaharika Schmid			3	}
E. tenellus Rambur			51	}
Amphipsyche indica Mart.		2	2	26
Hydropsychodes sp.				1
Trichosetodes argentolineata Ulmer			1	
Setodes inlensis Mart.			10	11
Setodes sp.				1
Setodellina nerviciliata Schmid			2	1
Oecetis meghadouta Schmid			1	1
Leptoceridae g. sp. no. 1				1
do. no. 2				1
do. no. 3				6

2. Discussion

The caddisfly fauna of Sri Lanka is comparatively well known, above all from the investigations of Schmid (1958), but also from contributions of other authors (Walker, Hagen, Brauer, Banks, Ulmer, Mosely, Kimmins: references see Schmid 1958; Malicky 1973). It is therefore not surprising that not one species in the samples was new to science. Only one species (*Ecnomus moselyi*) had not previously been recorded from Sri Lanka, although it was known from the Indian mainland (Martynov 1935: 142).

It may be asked whether all the caddisfly specimens in the light trap material came from the lake, i.e., whether they developed in it. The average flight range of adult caddisflies is about several hundred meters which means the distance between the breeding place and the light trap (Malicky, in press). Since there are no other bodies of water in the immediate surroundings, it is evident that practically all individuals came from the lake itself, and only

Schiemer, F. (ed.), Limnology of Parakrama Samudra - Sri Lanka
 ISBN 90 6193 763 9

single specimens may have come from other waters over greater distances. It is therefore certain that the light-trap catches represent the true trichopterous fauna of the lake, with the possible exception of species not attracted by artificial light. Continued trap operation would certainly reveal the presence of further species, present in the lake in low numbers. The ecological characteristics also confirm that the animals originate from the lake. According to Schmid (1958: 25), the species named in Table 1 are characteristic of stagnant waters in the lowlands of Sri Lanka, except *Ecnomus dutthagamani* of which only one specimen was caught and which, according to Schmid, is characteristic for large rivers. Since the trichopterous faunae of lakes and large rivers contain well known convergences, the presence of this species in Parakrama Samudra is not surprising. It is noteworthy that by far the most abundant species in this lake, *Ecnomus moselyi,* had not so far been recorded from the island. Regional differences in the fauna of lakes may therefore exist.

A few females in the material could not be identified as to species, but they certainly also belong to the typical lake fauna. I did not try to separate the *Ecnomus* females into species since this is impossible at our present state of knowledge. The females which are listed under *Orthotrichia* spp. may belong to other genera in the sense of Schmid, but these genera were united with *Orthotrichia* by Marshall (1979: 214–216). From the zoogeographical point of view, the material represents a typical palaeotropical lake fauna in which *Ecnomidae* and *Leptoceridae* dominate. Several species are known to be scattered over the Indian sub-continent, others were recorded from Sri Lanka only. *Ecnomus tenellus* has an extraordinarily large range and is also common in European lakes.

It should be emphasized here that adult material is indispensable for studying the trichopterous fauna of a body of water. Mature pupae are also useful, but a reliable identification of species is not possible with larval material alone. This is not even possible in much better investigated regions such as Europe or North America. Larval collections give valuable information, but only when accompanied by sufficient adult material.

References

Malicky, H., 1973. The Ceylonese Trichoptera. Bull. Fish. Res. Stn., Sri Lanka (Ceylon) 24: 153–177.

Malicky, H., in press. Anflugdistanz und Fallenfangbarkeit von Köcherfliegen bei Lichtfallen. Acta Biol. Debrecina.

Marshall, J. E., 1979. A review of the genera of the Hydroptilidae (Trichoptera). Bull. Brit. Mus. (Nat. Hist.) Ent. 39: 135–239.

Martynov, A. V., 1935. On a collection of Trichoptera from the Indian Museum. Part I. Annulipalpia. Rec. Ind. Mus. 37: 93–306.

Schiemer, F., 1981. Parakrama Samudra (Sri Lanka) Project, a study of a tropical lake ecosystem. I. An interim review. Verh. Internat. Verein. Limnol. 21: 993–999.

Schmid, F., 1958. Trichoptères de Ceylan. Arch. Hydrobiol. 54: 1–173.

Author's address:
Hans Malicky
Biological Station Lunz
Institute of Limnology
Austrian Academy of Science
A-3293 Lunz am See
Austria

Subject index

Species index
(species recorded from Parakrama Samudra)

ALGAE

CHLOROPHYCEAE

CRYPTOPHYCEAE

CYANOPHYCEAE

CYPRINIFORMES

MASTACEMBELIFORMES

OPHIOCEPHALIFORMES

PERCIFORMES

PROTOZOA

ROTATORIA

TENTACULATA

GPSR Compliance

The European Union's (EU) General Product Safety Regulation (GPSR) is a set of rules that requires consumer products to be safe and our obligations to ensure this.

If you have any concerns about our products, you can contact us on ProductSafety@springernature.com

In case Publisher is established outside the EU, the EU authorized representative is:

Springer Nature Customer Service Center GmbH
Europaplatz 3
69115 Heidelberg, Germany

Batch number: 10762209

Printed by Printforce, the Netherlands